THEORY OF SCIENCE

Dr. B. Bolzanos

Wissenschaftslehre.

Versuch
einer ausführlichen und größtentheils neuen Darstellung
der

Logik

mit steter Rücksicht auf deren bisherige Bearbeiter.

Herausgegeben
von
mehren seiner Freunde.

Mit einer Vorrede
des
Dr. J. Ch. A. Heinroth.

Erster Band.

Sulzbach,
in der J. E. v. Seidelschen Buchhandlung,
1837.

BERNARD BOLZANO

THEORY OF SCIENCE

Attempt at a
Detailed and in the main Novel Exposition
of

LOGIC

With Constant Attention to Earlier Authors

Edited and translated by

ROLF GEORGE

UNIVERSITY OF CALIFORNIA PRESS
Berkeley and Los Angeles 1972

UNIVERSITY OF CALIFORNIA PRESS
Berkeley and Los Angeles, California

ISBN 0-520-01787-0
Library of Congress Catalog Card Number: 71-126765

TO THE MEMORY OF

HENRY S. LEONARD

ACKNOWLEDGMENTS

During the earliest stages of my work on this translation, I enjoyed the co-operation and advice of my teacher, the late Henry S. Leonard. A preliminary draft of the first volume was finished in 1958, but at that time I was still thinking of a complete translation of all four volumes. I was eventually persuaded that early complaints about the unnecessary bulk of the work had their point. Kambartel's very successful attempt at shortening the first two volumes (*Bernard Bolzano's Grundlegung der Logik*, Hamburg, 1963) finally convinced me that an abbreviated version was not only feasible, but desirable.

With the help of a Summer Research Grant from the University of Waterloo I could resume the work in the summer of 1967. A grant from the Canada Council covered most of the cost of manuscript preparation.

I am deeply indebted to Professor Roderick Chisholm for reading the manuscript and making many suggestions concerning selection and translation, and to Mr. Craig Townson for his help in the preparation of the final manuscript.

Thanks are also due to the Macmillan Company for allowing me to quote from N. K. Smith's translation of Kant's *Critique of Pure Reason*, and to the Library of the University of Manitoba for lending me their copy of the *Wissenschaftslehre* for almost a year.

Waterloo, September 1970 ROLF GEORGE

CONTENTS

This table lists the complete contents of the first three books of the *Wissenschaftslehre*. Sections with * are translated either completely or in substantial part. Unstarred sections without page numbers are omitted altogether.

INTRODUCTION

BOOK ONE

THEORY OF FUNDAMENTALS

PART I

Of the Existence of Truths in Themselves

PART II

Of the Recognizability of Truth

BOOK TWO

THEORY OF ELEMENTS

PART I

Of Ideas in Themselves

CHAPTER I

Of the Concept of an Idea in Itself

CHAPTER 2

Internal Attributes of Ideas in Themselves

CHAPTER 3

Distinctions between Ideas that Stem from their Relation to each other

CHAPTER 4

Distinctions among Ideas that Result from their Relations to other Objects

APPENDIX

Earlier Treatments of the Subject Matter of this Part

CHAPTER 3

Distinctions among Propositions which are Based upon their Relations to each other

CHAPTER 4

Several Types of Propositions which State Relations between other Propositions

CHAPTER 5

Some Propositions whose Linguistic Expression Warrants Special Comment

APPENDIX

Earlier Treatments of the Subject Matter of this Part

PART III

Of True Propositions

PART IV

Of Arguments

APPENDIX

Earlier Treatments of the Subject Matter of this Part

BOOK THREE

THEORY OF KNOWLEDGE

PART I

Of Ideas

PART II

Of Judgments

PART III

Of the Relation between our Judgments and Truth

PART IV

Of Certainty, Probability, and Confidence in Judgments

BOOK FOUR

HEURETIC

BOOK FIVE

THEORY OF SCIENCE PROPER

EDITOR'S INTRODUCTION

Bolzano's Life

Bernard Bolzano was born in Prague on October 5, 1781. His early education, by his father and at the Piarist Gymnasium, was conducted in the spirit of the Josephinian Enlightenment (named after the second Austrian Emperor of that name) which emphasized utility, practical morality, and a somewhat pedantic concern with the common good. Balzano later wrote of his father that he was a man with "reverence for God free of all superstition, courage, compassion and patriotism born of a well organized love for mankind, whose diligence did not allow him to spend even a single hour without useful occupation."* He could have used these words to describe himself.

In 1796, he began the study of philosophy, mathematics and theology at the University of Prague. He was attracted by the "purely speculative" part of mathematics, "that part of it which is also philosophy". By this he meant proofs for opinions which everyone already holds, "seeking out the grounds on which our judgments rest".† His interest, in short, lay with the foundations of mathematics, and his dissertation appropriately dealt with some aspects of *Elementargeometrie*. In philosophy, his preferred subject was logic. As was then the vogue, he also spent a good deal of time studying Kant, and though he acknowledged his debt on several occasions, he remained critical of the Kantian system and opposed to many of its tenets. It seems that none of his teachers had any profound or lasting influence upon his opinions nor, according to his own testimony, was there any philosophical system which he took to be the only true one, or for which he harboured much admiration.‡

In 1804 he competed unsuccessfully for a chair in mathematics, and then accepted an appointment for a newly established position as professor of religious instruction in Prague. His views on the nature of religion and religious commitment made it quite impossible to perform the duties officially expected of holders of these new chairs. They had been established in the course of the so-called Austrian

* *Lebensbeschreibung des Dr. B. Bolzano* (ed. M. Fesl), Sulzbach 1836, p. 5.

† *Ibid.*, p. 19.

‡ From a manuscript, quoted in Edward Winter, *Leben und geistige Entwicklung des Solzialethikers und Mathematikers Bernard Bolzano*, Halle 1949, p. 20.

Catholic Restoration with the purpose of providing religious instruction for non-theologians, and to reverse deistic and atheistic tendencies among the lay students. Accordingly, Bolzano was expected not only to give orthodox interpretations of religious dogma in his lectures (which he was to base upon a book by the emperor's confessor, Frint), but also to read a homily each Sunday, to hear confession, etc. It must be understood that Bolzano was a devout man, and that his opposition to irreligion was as unfaltering as anyone could hope. At the same time, the contrast between the simplistic and pious Catholicism which was officially enforced, and Bolzano's rational faith must sooner or later lead to a conflict.

The foundation of Bolzano's religious faith was the principle of utility. He thought of himself as being in narrow agreement with Bentham, and affirmations of utilitarian tenets are found in several of his writings: "I am of the opinion that the supreme moral law demands nothing but the advancement of the common good."* He had adopted three maxims: "Advance the common good", "It behooves us to be happy and to make happy", and "I must progress".† His utilitarian convictions were coupled with extreme readiness for personal sacrifice. When he was eventually dismissed from office with a pension of only 300 *Gulden*, he found great comfort in the thought that this was not more than would be his share if all goods were equally divided. More to the point, he rigorously measured all activities, including religious pursuits, against the standard of public utility. Religion he claimed to be "the sum of such doctrines or opinions which have an either detrimental or beneficial influence upon the virtue and happiness of a man",‡ and a proposition is to be called religious if its consideration "not only moves us in our heart to declare either for or against it, but if through the acceptance or rejection of this proposition our virtue or happiness is altered."§ By virtue, Bolzano means "the persistent striving to make the sum of pain in this world as small as possible, and to enlarge the sum of well being as much as possible."¶ Though some of these quotations stem from a later date, his utilitarian convictions were well developed when he was offered the professorship. The choice to accept this position or even the priesthood did not come easily. Bolzano tells in his autobiography that he filled reams of paper with deductions, starting with the supreme moral law and determining the utility of

* *Wissenschaftslehre*, Vol. IV, p. 27. † Cf. Winter *op. cit.*, p. 60.

‡ *Lebensbeschreibung*, p. 199.

§ Bolzano, *Lehrbuch der Religionswissenschaft*, Sulzbach 1834, p. 60.

¶ Homily on the first Sunday of Advent, 1810, in *Erbauungsreden*, Vol. IV, Prague–Vienna 1852, p. 19.

each profession. He remarks resignedly that his decision was still not as detached as might be desired, since he was influenced by his mother's wishes. In the end, a chance remark of one of his professors that a doctrine is justified if faith in it leads to moral improvement convinced him that the priesthood was the correct choice.*

Bolzano was professor of religious instruction from 1805 to the end of 1819. His weekly sermons became immensely popular, frequently drawing as many as 1000 listeners, and resulting in a kind of movement, sometimes called the "Bohemian Enlightenment", which combined a rationally clarified catholic faith with a programme for social and political reform. It was partly this popularity, partly the general ferment of the Napoleonic wars, which kept him in this position for such a long time.

Bolzano's dismissal was part of a purge of unreliable elements, freethinkers, nationalists and progressives, which took place in Germany and Austria after the assassination of the conservative playwright and diplomat Kotzebue. Charges of heterodoxy and political unreliability had been placed against Bolzano much earlier, and personal grievances also seem to have played a role. As early as 1806, Frint had complained that his book did not sell well in Prague, and later Bolzano was expressly asked to justify himself for lecturing from his own notes rather than from Frint's book. Eventually presentations were made to the emperor, and objectionable passages were excerpted from his writings. The most offensive of these came from a volume of sermons of 1813: "There will be a time when the thousandfold distinctions of rank among men, which cause so much harm, will be reduced to their proper degree, when each will treat the other as a brother. There will be a time when constitutions will be introduced which are not subject to the same terrible abuse as the present one."† Saurau, then chancellor, pointed out that Bolzano's "innovations" cannot be justified. He held that in German universities, where professors must live on students' fees, new doctrines are a necessity; but in Austria, professors are paid by the state "so that they teach propositions that are approved by the church and the civil administration. It is a dangerous error for a professor to think that he can instruct the youth entrusted to his care according to the drift of his individual convictions or according to his own views."‡

An imperial decree dismissing Bolzano was issued on December 24, 1819; it forbade him to teach or preach in public; ecclesiastic

* *Lebensbeschreibung*, p. 27.

† Cf. Eduard Winter, *Der Bolzanoprozess*, Munich–Vienna 1944, p. 29.

‡ *Ibid.*, p. 35 f.

charges against him were ordered to be laid at once. The resulting proceedings did not come to a conclusion until 1825, when Bolzano himself wrote the final verdict in behalf of his ecclesiastic judges, withdrawing none of his earlier utterances, but expressing regret about any evil consequences that might have resulted from their being misunderstood.

From 1823 until 1841 Bolzano spent most of his time near Techobuz in the care of his friends Anna and Joseph Hoffmann. It seems that he moved away from Prague partly because the Hoffmanns could give him the care which his poor state of health demanded (he suffered from tuberculosis), and partly because the threat of further persecution never quite abated until one of his pupils became archbishop in Prague.

During that period he wrote a monadological essay *Athanasia, oder Gründe fur die Unsterblichkeit der Seele* (1827) to console Anna Hoffmann after her last child had died. Most of his time was spent in carrying out a plan conceived some years earlier, of writing a logic. He had concerned himself with the subject from the beginning of his academic career and his mathematical studies only deepened this interest. In one of his early publications* he had said that "a discussion of mathematical method is basically nothing but logic, and hence does not belong to mathematics". He must already have had in mind logic as a methodology of the sciences, a *Wissenschaftslehre.* A few years later, he had entered in his *miscellanea* the remark "I have decided, March 1812, to publish a logic under the title *Essay Concerning a New Logic, Which Would Necessitate a Restructuring of All Sciences; Offered for Examination to All Friends.* The first chapter should be: there are truths (concepts of truth); the sense ordinary people attach to it; not what philosophers improperly call subjective, but what it is objectively. Chapter two: We know several truths. Chapter three: sometimes we commit errors. Chapter four: making certain (purpose of ordinary, not of scientific, method). Chapter five: There is an objective connection between truths. Chapter six: it is sometimes possible to discover this objective connection. Chapter seven: Scientific method. Chapter eight: Different kinds of truths or judgments."† He appended the remark that probability judgments must be discussed, a proof that there are synthetic judgments must be included, and plenty of examples are to be provided.

* *Beiträge zu einer begründetcren Darstellung der Mathematik*, Prague 1810, part II, § 1; cf. Winter, *Leben*, etc., p. 26.

† Quoted from E. Winter, *op. cit.*, p. 26. Winter transcribes: "6 *Kapitel: Es gibt unter den Wahrheiten eine objektive*", but clearly, '*Verbindung*' or some such word must be added after '*objektive*'.

Thus many of the central thoughts of the *Wissenschaftslehre* were already present; especially noticeable, aside from his recognition of Propositions in Themselves, is his resolute opposition to the Kantian maxim that logic must under no circumstances be envisaged as an organon. From the very beginning, Bolzano made no distinction between logic and scientific method; his aim was to provide a method that would lead to a more thoroughly founded presentation of the sciences.

The work on the *Wissenschaftslehre* occupied most of the decade 1820–1830. In May 1830 he could finally write his student Přihonsky that the manuscript was finished. During the next several years only a few minor alterations were made, and the work was published in 1837. In 1840, upon the death of Anna Hoffmann, Bolzano returned to Prague and published in a variety of subjects. After the completion of the *Wissenschaftslehre*, he had continued his mathematical work. He was going to write a "theory of magnitudes" (*Grössenlehre*), and a theory of functions. He finished neither, but the libraries in Prague and Vienna contain considerable manuscript remains.* His theory of infinite sets, which he discussed at length with Přihonsky, was published by the latter in 1851.† Although Bolzano never had much opportunity to discuss mathematical problems, nor the benefit of professional contacts with other mathematicians of rank, his contributions are impressive. The most outstanding are perhaps the following: he defined convergence and indicated convergence criteria several years before Cauchy. He described a function continuous but not differentiable in an interval. He did this several decades before Weierstrass, though the discovery remained unpublished. Finally, he realized that any infinite set contains a subset that stands in biunivocal correspondence to it; more important, he realized that this is not a contradiction. Bolzano's work in mathematics is outstanding for its conceptual precision. Here as everywhere else he is remarkable not because of the imaginative sweep of his thought, but because he refused to accept what he had not carefully proved, and because of his close critical examination of accepted theories.

During the late 1830s and the 1840s efforts were made to publicize

* His theory of functions was published in 1930, and the theory of numbers in 1931. (*Funktionenlehre*, ed. Rychlik, Prague 1930; *Zahlenlehre*, ed. Rychlik, Prague 1931.)

† B. Bolzano, *Paradoxien des Unendlichen*, ed. Fritz Přihonsky, Leipzig 1851. New edition, ed. Alois Höfler, Meiner, Leipzig 1921, Hamburg 1955. Translated as *Paradoxes of the Infinite*, ed. Donald A. Steele, London 1950. For a discussion of Bolzano's mathematical contributions cf. Steele's excellent introduction, also Winter, *Leben und geistige Entwicklung*, etc., ch. 4; J. L. Coolidge, *The Mathematics of Great Amateurs*, Oxford 1949, ch. 16; and Jan Berg, *Bolzano's Logic*, Stockholm Studies in Philosophy 2, Stockholm 1962, chs. I and VII.

Bolzano's views, especially in Germany. In September 1838 several of his disciples gathered with him in Techobuz to work out a plan of action: a discussion of Bolzano's philosophy was to be stimulated in the leading journals, important philosophers should be asked to review his books, a prize was to be awarded for the best critical discussion of the *Wissenschaftslehre*, and books summarizing his views were to be published.* The efforts were not successful. Though some of his students assumed important posts, no philosophical school or tradition formed.

Bolzano lived to see the revolution of 1848. He was in sympathy with the goals of the revolutionaries, but stayed aloof. He had little in common with the nationalist, romantic and liberal forces which now carried the opposition to the regime. He died on December 18, 1848 after a life "full of physical and mental suffering", and was buried in the *Wolschaner Friedhof*.

Bolzano's work never became widely known, but it profoundly influenced some important thinkers. Georg Cantor, for example, knew the *Paradoxes* and thought very highly of the book. Later on, Franz Brentano alerted Husserl to the *Wissenschaftslehre*, and Bolzano's influence, most obvious in the first volume of the *Logische Untersuchungen*, fortified, if it did not engender, Husserl's anti-psychologistic standpoint. Husserl's praise is almost extravagant: (Bolzano's theory of elements) "far surpasses everything else world literature has to offer as a systematic exposition of logic." "It contains such an abundance of original, scientifically secured and fruitful thoughts, that he must be considered one of the greatest logicians of all time . . . logic as a science must be based upon Bolzano's work."†

In more recent years, interest in Bolzano was revived through the work of Eduard Winter and Heinrich Scholz and, among English speaking philosophers, William and Martha Kneale, and Ian Berg. At present, a complete edition of Bolzano's work in thirty-nine volumes is in preparation under the general editorship of E. Winter.

General Outline of the Wissenschaftslehre

In an introductory chapter, Bolzano defines a science as an "aggregate of truths whose known portion is important enough to be set

* B. Bolzano, *Dr. Bolzano und seine Gegner*, Sulzbach 1839, and *Bolzano's Wissenschaftslehre und Religionswissenschaft in einer beurteilenden Uebersicht*, Sulzbach 1841.

† *Logische Untersuchungen*, Halle 1900, Vol. I, p. 225 f.

forth in a special book" and logic as the science which deals with the division of the domain of all truths into suitable parts, and supplies the rules for the composition of the respective treatises. These rules and the division of the domain of all truths are discussed in the final, fourth volume of the German edition. But before the domain of truths can be divided into sections, and treatises written, a sufficient number of truths must first be discovered. Accordingly, the theory of science proper is preceded by a book entitled *Erfindungskunst* (Heuretic), which is concerned with the discovery of truths. This section, in turn, presupposes a discussion of the conditions of human knowledge in general. But epistemology can be transacted only if it is preceded by a theory concerning the entities which are known, namely propositions in themselves and their terms (Theory of Elements). Finally, the first section of the work is the Theory of Fundamentals, in which Bolzano undertakes to prove that there are truths in themselves and that some of them can be known. I shall briefly comment on each of the five major divisions of W.L.

Propositions and Ideas in Themselves

One of the reviewers of W.L., a Dr. Menelaos, remarks that "throughout, the author assumes the old, strictly objective or dogmatic, viewpoint, in contrast to the contemporary one, which is based on the psychological selfconsciousness of the thinking mind."* It is, of course, one of the main tenets of W.L. that there are objective propositions and ideas [*Vorstellungen*], which need to be manifested neither as written or spoken statements, nor as thoughts in a mind. Bolzano introduces these concepts by way of a *Verständigungssatz*, i.e. a sentence which conveys the meaning of a term without being a definition. (He thought that definitions were often not the best method for reaching an agreement about meaning.)

By a proposition [*Satz an sich*] he means "any assertion that something is or is not the case, regardless whether somebody has put it into words, and regardless even whether it has been thought."†

This introduction of propositions and ideas in themselves has earned him the name of a "logical Plato",‡ and commentators have

* Review of W.L. by a Dr. P. Menelaos (probably a pseudonym) *Zeitschrift für katholische Theologie*, 25. Cf. *Dr. Bolzano und seine Gegener*, Sulzbach 1839, pp. 157 f.

† W.L. § 19.

‡ Cf. Friedrich Kambartel, *Bolzano's Grundlegung der Logik*, Hamburg (Meiner) 1963, p. xxi. K. makes certain qualifications, but takes the appelation to be essentially justified.

been moved to say that he postulated a *supersensible world* of propositions and ideas. Friedrich Kambartel, in the introduction to his edition of W.L., writes that "Bolzano's Platonism puts him into fundamental opposition to the . . . viewpoint of logical empiricism, as well as to all theories of science which have developed from Kant's transcendental philosophy . . .". And . . . "Rudolf Carnap, not Bolzano, has become the anti-Kant of this time . . ."* But since Carnap, too, has been called a Platonic Realist,† the contrast is not very enlightening. Indeed it can be shown, and I intend to do so, that Bolzano's postulation of propositions in themselves does not much differ from Carnap's position as expressed, e.g., in *Empiricism, Semantics, and Ontology*. A brief account of Bolzano's position will show why we should consider either both or neither of them to be Platonists.

Bolzano denied emphatically that propositions and ideas in themselves exist, or have reality. In the course of his argument he claims that none of the ordinary nouns indicating existence are applicable to propositions: they have neither *Sein*, nor *Dasein*, nor *Existenz*, nor *Wirklichkeit*.‡ He wishes to be committed only to the assertion that *there are* propositions. The distinction between saying '*A*'s exist' and 'there are *A*'s' is considered in W.L. §§ 142 and 137. Formally, the two kinds of proposition differ in that 'exists' is construed as a predicate, hence existence as an attribute, while propositions of the form 'there are *A*'s' indicate that the idea '*A*' has a referent [*einen Gegenstand*] or reference [*Gegenständlichkeit*]. It is not unreasonable to see the distinction between these two types of proposition as analogous to the distinction between existentially quantified expressions, and expressions in which 'exists' occurs as a predicate. In any case, Bolzano's "ontological commitment" to propositions and ideas in themselves does not seem to be any heavier than the commitment of a philosopher who, like Carnap, quantifies over variables of these types. Bolzano states repeatedly, especially in a series of letters to Exner, that he does not commit himself more than, e.g., a mathematician does who speaks about a formula that generates all prime numbers (B. thinks that such a formula must be a proposition in itself and not a proposition thought by somebody),§ or to more than a man who says that there are truths which are not yet known.¶ W.L. contains a number of further examples which make the same point,

* Kambartel, *loc. cit.* † By Quine. ‡ W.L. § 19.

§ *Der Briefwechsel Bolzano's mit Exner*, ed. E. Winter, Bernard Bolzano's Schriften, Vol. 4, Prague 1935, p. 83.

¶ *Ibid.* p. 24.

and in his defence of W.L. he wrote: "We hope that it will suffice to say that B. means by propositions and truths in themselves nothing but what we all mean by these words . . . when we ask, for example, whether every *truth* is recognized by some being . . . [or when we say that] if there were no thinking being, then the proposition that there is no thinking being would be a *truth*."* In the end, he claims that usefulness alone could be a sufficient ground for the "introduction" of propositions, truths, etc. into logic: "Once it is agreed that it is necessary or even simply useful to speak of truths in themselves, i.e. of truths irrespective of whether or not they have been recognized by anybody, and especially of the connection between them, it will not be denied that the concept of propositions in themselves in the indicated sense deserves to be introduced into logic."† Notice that Bolzano's concern is whether a certain concept "deserves to be introduced into logic". This locution itself shows that the point at issue is not an existence proof, but a pragmatic decision. We must realize that it is *worth while* to use this concept. Bolzano's claim that there are truths and propositions in themselves can be presented as a decision rather than a discovery, and it is fair to say that he himself saw the matter in this way. Further on in W.L., he claims that "the logician must have as much right to speak of truths in themselves as the geometer has to speak of spaces as such (i.e. of mere possibilities of certain locations) without thinking of them as filled with matter, although it is perhaps possible to give metaphysical reasons why there is no, and cannot be any, empty space."‡

It is worth dwelling a moment longer on the distinction Bolzano makes between 'There are *A*'s' and '*A*'s exist', especially since he rejects the view that there are several kinds or levels of existence, and that propositions and truths partake in an attenuated, ghost-like, or spiritual, sort of existence.§ At one point he remarks that a proposition of the form 'there is an *A*' is equivalent to one of the form '*A*'s exist' (or *A* exists) "only in case it lies in the concept *A* itself that the corresponding referent exists, as with the concept 'God'."¶ This passage should not be construed as affirming the validity of ontological arguments. Rather, Bolzano is here saying that 'there is an *A*' follows from '*A*'s exist' in any case, but the converse holds only if *A* is an existence-entailing concept and this last expression can be given an innocent interpretation by referring back to an

* *Bolzano's Wissenschaftslehre und Religionswissenschaft in einer beurteilenden Uebersicht*, Sulzbach 1841, p. 28. The book was published anonymously. In it, B. refers to himself in the third person.

† W.L. § 20.1.

‡ W.L. § 25.c.

§ Cf. Letters to Exner, p. 74 and p. 29.

¶ W.L. § 137.

earlier passage: "a proposition in itself does not exist. It is therefore as absurd to say that a proposition has eternal being, as it is to say that it began at a certain moment and ceased at another."* I don't take Bolzano to make the utterly trivial point that since propositions do not exist, they also don't exist at a time, or at all times. Rather, I take his words to imply that being-at-a-time is one, though perhaps not the only, sufficient condition for existence, or, to put it into Bolzanese: if the idea of an *A* at time *t* (for some *t* or for all *t*) has reference, then *A* exists (or existed). Thus if it followed from a concept that its referents, if it had any, would have a position in time, then this concept would be existence-entailing. For example, the concept of a winged horse is existence-entailing since its referent (if there were one) would have a certain temporal position. Bolzano would consider it a conceptual truth that horses begin and end in time. Hence it follows that if there is a winged horse, then a winged horse exists, but it does not similarly follow that if there is a proposition in itself, then propositions exist. Unfortunately, Bolzano does not make clear just what concepts are existence-entailing; he probably took the matter to be intuitively obvious. But this is a serious omission if, as I think, he meant to restrict pragmatically motivated ontological commitments to concepts that are not existence-entailing. We are not free to postulate a winged horse, but we may postulate propositions in themselves.

The second part of the Theory of Fundamentals consists of a refutation of scepticism. Perhaps its function is best described by saying that it should make the exposition of logic accessible even to the sceptic: since a good scientific presentation should generate conviction as it progresses, a refutation of scepticism seems required as early as possible. Scepticism is here treated as an aberrant view: the sceptic must be *healed*, he suffers from a delusion. Bolzano held that doubt and belief are psychological states that normally cannot be voluntarily induced. For example, he considered Descartes' rule that we should doubt everything once in our lives to be absurd, since we cannot doubt at will. Descartes, he thought, should have said that we ought to *examine* everything. But in the refutation of scepticism, B. is concerned with doubt as a psychological state, and universal doubt as a psychological deviation. In view of the structure of W.L. the passages on scepticism can be treated as an aside.

Deducibility; Ground and Consequence

Bolzano takes propositions to be composed of terms (ideas, *Vor-*

* W.L. § 19.

stellungen) much in the same way in which sentences are composed of words; indeed the composition of sentences is often taken as clue to the composition of the corresponding propositions. Bolzano held, though he did not claim to have a proof, that all propositions have the same basic structure, namely '*A* has *b*': they all assert that a subject has a certain character. (Bolzano generally uses capitals for denotative, and lower case letters for attributive expressions.) He expands not a little ingenuity on fitting various expressions into this pattern: 'some *A* are *b*' becomes 'The idea of an *A* which has *b* has reference'; '*p* v *q*' becomes 'The idea of a true proposition among *p* and *q* has reference'. Without doubt, Bolzano's insistence on a common form for all propositions was detrimental to the development of several aspects of the system. W. and M. Kneale have pointed out that the development of argument patterns, the calculational aspect of logic, was not notably advanced by Bolzano, and they justly cite his rigidity in matters of logical form as one of the reasons.* In particular, since the logical connectives are absorbed into the predicate of standard form propositions, Bolzano "has made it necessary to express notions which are usually regarded as formal (e.g. those of negation and particularity), by signs which enter his reductive formulae in the same way as signs which would ordinarily be said to express material notions (e.g. that of wisdom)."†

Every part of a proposition that is not itself a proposition is called an idea in itself. The first part of the *Theory of Elements* consists of a discussion of these ideas in themselves and their relations to each other and to other things. This is followed by the theory of propositions, perhaps the most important and interesting part of W.L. After several sections dealing with various properties of propositions, Bolzano, in § 147, introduces the notion of 'degree of satisfiability' [*Gültigkeitsgrad*]. He envisages certain ideas of a given proposition as replaceable by other appropriate ideas. In this way new propositions are generated, and the class of propositions so generated will have certain properties; for example, it will contain none, one, several, or only, true propositions; the original proposition will be said to have a higher degree of satisfiability the larger the number of true propositions in that class.

This procedure of Bolzano's calls for some comment. Bolzano, although he frequently speaks of variable ideas, did not have the concept of a propositional function or a variable in the contemporary sense. A language which allows for propositional functions must

* W. and M. Kneale, *The Development of Logic*, Oxford 1964, p. 370 f.
† *Ibid.*

contain certain symbols (constants) which "designate" objects, and other symbols (variables) which "range over" the same objects. Bolzano's propositions do not contain parts which exhibit these different semantic relations; his ideas can at best designate, but cannot range over, a class of objects. He gets by without using propositional functions by introducing appropriate classes of propositions instead. Whenever he appoints certain ideas as "variables", he has in mind parts of fully articulated propositions. For example, in the proposition 'Caius is a friend of Titus', we may consider 'Caius' as variable, but this means that we are to consider the class of propositions which result from 'Caius is a friend of Titus' when we substitute appropriate other ideas for 'Caius'. This does not turn 'Caius' into a variable; it remains a proper name.

Just as contemporary logic relies on propositional functions to exhibit the "forms" of propositions, so Bolzano took his classes of propositions to be the formal element in logic. He reflects that '*forma*' was used by Cicero in the same sense as '*species*'.* Hence the form of a proposition is a class of propositions that may be generated from it by replacing a certain constituent idea by other ideas. Most propositions will have many forms, because we may vary one or the other constituent, and since the parts of complex ideas may also be varied, a moderately complex proposition will have a considerable number of "forms". Some of them may be unknown if the proposition contains ideas which we have not been able to analyse fully.

Bolzano's theory of deducibility forms the core of the *Theory of Elements*. He was the first to give a formal definition of the notion of consequence. It is akin to that given a century later by Tarski.†

Bolzano defines deducibility as follows: "Propositions *M*, *N*, *O*, . . . are deducible from propositions *A*, *B*, *C*, *D*, . . . if every class of ideas whose substitution for *i*, *j*, . . . makes all of *A*, *B*, *C*, *D*, . . . true also makes all of *M*, *N*, *O*, . . . true."‡

Tarski's definition of consequence (actually a preliminary version) is: "The sentence *X follows logically* from the sentences of class *K* if, and only if, every model of the class *K* is also a model of the sentence *X*."§ A model for class *K* of sentences is any set of objects

* W.L. § 81, note.

† Heinrich Scholz was the first to call attention to this kinship. Cf. "*Die Wissenschaftslehre Bolzano's*", Abhandlungen der Friesschen Schule, N.F. 6, 1937, pp. 470 ff.

‡ W.L. § 155.

§ "*On the Concept of Logical Consequence*" (1937) in A. Tarski, *Logic, Semantics, Meta-Mathematics*, Oxford 1956, p. 417.

which satisfies the propositional functions generated from K by replacing all extralogical constants of K by variables.

At first sight, the principal difference between the two definitions is the presence of the term 'truth' in Bolzano's definition, but since he says that "our judgments are true if we combine with our idea of a certain object an idea which this object really has,"* the word 'truth' can be eliminated from the definition. One can then see that the main difference lies in the absence of the notion of a function in Bolzano and the fact that he does not draw a sharp dividing line between logical and extralogical parts of propositions. Thus, when he speaks of variable ideas, he may mean any parts or particles whatever of the propositions in question. We have already seen that the problem of effective identification of constituent ideas is not considered in W.L., and that it is therefore not made clear what "parts" of propositions may be substituted upon. This problem does not arise in the constructed languages of modern logic.

Another major difference is that in Bolzano deducibility is always relative to a set of variable (in his sense) ideas. Thus, the concept of deducibility becomes much wider in Bolzano than it is in Tarski. Consider the argument 'Socrates was a man, therefore Socrates was mortal'. It is valid in Bolzano's sense relative to the idea 'Socrates' since every substitution on 'Socrates' which makes the premises true also makes the conclusion true. But we can know this only if we first know that all men are mortal. Hence to assess the validity of an argument it will generally be necessary to have a good deal of extralogical knowledge.†

Tarski, by contrast, asks us to turn *all* extralogical constants into variables, hence knowledge of mortality or humankind will not be required to assess validity. On the other hand, since Bolzano does not distinguish logical and extralogical parts of propositions, he would presumably also permit substitutions on the former. This would make some arguments that are valid in Tarski's sense Bolzano-invalid relative to their logical particles; e.g. *modus ponens* is invalid with respect to 'if-then'.

A further important distinction between the two concepts of deducibility derives from the fact that in both cases the definiens is a universal affirmative proposition, i.e. a proposition of the form 'All S are P'. Most contemporary logicians, certainly Tarski, regard such propositions as true when there is no S. Bolzano, on the other hand,

* W.L. § 42.
† Cf. W.L. § 223.

regarded them as false under these circumstances. It follows that if there is no model for the class *K* in Tarski's definition, then the definiens is true, hence *X* follows from the class *K*. This simply means that if the set of premises is inconsistent, anything follows from it. By contrast, Bolzano's interpretation of universal affirmative propositions makes his definiens false if there is no class of ideas whose substitution for *i*, *j*, . . . makes all of *A*, *B*, *C*, *D*, . . . true. Hence for Bolzano nothing follows from an inconsistent premise set. It is important to notice that Bolzano did not hold the truth of universal affirmative propositions to be *undetermined* if their subject terms did not have a referent. He took them to be *false* under this condition. This was a considered position which is sustained throughout W.L. For instance, in order to maintain it, he abandons the view that *A* and *O* propositions are contradictory.*

The discussion of deducibility and the related topics of consistency, equivalence, etc. are followed by an investigation of the concept of probability. There is a noteworthy similarity between Bolzano's and Wittgenstein's treatment of that relation,† as well as Carnap's conception of it. I have attempted to include most of what Bolzano had to say about probability and confidence, though I had to leave out a number of occasional remarks. It is his view that we tend to accept judgments whose probability on the basis of accepted judgments is greater than 1/2, i.e. if the probability of *p* exceeds 1/2 we tend to *form* the judgment that *p*. But the judgment that *p* is the case can be formed with more or less *confidence*. If with greater confidence, the judgment will have more *force*. Confidence is said to be a function of both the degree of vividness of the constituent ideas and of the degree of assent,‡ though he states later§ that nothing but the degree of probability can have an influence upon the degree of confidence. It seems that Bolzano generally considers confidence to be a feeling of some sort, but he occasionally seems to allow for external manifestations, e.g. through wagers. He also develops some aspects of the notion of expected value¶ and seems to employ a principle of minimizing regret. Unfortunately, the suggestions that confidence is somehow exhibited in action are not followed up, and his discussion of the relation between probability and confidence is disappointingly short in view of his breadth in other matters. Still, it is remarkable that he clearly distinguished between objective measures of the probability of a proposition, and subjective strength of conviction. His suggestion that the latter might be influenced by the vividness

* W.L. § 230.
† *Tractatus* 5.15.
‡ W.L. 293.
§ W.L. § 318.
¶ W.L. § 317.

of the ideas in question has not been followed up in contemporary investigations.

After the concept of probability, Bolzano introduces another relation which he calls *Abfolge* (ground-consequence). This relation is said to hold between truths, but no definition of *Abfolge* is attempted; the relation is introduced only by way of examples. Consider the two propositions 'the temperature is high', and 'the thermometer registers high'. According to Bolzano, we can *deduce* the first from the second, but the second is not the ground of the first. Conversely, the second is deducible from the first and the first is also the ground of the second. Bolzano refers to the authority of Aristotle and his distinction of explanations into explanations ὅτι and διότι. (The schoolmen similarly distinguished *demonstrationes quia* from *demonstrationes propter quid.*) Thus we can *infer* that the temperature is high from the high thermometer reading, but the high reading does not *explain why* the temperature is high.

From Bolzano's discussion we gather that the formal differences between deducibility and ground–consequence are that the former is transitive, non-symmetrical and simply reflexive, while the latter is intransitive, asymmetrical, and irreflexive. More importantly, the ground–consequence relation among empirical propositions corresponds to the causal relation among facts: if p is the ground of q, then the fact that p is the cause of the fact that q. This parallelism allows Bolzano to provide rather persuasive examples of the relation among empirical propositions. He is much less successful in conveying his intuition that the relation of ground and consequence also holds between purely conceptual and mathematical truths. For example, he writes in connection with the problem of constructing an equilateral triangle with a given side: "Without doubt, the proposition that for any two points a and b there is a third point c, such that the distances $ac = ab = bc$, can be *inferred* or deduced from the proposition that two coplanar circles around a and b with radius ab will somewhere intersect . . ., but nobody who has a clear concept of ground and consequence will deny that the first proposition is not objectively grounded in the second, but the second in the first; it is not the case that there is a third point for the given two because the two circles intersect, but they intersect because there is such a point."* It seems absurd to try to decide whether two lines intersect because they have a point in common, or have a point in common because they intersect. Bolzano seems to have held that among two mutually deducible

* *Versuch einer objektiven Begründung der Lehre von der Zusammensetzung der Kräfte*, Königlich Böhmische Gesellschaft der Wissenschaften, 2, Prague 1843, p. 433.

truths exactly one must be the ground of the other; but this assumption is nowhere established.

While Bolzano attempts to establish a number of theorems about the relation of ground and consequence, he does not tell us very much about the relation itself; in fact, he does not quite seem to have made up his mind about its nature. He emphasizes that the ground must always be simpler than its consequence, but he is not sure if 'ground' and 'consequence' are simple (hence undefinable), or complex concepts. On one occasion he holds that not every case of a ground–consequence relation is also a case of deducibility,* but later he conjectures that the concept (ground) "which I have claimed to be simple above, might be complex after all; it may turn out to be none other than the concept of an ordering of truths which allows us to deduce from the smallest number of simple premises the largest possible number of the remaining truths as conclusions."†

Bolzano's insistence that truths are ordered in themselves according to the relation of ground and consequence, and that this ordering is independent of our order of recognition has tended to re-enforce the Platonizing interpretation of Bolzano. Against this it is well to notice that Bolzano thought that the presence of an objective relation of ground and consequence can also be admitted by somebody who does not believe that there are truths in themselves at all. Bolzano writes that such a person could admit that there is a certain *relation* between truths which deserves the name of an *objective connection* since it rests upon circumstances which do not depend on how we happened to recognize these truths. "If he were to lay down the rule that only the simplest among all the classes of mental [occurrences of] truths from which a given mental truth can be deduced is to be called its objective ground; furthermore, if he were to use the criteria for this relation set forth in § 221, his views would be very similar to B.'s doctrine, except for the circumstance that he speaks of true *thoughts*, while B. speaks of truths in themselves."‡

Subjective Propositions and Ideas

Bolzano next considers propositions and ideas as they are manifested in the mind. So far, not much attention has been paid to this part of his work; interest has generally centred around his logical theories and the doctrine of propositions and ideas in themselves.

Bolzano assumes, perhaps naively, that a judgment is the manifest

* W.L. § 200.

† W.L. § 221. Interesting comments on this definition can be found in Kambartel and Buhl.

‡ Ueberricht p. 68 f.

presence of a proposition in itself in the mind. The proposition is said to be the matter, or stuff, of the judgment. Since propositions are composed of ideas in themselves, judgments are said to consist of subjective ideas which pass through the mind, one after the other, though the judgment is not merely the presence of a series of ideas, but also contains an element of affirmation or acceptance. Being actual affections of the mind, judgments and subjective ideas are actual, occur in time and have duration. Unlike objective ideas, subjective ones can be clear or abscure, distinct or confused, and more or less vivid; judgments can be formed with more or less confidence.

The view that subjective ideas are parts of judgments was not new, but Bolzano's theory of objective contents allowed him to avoid a certain widespread confusion. It was generally acknowledged that ideas [*Vorstellungen*] pass through a man's mind when he thinks, i.e. judges. At the same time it was claimed that ideas are the kind of thing that occurs in the mind when a man hears or sees (sounds and visible shapes). In other words, ideas were seen to be both, the elements of mental judgments, as well as the elements of sensation. But it is not clear how anything can function in both these capacities. Ideas were often described in terms not consistent with their roles as terms of judgments: they were said to be divisible, round, moving, receding, semblances of their objects, etc. (Hume, for example, said of ideas that they are not infinitely divisible, also that a man without certain organs of sense will be deprived of certain ideas.) It is clear that mental occurrences of this sort, whatever we want to call them, are not what is required to fill the role of terms of judgments. If judgments are indeed mentally articulated propositions, their terms will have to be something other than sense-data and the like. Sensations, etc. could be seen to be parts of judgments only under very peculiar circumstances; they are, in any case, not *eo ipso* what is here wanted. It seems, then, that any theory which envisages judgments as manifest mental events, and which also holds sensations to be mental events, must divide subjective ideas into two classes, or at least must explain how one kind of entity can fulfil both of these functions. (Hume, for example, did not separate these kinds. He claims reasoning to be the operation of our thoughts and *ideas*, but it is not very plausible to call reasoning the operation of something that can be divided, or pointlike, etc.)

Bolzano did not become a victim of that confusion. For him a subjective idea is either part of a mental proposition or at least could be such a part, i.e. he concentrates on the *logical* functions of ideas; he never envisages them as something that is at all like a sensation.

This concentration on the logical aspects of mental activity, one-sided though it may be, allows him to avoid certain classical mistakes. In particular the view that knowledge consists in the similarity or resemblance between our ideas and their objects is exposed as fallacious. Terms of propositions *refer* to their objects. They do not need to resemble them. The truth of a proposition, and hence our knowledge of an object, does not depend upon the similarity between idea and object, rather, "a proposition is true if we connect with the idea of an object the idea of an attribute which this object actually has."* Notice that this rejection of the resemblance theory is not based on the classical argument that we can never know whether our ideas resemble their objects since we can never compare the idea with its object (the object being altogether inaccessible—all we have are ideas). Rather, the critical point is that it is of no consequence whether the idea resembles its object, since resemblance is of no importance in connection with the idea's logical function.

If we understand by subjective ideas the terms and particles of subjective propositions, the origin and "adequacy" of ideas ceases to be a central problem of epistemology. The important problem is no longer whether our *ideas* match or resemble their objects, but whether our *judgments* are true, and how we come to make true judgments. The task of the theory of knowledge becomes to explain how true and false judgments arise in the mind; since Bolzano took a judgment to be composed not of sensations and similar entities (though it can of course be about sensations), but of mental *terms*, i.e. the mental counterparts of *logical* entities, he took the clues for his epistemology more from logic than from psychology. Given a subjective idea or proposition, it must be possible to do two things; namely to describe in logical terms the corresponding objective idea or proposition, and then to inquire why this mental phenomenon arose in the mind at that time. For example, if some of our ideas are *intuitions* [*Anschauungen*], and if others are *concepts* [*Begriffe*], and if these play a different role in our judgments, then it must be possible to indicate the *logical* difference between them. Of course, it sounds strange to speak of an intuition in itself, i.e. an intuition irrespective of whether or not somebody has it, but this is only Bolzano's manner of facing the logical issue involved. If intuitions differ logically from concepts, then it must be possible to describe their different logical character without reference to the mode of their psychological origin, i.e. without reference to impressions, etc. Bolzano's view is that intuitions are ideas that are both simple and singular. By this he

* W.L. § 42.

means that intuitions do not consist of further ideas, and that they have precisely one referent. It must be understood that the definition of 'intuition' without reference to sensation is not a quaint rationalist aberration, but is absolutely required if intuitions are to be envisaged as parts of propositions, mental or otherwise.

An intuition is said to be commonly designated by the word 'this', and to have as its object a change that "just now takes place is us". This change is also said to be the "immediate hence unanalysable cause" of the idea.* It seems to me that the "changes" here alluded to are the nearest thing in Bolzano to the sensation-like ideas of the Empiricists. They hold almost no interest for Bolzano and are barely mentioned as causes and objects of intuitions.

Concerning the origin of judgments, Bolzano distinguishes immediate and mediated judgments. He makes the claim that all immediate judgments are infallible, but, characteristically, attempts to show that this is a purely conceptual truth which can be established without even citing an example of an immediate judgment.† Bolzano does not state the complete argument, but it can be reconstructed in the following manner: In the *Theory of Fundamentals* it was established that we know at least one truth. This knowledge must be either immediate or derived. In either case we must have some immediate knowledge. Assume that we know that not all but only some of our immediate judgments are true. But if we know this, then we must have *reasons* why some of them are true, and others false, and if we have such reasons, then the judgments are *not immediate*. Bolzano claims that if we hold some of them to be false, we must hold all to be false, since they all originate in the same way (namely without further reasons for their truth or falsity). If valid, the argument would establish only the fact that we are in the possession of a class of judgments that are infallible, but it does not identify these judgments. In casting about for likely candidates for that role, Bolzano, not surprisingly, settles upon judgments of the form either 'I have appearance *A*' or 'this (what I now see) is an *A*'. There must be at least some immediate judgments of this form, Bolzano argues, because "every derived judgment presupposes another of the same kind." Hence "if 'the intuition *X* is an *A*' is not an immediate judgment, then it must be derived from a pair of others, namely 'the intuition *X* is a *B*', and 'all *B* are *A*.' "‡ The judgment 'the intuition *X* is a *B*' is either immediate, or derives in the indicated fashion from another judgment of the same form. Eventually, we must arrive

* W.L. 286. † Cf. W.L. 42, 311. ‡ W.L. 300.12.

at an immediate judgment. What is remarkable about this argument is that Bolzano does not argue from the phenomenal character of some judgments to their immediacy and infallibility, but argues for the infallibility of some forms of judgment from the fact that we know anything at all.

All mediate judgments are based on, i.e. mentally caused by, immediate judgments. B. held that a given judgment p can be mediated by another, q, only if it is deducible from q, or q probabilifies p. He obviously envisaged the mind as some sort of machine which produces judgment according to certain rules.* (B. held the production of judgments to be involuntary. They are caused either by something that is presented to me, as in the case of immediate judgments, or by other judgments already present in the mind. I can only exercise a modest amount of control by the direction of my attention.) The rules which the mind follows in the production of judgments belong to, and constitute much of, the objective part of logic. In other words, B. assumed that judgments stimulate and produce one another in the mind in conformity with the objective rules of deducibility and probability. Errors can, of course, occur, but only if a judgment which is merely probable on the basis of previously accepted judgments is accepted without qualification and turns out to be false. I.e. we err when we forget that we derived a certain proposition through a probability argument, and proceed from 'the probability of p is m/n' to 'p' itself.

With this exception, thought processes and ratiocination are claimed to conform to objectively valid arguments, so that neglecting probability-riders is the only possible source of error. But what could be the evidence that if p and q are present in the mind, and r is stimulated by them (how do we know that r was stimulated by them?) then r is deducible from them? B. claims in effect that there is no evidence *against* this assumption, but it is fair to ask what such contrary evidence could be. It seems that evidence *against* B.'s claim would have to consist of an argument which is everywhere accepted as valid but which is objectively invalid, or, conversely, by an argument which is objectively valid but is generally rejected. In the nature of the case, this evidence, cannot be *produced.* Thus while there might be such arguments, they cannot be brought forth as *evidence.* Hence, Bolzano's claim, if it makes any sense at all, is not the kind of proposition which can be confirmed or disconfirmed.

It is important to notice in this connection that not every proposition that appears in the mind is a judgment, i.e. an *accepted* proposi-

* Cf. Jan Berg, *op. cit.*, p. 67.

tion. We can have the mere idea of a proposition in our mind without affirming it. Bolzano's theory as I sketched it above was concerned only with the origin of *judgments*; ideas, even ideas of propositions, can be generated in quite different ways, for example by association.

In many ways, Bolzano's theory of knowledge holds more interest than his logic, since the latter covers ground that has been gone over many times since, while the former contains a wealth of original insights which have not been closely scrutinized. Nevertheless, I am persuaded that Bolzano's basic epistemological assumption is unsound, the assumption namely that judgments must be considered to be manifest mental occurrences of a certain sort. But he has worked out the consequences of this assumption in greater detail than any other writer.

Heuretic and Theory of Science Proper

The fourth part of W.L. consists of a set of definitions and rules useful for the discovery of truths. It is only of minor interest, and I have included only the section on the discovery of causes to given effects because of its similarity to Mill's Canons. It can serve as an example for the rest of the Heuretic.

The fifth part of W.L. is the theory of science proper. The section of it included in this edition is not at all typical of that part, but is similar to the logical investigations of the Theory of Elements and forms an interesting extension of them.

The *Theory of Science in the Narrow Sense* describes in detail how a good treatise [*Lehrbuch*] is to be put together; it is little more than a manual of style. According to the definition of '*Wissenschaftslehre*' it is the actual purpose and consummation of the whole logical enterprise: Bolzano had defined Logic as that science "Which teaches us to compose other sciences (actually only their treatises)".* Yet in a letter to Romang, Bolzano advises that "the whole fourth volume can safely be laid aside . . .".† Bolzano's attitude will not appear paradoxical when we reflect upon the character of almost all of his work: most of his intellectual endeavours were directed not toward the discovery of new truths, but toward testing and analysing of what appears to be known already, and for seeking a solid foundation for accepted positions. In the same letter, Bolzano writes, "If I am asked to indicate briefly the essential difference between my own philosophical and theological concepts and those of others, I would say that it is the fact that I have tried with especial care to raise

* W.L. 1.   † From Winter, *Leben*, etc., p. 51.

everything I think to the level of clarity and distinctness. It is only in this way, only through precise definitions, that I arrived at the particular opinions and doctrines you find in my writings (even the mathematical writings)." Some of Bolzano's titles reflect the same tendency. One of his earliest writings bears the title *Contributions to a more Well-Founded Presentation of Mathematics*, and one of the last papers he projected was to have borne the title *Concepts of Geometry which Everybody Knows and does not Know*. His monumental *Religionswissenschaft* is not aimed at discovery, but toward more secure and rational proofs for already accepted tenets of Catholicism.

Similarly, W.L. ends by providing a set of rather uncontroversial rules for the construction of proper treatises, but these rules, though they may be generally accepted, require a justification which has now been provided. It is quite consistent with Bolzano's concept of a science that in some cases the supporting truths are more important than the final conclusions: There is no particular reason why, in the progression from grounds to consequences, the most important truths should occur last. The theory of science is itself a science in which the most important truths occur early in the development. Hence the preparatory studies are the truly noteworthy part of W.L.

The Text

W.L. was first published in 1837 by Seidel in Sulzbach, Bavaria. In 1882 Braumüller in Wien re-issued, in 12 volumes, all of Bolzano's works published earlier by Seidel. Winter* calls this a new printing, but it looks as if Braumüller had obtained the unsold signatures from Seidel and put his own wrappers on them. In any case, the 1882 edition of W.L. is precisely like that of 1837. The first two volumes were reprinted in 1914, and another corrected reprinting of all four volumes was made in 1929–31. The present edition is based on that printing.

Principles of selection. In Bolzano's letter to Romang, mentioned earlier, he gives advice concerning the reading of W.L. After advising Romang that he can lay aside the fourth volume, he continues† "You can omit the entire book titled *Heuretic* in the third volume, and the lengthy part on arguments in the second book. It will perhaps be sufficient if you read in the first volume §§ 19, 25, and 26 (on the

* *Leben*, etc., p. 96.

† *Ibid.*, p. 51.

concept of propositions and truths in themselves) 48, 49, 50 (about ideas in themselves) 55, 56, 57, 58, 63, 64, 66, 67, 68, 70, 72, 73, 79 (time and space), 85 (sequence) 87 (where you find the concept of infinity which I put in the place of the self-contradictory Hegelian absolute). From the second volume §§ 125, 127, 133, 137, 148 (analytic and synthetic propositions), 154–58 (propositions with variable parts, where I define the relation of deducibility: whenever *A*, *B*, *C*, . . . are true, then *M* is also true), 170 (propositions of the form 'a certain *a* has *b*'), 179 (propositions with if–then), 182 (the important concepts of necessity and possibility), 183 (time determinations), 197 (analytic and synthetic truths), 198 (concept of ground and consequence between truths), 201, 202, 214, 221 (basic truths). From the third volume all you need to read is what I say about the concepts of clarity and distinctness (280, 281), and then you may lay the entire logic aside, provided only that you look up one or the other item as the occasion requires; the index in the fourth volume will simplify this considerably."

With very few exceptions, these sections are fully contained in the present edition. I have included a good deal of material in order to give a reasonably complete survey of Bolzano's logical and epistemological doctrines; I have also made sure that all important definitions are present. Some sections were added because of their individual merit, for example, § 42 because of its general philosophical interest, and § 303 because it is an outline of a constructional system; there are only very few detailed attempts of this sort in the history of philosophy.

Many of the historical notes were included, but I could not aim for complete or even adequate coverage in that department, even though Bolzano's critical comments on his predecessors are a very prominent part of W.L. I have chosen the most important remarks, especially those dealing with Kant and the Kantians.

The Text. Two sizes of type have been employed. Any long stretch of literal translation, including Bolzano's notes, has been set in the larger type, while the smaller has been used for the paraphrases, including interspersed quotations.

Some of the paraphrases are quite detailed, containing lengthy literal quotations, and should give the reader an accurate impression of Bolzano's argument. Others give the merest hint of the contents of their sections. Still, I thought it appropriate to include them; this makes it possible for the present edition to serve as a guide to the complete German text.

The decision whether to paraphrase or give a literal translation was not always wholly determined by my estimation of the importance of a given passage. Some sections lent themselves more readily to abridgement than others. For example some of the sections in which B. develops his theory of deducibility (§ 155, etc.) are already so terse that abridgment would have been tantamount to omission, while other equally important sections were adequately represented by paraphrases liberally laced with quotations. Thus size of type is not always a reliable guide to the importance of a section.

Spacing for emphasis was replaced by italics, though in many cases the emphasis was deleted. Single quotes were used in order to indicate that reference is made to a word or an idea. Thus, where B. writes "*die Vorstellung: Gott*" I write "the idea 'God' ". Actually, B. does not have a uniform way of writing such expressions; accordingly I have supplied the marks as I saw fit. Double quotation marks are used in the paraphrases to indicate quotations from W.L. or from other authors.

The numbers in the headings refer to the original pagination, which is identical in all previous editions.

I provided the complete table of contents of the original through the book *Theory of Knowledge*. Sections reproduced in full or in substantial part are marked by an *; sections completely omitted do not have page numbers following them.

The subject index is based upon Bolzano's own index. The compilation of the author index was made much easier by the use of Schultz' index of authors in the edition of 1929–31.

Unfortunately, I did not have an opportunity to check Bolzano's quotations. Many of the books are quite rare and are accessible only in Europe. In the case of major authors, I have followed Kambartel and changed Bolzano's references to standard editions; e.g. the references to Aristotle are to the Bekker edition.

I have made several changes in the text in addition to the corrections made in the 1914 and 1929 editions, the most important of these are listed on page xlvii.

Page	*Line*	*Original version*	*Revised version*
100	33	ideas	objects
142	12	subjective	objective
206	20	*M*, *N*, *O*, . . .	*G*, *H*, *I*, *K*, . . .
222	8	ideas *A*, *B*, *C*, . . .	propositions *A′*, *B′*, *C′*,
222	18	*A′*, *B′*, *C′*, . . .	*M*, *N*, *O*, . . .
226	8	'every Y has z'	'every Y has x'
252	25	between 'should' and 'which' I have inserted: "what is meant is that Titus should not accept Gifts"	
262	14	*A* is . . . from *A* . . .	*M* is . . . from *A* . . .
284	28	*a* more complex than *b*	*b* more complex than *a*
351	5	immediate	mediated

BIBLIOGRAPHY

The most complete Bolzano bibliography is contained in Jan Berg, *Bolzano's Logic* (see below). I only list some recent studies and editions.

Bar-Hillel, "Bolzano's Definition of Analytic Propositions", *Theoria* 16 Lund 1950.

— "Bolzano's Propositional Logic", *Archiv für mathematische Logik und Grundlagenforschung* 1, 1952.

Berg, Jan, *Bolzano's Logic*, Stockholm 1962.

Bolzano, B., *Bernard Bolzano's Grundlegung der Logik* (*Wissenschaftslehre I/II*) Edited and with an Introduction by Friedrich Kambartel, Hamburg 1963.

— *Paradoxes of the Infinite*, ed. and with an introduction by Donald Steele, London 1950.

(A new complete edition of Bolzano's work in 39 volumes is shortly to commence publication under the editorship of Eduard Winter.)

Buhl, G., *Ableitbarkeit und Abfolge in der Wissenschaftstheorie Bolzanos.* Kantstudien, Ergänzungsheft 83, Cologne 1961.

Kneale, W. and M., *The Development of Logic*, Oxford 1962.

Scholz, Heinrich, "Die Wissenschaftslehre Bolzanos", *Abhandlungen der Friesschen Schule*, Neue Folge 6, 1937, pp. 401–72.

Winter, E., *Bernard Bolzano und sein Kreis*, Leipzig 1933, (2nd ed. forthcoming).

— *Leben und geistige Entwicklung des Sozialethikers und Mathematikers Bernard Bolzano*, Halle 1949.

BERNARD BOLZANO

Engraving by Schulz after a drawing by Kriehuber, published in *Starý Světozor*, 1868. (Picture by courtesy of the Museum of Czech Literature.)

INTRODUCTION

§ 1. *What the Author Means by Theory of Science*

1. Suppose that all truths which are now, or ever were, known to any man were somehow collected together, e.g. compiled in a single book; I would call such an aggregate the sum of all human knowledge. Compared to the immense domain of truths in themselves, most of which are altogether unknown, this sum is very small; but it is a large, even too large a sum for the mental capacity of any single man. For even the most capable man, his perseverance and favourable circumstances notwithstanding, would be unable to comprehend as much as the most important part of this sum, let alone everything that has to date been discovered by the common efforts of mankind. We must therefore agree to some sort of division. Since none of us can come close to learning everything that seems worth knowing from some point of view, each must concentrate upon some particular subject and must learn whatever is most necessary or useful for us under the circumstances. In order to make it easier to select and acquire what we need to know most (and for various other reasons as well) it should prove advantageous first of all to divide the total domain of human knowledge, or rather the domain of truths in general, into several areas; and then to take the most important of these truths and compile them in special books in the most comprehensible and convincing manner possible, adding, where necessary, further propositions as they are required for purposes of clarification or proof. I therefore take the liberty to call an aggregate of truths of a certain kind a *science* [*Wissenschaft*], if what is known of it is important enough to be set forth in a special book, in the way just mentioned. Any book which appears to have been written with the definite intent of recording all known and important truths of a science in a manner that makes it easy to comprehend and accept them with conviction I shall call a *treatise* [*Lehrbuch*] of that science. For instance, I shall call the class of all truths which assert something about the constitution of space the science of space (geometry), because these propositions form a separate species of truths; their known and important part unquestionably deserves to be recorded in special books, and ought to be accompanied by proofs which make them as comprehensible and convincing as possible. Such books I shall call treatises of geometry.

2. I admit that the meanings which I have given to the two words *science* and *treatise* are rather unusual. Let me point out that there

is no commonly accepted meaning at all for these two words, and that I shall not fail to justify my definitions later on. . . .

3. It is obviously not a matter of indifference how we proceed in dividing the totality of human knowledge, or rather the entire domain of truth in general, into the parts I have called sciences in (1), nor how these individual sciences are to be represented in particular treatises. One need not overrate in the least the value of mere knowledge to see that mankind is beset with innumerable evils simply because of ignorance and error, and that we would be incomparably better off and happier on earth if only each of us could acquire exactly that information which would benefit us most under the circumstances. This goal could of course not be achieved simply by dividing the domain of all truths into individual sciences and by making available a good treatise for each of them, but we would come significantly closer to it, especially if some additional measures were taken. (a) Everyone who has the necessary preliminary knowledge could instruct himself in the most reliable and complete manner about any subject in which he needs information, and could learn everything that is known on that topic. (b) If everything contained in these treatises were represented as clearly and convincingly as possible, doubt and error could be expected to diminish even in disciplines where passion makes it difficult to accept the truth: in religion and morals, especially since (c) the more widespread study of certain sciences from well written treatises would engender greater skill in right reasoning. (d) The discoveries which have so far been made would certainly lead to many new discoveries, if they were only more widely known among us; hence it is obvious that the benefits from these measures will spread in the course of time rather than diminish.

4. It should be possible through some reflection to find the rules which we must follow in dividing the total domain of truth into individual sciences and which must govern the writing of the respective treatises. There can also be no doubt that the sum of these rules itself deserves to be called a science, since it is clearly worth while to collect the most important part of them in a special book, and to order them and provide proofs for them so that everyone can understand and accept them with conviction. I allow myself to call it the *theory of science* [*Wissenschaftslehre*], since it is the science which teaches us to represent other sciences (actually only their treatises). Thus, by theory of science I mean the aggregate of all rules which we must follow when we divide the total domain of truths into individual sciences, and represent them in their respective

treatises, if we want to do a competent piece of work. Actually, it is evident that a science which is to teach us how we should represent the sciences in treatises must also teach us how to divide the total domain of truths into individual sciences, since a science can be properly represented in a treatise only if its boundaries are well defined. Hence we can express our definition of the theory of science more briefly by saying that it is the science which instructs us in the representation of sciences in adequate treatises.

§ 2. *Justification of this Concept and its Name*

Since, according to § 1 the introduction of a new science is not a slight matter, B. wishes to give a justification for the theory of science itself. At this stage he can give only common-sense reasons, since only the theory of science itself will indicate how the sum of all truths is to be divided.

B. takes it to be evident that there are at least some truths concerning the proper presentation of sciences in general, and that it would be of benefit to have them collected together. "One may perhaps doubt that the proposed theory of science is *possible*. Since, according to the above definition, the theory of science is to teach us how the sciences ought to be represented, and since it is itself a science, one might ask how it can be produced, since one does not know how any science is to be represented as long as there is no theory of science. This doubt can easily be removed. One can proceed according to the rules of the theory of science and thus generate any number of sciences (or rather their written expositions), among others the theory of science itself, without being clearly aware of these rules; one can find through reflection many or all of these rules without ordering and connecting them, as it must be done in the treatise of a science. Once these rules are known, every science, including the theory of science itself, can be further elaborated and represented in writing. This amounts to no more than arranging certain known truths in an order and connection which they themselves prescribe." But even though the theory of science is possible, it may not be worth while; the concept could be too narrow: should it also cover instruction in the *discovery* of truth, and its *oral presentation*? The latter concern is said to belong properly to didactics, an already existing science, and the former to be quite compatible with the given concept, "since one cannot present a truth until he has first found it".

Finally, B. justifies his use of '*Wissenschaftslehre*' against Fichte and Bouterweck by claiming that his use of the word seems a natural one.

§ 3. *The Author's Theory of Science is a Science which has Long been Known and Pursued under Different Names*

"If the science which I have just described, and upon whose independent recognition I have insisted, is really as useful and necessary as I maintain, then it is hard to believe that it should so far have been ignored. In my view this has indeed not been the case; rather, I believe that this science has constantly engaged the attention of all philosophers, starting from the age of the Zeno the Eleatic, or at least from the time of Parmenides, to this day. In my opinion, the subject of all the countless writings with titles such as *Canonic*, *Dialetic*, *Topic*, *Logic*, *Heuristic*, *Organon*, *Dianoiology*, *Ideology*, . . . *Way to Truth*, *Way to Certainty*, *Therapy of the Understanding*, and many others was the just discussed theory of science."

In the remainder of § 3, B. gives a wealth of quotations to establish his point; for example, he claims that "Aristotle speaks explicitly of the need for a special science which would teach us how to proceed in the presentation of sciences in general 'since it is absurd to study method and subject matter at the same time' (*Met.*, 995^{a}13). Hence must he not have thought that he had put forth the science of methodology, and where could he have done so but in the *Organon*?"

§ 4. *Why this Definition has so far not been Given*

B. gives four reasons why logic has so far not been defined as the theory of science:

1. The actual instruction how a science is to be presented must be preceded by a great many doctrines of diverse kinds. It was generally overlooked that it is the purpose of the entire enterprise, though it forms but a small part of the whole.

2. It has become an almost universal practice to teach logic only to adolescents in order to instruct them in correct thinking. Although this is meritorious, it detracts from the correct definition of logic, since its most essential part, namely the doctrine of scientific exposition, is generally omitted under these circumstances.

3. Since the definition of 'logic' requires a prior definition of 'science', and since this latter concept is controversial, other, less perspicuous, definitions of 'logic' tended to be favoured, e.g. *logica est ars cogitandi*.

4. In more recent times, Kant has emphasized that logic should not be thought of as an organon. "The definition that logic is a theory of science has much similarity with the decried doctrine that logic is an organon . . . and if now and then somebody dares to imply that logic is a theory of science, he does it in the foreword or some other place where it is less conspicuous than in the definition of logic itself."

§ 5. *What the Author Thinks of these Reasons*

Against 1 and 2 of § 4, B. maintains that the extensiveness and importance of the preliminary investigations do not militate against defining logic as the theory of science.

Against 3 he points out that there are many other sciences whose concept cannot be developed except within these sciences themselves, e.g. jurisprudence is the science of law, but the definition of law must be given within jurisprudence itself.

Finally, against 4, B. remarks that "the assertion that logic is a theory of science, or even an organon, is not in the least objectionable, if it is taken to mean that it establishes the rules according to which the total domain of truth is to be divided, and each science has to be treated. It would only be false if one were to imagine that this science had to contain the first principles upon which the edifice of all the other sciences is to be erected. Logic does not treat of the principles which lie at the basis of each science, but it deals with the procedure which has to be followed in their exposition."

§ 6. *The Author will Generally Call his Theory of Science Logic*

After what has been said, I take it to be established that the science whose concept I have introduced in § 1 under the name *theory of science* is essentially the same as that which has been known and pursued under different names, but most commonly under the name *logic*. Because of this, I shall from now on only rarely use the term *theory of science*, though it is very descriptive, but rather employ the name *logic*, which recommends itself through its shortness and flexibility.

§ 7. *Examination of other Definitions*

Since the definition of logic which I here give differs from the definitions of others, it is only just that I should mention them, and indicate briefly why I am not satisfied with any of them.

1. One of the most common definitions states that logic is the doctrine or *science of thinking*. Thus Kant says in his *Logik* (A 4): "The science of the necessary laws of understanding and reason in general, or of the mere form of thinking, is logic." A similar definition is found also in Kiesewetter, Krug, Tieftrunk, Calker, Esser, Roesling, Sigwart, and many others. It seems to me that all these definitions, if they are to be interpreted literally, are too wide. For no matter how we proceed with the business of thinking, at whatever conclusion we may arrive, or merely intend to arrive, whether we find truth, or entangle ourselves in errors; whether we seek truth or take pains to deceive ourselves, or do neither of them but entertain merely for pleasure some idea or another without believing that anything corresponds to it: do we not in all these cases follow certain *laws* or *rules*? For example, if we want to deceive ourselves, do we not have to follow the rule that we should withdraw our attention from the principles of truth and direct them toward the pseudo principles which belong to the opposite error, etc? Could not any description of rules of this kind be called a doctrine of thinking, of the laws and rules of thinking? Could it not even be called, if you will, a science of the law-governed use of reason and understanding?

Yet such a collection of rules would be quite unlike what we all think by logic. If one were to attempt to save the above definitions by restricting the laws in question to those which agree with the purpose of our faculty of understanding, then I should have to demand a closer interpretation of these words: if we restrict ourselves to laws which hold absolutely, i.e. without presupposing an arbitrarily adopted purpose, then we are concerned with moral laws whose development belongs to ethics. As a consequence of this definition, all of logic would be changed into a chapter of ethics concerning the dutiful use of the faculty of understanding. This is M. Damiron's serious intention (in his *Cours de Philosophie*, 3 vols., Paris 1831–1836). But how many rules of which so far in logic nobody has dreamed would he not have to admit into his exposition! If, on the other hand, we were to admit only laws which spring from a chosen purpose, then a statement of this purpose is necessary for the completeness of the definition. If this purpose were the recognition of truth, then the definition would be changed into one which I shall examine in the sequel.

2. Some authors who seem to have felt the great vagueness of the above definition declare logic to be a *doctrine concerning the education of our faculty of understanding*. Thus, Reusch says in his *Systema Logicum* (ed. Polzius, Jena 1760, § 99): "Logic is the science of the perfection of the cognitive faculty by appropriate means." The same concept was already expressed by Clauberg, though less clearly (*Logica Vetus et Nova*, 3rd ed., Sulzbach 1685): "Logic is the art of shaping our reasoning", etc. Through this definition all those rules for the use of the understanding which would impede its perfection are banished from logic; on the other hand, a great many subjects which are quite alien to logic will now be drawn into its field. There are very many different ways of improving the faculty of understanding which are discussed, or at any rate should be discussed, e.g. in educational science, in medicine, morals, political science, and many others, but which would be out of place in logic. Does, for instance, the question whether coriander improves one's memory belong to logic? And yet, it most certainly would, if logic were to be an *ars rationis formandae* in the full sense of the word.

3. More precision was achieved when logic was defined as *the science of those laws which we have to follow in thinking if we want to find the truth*. This was roughly the sense of Keckermann's definition: "Logic is the art of ordering and directing the operations of the human intellect or human cogitation toward the cognition of things" (*Praecognitorum Logicorum Tractatus*, Hanover 1606, I, ch. 2).

Others have expressed this even more clearly, e.g. Wolff: "Logic is the science of directing a faculty toward true cognition" (*Logica*, 3rd ed., Frankfurt and Leipzig 1740, § 61), likewise Gaudin, Crusius, Darjes, Miotti, Ulrich, and many others. I admit that the rules which one should observe in thinking in order to arrive at a recognition of truth, can be, and in a sense even should be, discussed in logic. For even though they do not essentially belong to the contents of logic, according to my definition, they still are so closely connected with it that their inclusion can in no case be condemned. Yet if the purpose of logic is to teach how sciences should be represented, then it would, strictly speaking, already fulfil its duty if it only gave the rules which tell us whether a given truth belongs in this or that science, and in which order and connection it should be represented there. However it would clearly be commendable if logic also gave instructions about the important problem of how those truths can be found in the first place. Indeed, since we do not have a special science, distinct from logic, which aims at solving this problem, it is reasonable to make it a duty for the logician that he should at least not remain silent about a matter of such importance. But no matter to what extent one may hold logicians responsible for the investigation of this question, the above definition of logic is still mistaken. In my opinion it is shown to be false by the mere fact that it does not indicate why the rules of scientific exposition should constitute such an extremely important part of logic, even its ultimate goal. For if logic is defined as the doctrine of the rules which our thinking must follow if we would find the truth, then its ultimate goal is obviously no other than—discovery of the truth. However, the synthesis of already discovered truths into a scientific whole contributes but little to this purpose. It may indeed help others a good deal in grasping the truth, but if logic only gives the rules which one has to observe with one's *own* thinking, in order to find the truth, then a description of the methods by which an already discovered truth should be presented to *others* lies outside its purpose.

It follows that logic should give the rules of scientific exposition only to the extent in which a scientific compilation of already discovered truths is a means which may occasionally lead us also to some new truths. Now I do not wish to deny that a truly scientific representation promises this advantage, but who could believe that this is the only reason why the art of scientific exposition should be taught at all?

4. Others have seen the importance of this art and its merit for all instruction, and have been prompted to define logic as the doc-

trine of exposition in general. It seems that Cicero already had this notion of logic, since he mentions it under the name "accomplished skill of discussion" (*Topica*, cf. also bk. II of *de Oratore*). In this context also belong the definitions of *Melanchthon*: "dialectic is the art of teaching" (*Dialectica*, 1536), George of Trebizond (*De Re Dialectica*, Cologne 1536) Fonseca (*Institutionum Dialecticarum*, Cologne 1623, bk. 8), Peter Ramus, of the authors of the *Ars Cogitandi* (new ed., Basel 1749), Hollmann (*Logik*, Goettingen 1746) Watt (*Logic*, 12th ed., London 1763), and many others. If these definitions were valid, then the content of logic would have to be much larger than it is. It would not only have to deal with concepts, judgments, inferences, and the order and connection in which truths must be represented in a book, but also with the problems of *oral* presentation of such truths; and not just one or the other kind of oral presentation, but every kind that might be useful under certain circumstances. Hence, it would have to contain, for example, rules for discourse with children, congenitally blind or mute persons, etc. But investigations of this sort have so far never been conducted in logic, nor is it desirable to unite into one science the rules that govern the written representation of sciences as well as those that govern instruction, i.e. theory of science and didactics.

5. Salomo Maimon (*Versuch einer neuen Logik oder Theorie des Denkens*, Berlin 1794, Foreword, p. xxxi) defines *philosophy* as the science whose subject is "the form of a science in general". In other words, it would seem to be precisely what I call theory of science and others call logic. But since he means by this form "the absolute first principles of human understanding", the matter is indeed quite different. Logic itself he defines in § 1 as the science of the thinking of an object in general, which is not determined through inner characteristics, and determined only in its relation to its conceivability. He holds that this is like the "general doctrine of quantities", which "investigates all possible forms in which quantities can be thought, irrespective of whether their application will lead to determinable, undeterminable or even impossible quantities." In my view, it is not quite precise to say that the general doctrine of quantities (arithmetic, algebra and analysis) is not concerned whether the forms of quantities which are here considered lead in their application to determinable, undeterminable or even impossible quantities. Rather, the investigation which is conducted with respect to a certain form of quantity, e.g. $\int \frac{dx}{\log x}$, consists mostly in investigating whether and how they are determinable. Moreover, one must

not confound the mere conceivability of the thing with its *possibility*, not even with its so-called *inner* possibility, which is opposed to the self-contradictory. For even the contradictory, for instance a square circle or $\sqrt{-1}$, is conceivable, and we do in fact think it whenever we speak of it. Something is inconceivable only if and in so far as we have no representation of it at all; as red colour may be inconceivable for somebody who was born blind. One can also see from this example that the mere conceivability or inconceivability of things is only very rarely considered in logic, and it certainly does not constitute its only content. But even if we replace the word conceivability by inner possibility (or consistency), it is quite wrong to say that logic does not concern itself with anything but the laws of inner possibility. It is true that there are several rules of logic which, if we observe them, keep our assertions from coming into an inner contradiction with each other; but rules of this sort do not constitute the total content of logic. Finally, I do not understand how it can be said that in logic we think the objects "altogether indeterminate as to their inner characteristics." For, if we think an object as altogether indeterminate, then we cannot assert anything of it. This whole definition probably stems only from the fact that in the examples which are used in logic, e.g. in the syllogism: all *A* are *B*, all *B* are *C*; therefore all *A* are *C*, the signs *A*, *B*, and *C* may mean, as we say, "anything". But this statement is not altogether precise. The signs *A*, *B*, and *C* can here mean very different things, but not quite anything we may choose. They must signify ideas such that *B* is an idea which can be predicated of all *A* and *C* one which can be predicated of all *B*. Thus it can be seen that the objects *A*, *B*, and *C* are not left indeterminate as to *all* their characteristics, but only as to some of them. There are many similarities between this definition by Maimon and that of Twesten, namely that logic is the theory of the application of the axioms of identity and contradiction (cf. Twesten, *Die Logik*, Schleswig 1825).

6. While, in Germany, philosophers attempt to eliminate all empirical content from the province of logic, in France they try to make logic into an altogether empirical and subjective science, into some kind of empirical science of the mind. One of the latest and most original students of this science in France, Count Destutt deTracy, says in the third volume of his *Elémens d' Idéologie* (2nd ed., Paris 1818, ch. 1, p. 124), "Logical science consists only in the study of our intellectual operations and their effects. The theory of logic is nothing other than the science concerning the formation of our ideas and of their expression, of their combination and deduction.

In a word, it is nothing but the study of our tools of knowledge."

According to this definition, logic would be the *science of the way in which we attain our knowledge*. Among the Germans, it seems that especially Ernst Platner thought of something similar when he called logic in the widest sense a pragmatic, i.e. critical, history of the human faculty of understanding (*Philosophische Aphorismen*, 2nd ed., Leipzig 1800, § 21). Beneke, too, maintains that logic, as the science of thinking, is given the task of describing completely and clearly the form and origin of the development of our thought (*Lehrbuch der Logik*, Berlin 1832).

If not even methods for separating truth and falsehood belong to the essential content of logic, it will be even less important that it should inform us how some piece of knowledge originated within us. It would be completely wrong to believe that the latter is a necessary prerequisite for the former; that is, that we could not ascertain at all whether any one of our opinions is true as long as we do not know how it originated. Krug says very aptly about this point (*System der theoretischen Philosophie*, Königsberg 1806–1810, part 1, § 8, note 1), "It is not at all necessary to know how thoughts originate in order to find out how they must be treated in their relation to one another." (To this treatment belongs obviously also the judgment of their truth and falsity.) "There are many things in this world whose origin we don't know, and yet are capable of handling very aptly." However, I admit that the investigation of the origin of our knowledge is a very deserving undertaking in logic, since it calls our attention to the source of ever so many errors and thus allows us to avoid them. . . .

In 7, B. rejects Reinhold's definition, that logic is the doctrine of the forms of thought. He will give a full discussion of this viewpoint in § 12.

In 8, B. rejects as too narrow Herbart's definition that logic is concerned with the clarity of concepts.

9. Quite different from the hitherto existing opinion was not only the *definition* but also the *concept* of our science which Hegel has brought forth, saying that it "is to be taken as the system of pure reason, as the realm of pure thought, in general, as the pure science which presupposes the liberation of the antithesis of consciousness, and which contains the thought in so far as it is identical with the thing and the thing inasmuch as it is identical with the pure thought." I confess that I have never succeeded in discovering any sense in this definition. For the thought of a thing and it, the thing itself, which is thought in such a thought, are in my view always different from one another, even in the case in which the thing about which we think is itself a thought. For even in this case, the thought of my

thought is not the same as the former, but another thought. Therefore, I do not understand how one can say that logic contains the thought inasmuch as it is identical with the thing, and the thing inasmuch as it is identical with the thought.

10. Twesten (*Die Logik*, Schleswig 1825) maintains that logic in the received sense, which he wants to uphold, "is the theory of the application of the principles of identity and contradiction." I am of the opinion that these two principles do not even allow us to deduce the few rules of syllogistics which Aristotle gave.

In 11 and 12, B. criticizes the definitions of Troxler and Umbreit as incomprehensible.

In a note, B. discusses Fichte's conception of logic. Fichte's *Wissenschaftslehre* was to be a doctrine of knowledge in general, consisting of intuition and thought, while logic was to be only part of this, namely the doctrine of thinking. B. objects to this on the grounds that such distinctions as that between intuition and concept, experience and *a priori* truth must be discussed within logic. He then takes the opportunity to confess "that I belong to those people on whom the *entirely new sense* which Fichte demands for the understanding of his *Wissenschaftslehre* (Berlin 1810, p. 4) has not dawned, although I have taken some pains to do everything that Fichte demands from his readers for that purpose. In other words, I believe that Fichte is mistaken, and I believe also that I understand what may have led him to make some of his erroneous assertions. However, I can do this only with very few of his statements, while a great many other utterances of his sound so extraordinary to me so that I doubt whether I even understand their correct meaning. Therefore, much is lacking to assure me that his system is false. As a tribute to truth, I must make a similar confession with regard to Schelling, Hegel and other thinkers who philosophize in a similar way. This holds even with regard to Herbart, which may here be said once and for all."

§ 8. *Some Concepts Related to Logic*

B. discusses the various concepts that are associated with the word 'logic', i.e. his own concept, logic book, logic as a class of opinions that somebody holds, innate and acquired logic. Furthermore, skill in following logical rules is distinguished from knowledge of these rules, and a distinction is made between logical talent and logical accomplishment.

§ 9. *The Uses of Logic*

B. states against Hegel that the knowledge of logical rules will make reasoning more certain than their mere intuitive application. This knowledge will be of especial aid in the avoidance of fallacies. He also points out that one cannot hope to be able to transact certain very difficult sciences, such as metaphysics, without being conscious of the rules of logic. The usefulness of logic is enhanced if it is thought capable of perfection. "It is one of Kant's literary sins that he attempted to deprive us of a wholesome faith in the perfectibility of logic through an assertion very welcome to human indolence, namely that *logic is a science which has been complete and closed since the time of Aristotle*. It seems to me that it would be much better to assert as a kind of practical postulate that faith in the constant perfection not only of logic but of all science should be maintained."

§ 10. *Time and Preparation for the Study of Logic*

B. suggests that the most appropriate age for the study of logic is the later years of adolescence.

§ 11. *Whether Logic is an Art or a Science*

A science that consists mostly of rules of conduct is called an art. An art in the narrower sense is one where the successful application of the rules requires special training. Logic could then be called an art in the narrower sense, but since every art is a science, it would also be a science.

§ 12. *Whether Logic is a Purely Formal Science*

1. Most contemporary treatises of logic state that "in logic not the *content*, but the mere *form* of thought is to be investigated". For this reason it is said to be a purely *formal* science. The meaning of this somewhat obscure expression is explained in different ways. Thus Jakob, in his *Grundriss der allgemeinen Logik* (2nd ed., Halle 1791, § 62) says "It [logic] abstracts from all the differences of objects and considers merely the ways in which the understanding thinks and must think them. Therefore it is a purely formal science." On the other hand, *Hoffbauer* says in his logic (*Anfangsgründe der Logik*, Halle 1794, § 11) that "the content of a thought is that in it which corresponds to the thing that is thought of. Its form is that in it which is created through thinking. The contents of thought are the ideas from which thoughts can be generated, the form of thought is the way in which this happens." "Pure logic", he says in § 17, "is the science of the form of thought." *Metz* says in his *Handbuch der Logik* (2nd ed., Würzburg 1802, p. 4) that in logic we must abstract from the differences between the objects of thought as well as from the thinking subject, and must concentrate only upon thought as such, *in abstracto*. Therefore only the form of thought can here come under consideration. This form, he says, is "that through which representing becomes thinking, and this is the determination of the ideas (matter of thought) through the unity of consciousness". Krug (*Fundamentalphilosophie*, Züllichau 1803, p. 332) says that formal

thought, which makes up the subject matter of logic, consists in the fact "that the ideas are related only to one another without any regard for the objects to which they might correspond"....

2. It seems to me that these definitions are either erroneous or too loose to settle the matter. Let me then develop my own opinion.

Scholars who make claims like those above tacitly assume that the objects forming the field of logical enquiry must all be thoughts; that is, if they are nothing else, they must at any rate be thoughts. But this almost universal assumption might very well be false. The history of science knows many examples where the field of a science was broadened, which amounts to an admission that it had been too narrow. For example, geometry began as the science of land measurement, but was gradually extended to become the science of space in general. Similarly, it might very well be the proper task of logic to discover more than the laws of *conceived* truths (or true thoughts), namely the laws that hold for truths as such. If it were the case that the validity of logical rules extends not merely over propositions that are thought, but over all propositions, irrespective of whether they have been thought by anyone, then the subject of logic would be too narrowly circumscribed if it were not extended beyond thoughts to propositions in general. I hope to demonstrate in the sequel that this is indeed the case, and that the source of most errors in logic has been the lack of distinction between thought truths and truths in themselves, and between thought propositions and thought concepts on one hand, and propositions and concepts in themselves on the other.

But even if this were not agreed upon, there would still be unanimity that logic should state only the general rules for the pursuit of a science and need not specifically consider any particular truth of a given science. Thus the truth that space has three dimensions will at most occur as an example. It is the task of logic to give rules which apply simultaneously to several truths or, what amounts to the same, to a whole class of truths.

For this reason, the theorems (though perhaps not the examples) of logic never concern a particular, fully determinate proposition, i.e. a proposition in which subject, copula and predicate are all given. Rather, theorems concern a whole class of propositions at once, i.e. propositions some of whose parts are determined, while the remainder is undetermined. Thus the proposition "some people have white skin" occurs in logic at best as an example, and not as the subject of a theorem, while a class of propositions, such as the class determined by the expression 'Some *A* are *B*' may well be the subject of a theorem. If these classes of propositions are to be called general *forms* of propositions, then it is permissible to say that logic is concerned with forms rather than with individual propositions. (Actually, only the written or oral expression 'Some *A* are *B*', and not the class itself, should be called a form.)

Furthermore, if we want to call what is left indeterminate in such a class of propositions the *content* of the propositions in the class, such as the *A* and *B* in the above example, then we may say that logic is concerned merely with the form, and not the content, of propositions. (Actually, the parts of these propositions which are determinate have, in certain respects, the same claim to be called content as do the parts left indeterminate.)

I should not object to calling logic a formal science, if what is intended is this feature. I don't think it is necessary to guard against the misunderstanding that logic itself, being called formal, has no contents, and hence, since only fully determinate propositions can be true, does not contain any truths. This would be so outlandish an error that it is unlikely to be committed. Although determinate propositions are not the *objects of* the theorems of logic, yet these theorems themselves *are* propositions. One must distinguish between the *objects* of a science (what it deals with or treats) and the *content* of a science (its theories). For example, the object of geometry is space, the contents of geometry are propositions about space.

3. However, it seems that some who have called logic merely a formal science had something different in mind, or else they did not draw the proper conclusions from their conceptions. Otherwise they could hardly have said that logic "has to abstract from all differences between objects". Logic must indeed *attend* to the differences between possible objects of thought in so far as they influence the rules for the investigation of these objects. For example, logic must attend to the difference between truths which can be discovered only by experience, and those that can be found independently of experience. Otherwise it could neither describe nor justify the difference between the two methods of investigation.

One sometimes has the impression that some scholars want to avoid certain difficult investigations by hiding behind the vague expression that logic has to abstract from all contents of thought. For example, in the theory of judgments, even more than in the theory of concepts, many a happy distinction and formulation which did not fit into the cherished framework of the categories was rejected on the grounds that it rested upon the content rather than the form of the objects in question. I merely mention the division of judgments into analytic and synthetic. But this criterion could not have been used, if the formal character of logic had been understood in the above-indicated way. For, what is distinguished in any such division is not individual propositions, but whole classes of propositions. Thus, logic does not cease to be a merely formal discipline in the above sense, even if it makes distinctions of this sort.

4. B. now suggests that the following could have been meant by the distinction of formal and material elements: a certain kind of proposition or concept was called formal if they could be characterized by attending only to some of their constituents, and not to others. Thus the distinction between affirmative and negative was said to concern only their form, since it depends on only one of their parts (according to most theories the copula), while the division of propositions into *a priori* and *posteriori* is said to depend upon the total content. If this is what is meant by the formal character of logic, then "the restriction of logic to mere form is arbitrary and detrimental to science".

5. B. criticizes the view (Fries) that logic is the system of analytic judgments. Analytic judgments, such as 'an equilateral triangle is a triangle' are too unimportant to be the peculiar contents of any science.

§ 13. *Whether Logic is an Independent Science*

Some logicians, among others Metz (*Handbuch der Logik*, Würzburg 1802, § 37), declare logic to be "an independent science . . . through which alone the systematic study of other sciences can be approached." This flattering view tends to draw applause from logicians. Yet, there is no scarcity of logicians who explicitly or implicitly assume the contrary. Krug asserts that logic, like all other philosophical disciplines, is founded on what he calls fundamental science (*Logik*, Königsberg 1806, § 8, n. 2). Others think it depends on psychology or even on metaphysics. My own view may be gathered from the following:

1. In § 1, no. 1, above, I have said that a science is nothing but a class of truths of a certain kind. It follows from this that a given science itself will not contain truths of another kind, but every handbook of that science must contain quite a few additional truths, for example, truths that are required as lemmas to prove the theorems of the science in question. Such lemmas may well belong to another, already existing science, and may be said to be *borrowed* from that other science. In such a case, the borrowing science will be said to be *dependent* on the other science. For example, the general theory of space (geometry) depends in this fashion on the general theory of magnitude (arithmetic, analysis) since treatises of geometry contain certain truths, indispensable for the proof of geometric

theorems, which deal with magnitude in general, and not specifically with space, e.g. the proposition that equals added to equals will be equal.

If one adheres to the above definition of science, it is obvious that only very few sciences are altogether independent, and that only those are independent whose essential doctrines can be proved from premises indigenous to that science. One can guess in advance that logic is not one of these. To demonstrate the rules which are its essential content we must use as premises many truths that are not rules, and are therefore not part of the essential content of logic. If any such premise is a theorem in some independently developed science, we must say that logic is dependent on that science.

To show how many sciences are auxiliary to logic, one would have to decide first how many sciences there are and what their boundaries are. It is premature to do this, so I shall here content myself with the following: Logic is to teach us rules by which our knowledge can be organized into a scientific whole. To do this, it must also teach us how truth may be found, error discovered, etc. It cannot do the latter without attending to the way in which the mind acquires its ideas and knowledge. The proofs of its rules and theories must therefore make reference to the faculty of representation, to memory, the association of ideas, imagination, etc. But the human mind and its faculties are the subject of an already existing science, namely empirical psychology. From this it follows that logic is dependent at least on psychology, and that it must forego the reputation of being a completely independent science.

2. B. does not think that Logic is dependent upon Krug's "fundamental science", whose task it is to stake out the areas of the special sciences, and which is to contain the first principles of all other sciences. The former task B. wants to award to logic itself; as concerns the latter, he does not think it reasonable to create a science of this sort; it would be "reminiscent of a building consisting of nothing but footings".

In a note he indicates that he is not inclined to share the concern of some logicians whether logic is a branch of philosophy. This can be decided only after we have an answer to the question 'what is philosophy?'.

§ 14. *General and Special Logic*

General logic is concerned with rules governing all sciences, special logic with rules governing only one or several special sciences. The present work is concerned with general logic only.

§ 15. *General Outline of this Treatise*

It is desirable that the theory of science proper should be preceded by a discussion of rules to be followed in the discovery of truths: heuretic. Heuretic seems to require an antecedent discussion of the general conditions or human knowledge: epistemology. Epistemology can be fruitfully developed only if it is preceded by the theory of ideas, propositions and deductions: the theory of elements. The latter will be preceded by a theory of fundamentals in which it is proved that there are truths and propositions in themselves.

§. 16. *Some Remarks about the Plans Followed in the most Important Modern Treatises*

Most current logic books, according to B., divide logic into a pure and an applied or empirical part. The former deals with rules that hold for all rational beings whatever, the latter with rules that hold specifically for man. In both parts, however, a proposition is always envisaged as somebody's judgment, and a truth as a true judgment. But if an exposition of logical theory starts out by defining "ideas, propositions and truths as phenomena in the mind of a thinking being, it cannot possibly arrive at a true notion of the connections that obtain between truths in themselves. It will continue to confuse this connection with relations between experiences." Where logicians have paid attention to the relations between objective truths, it was the result of a happy inconsistency.

But even if logicians had consistently talked about mental phenomena only, their procedure would nonetheless have been incoherent. For if pure logic deals with the laws of thought that hold for *all* beings (even God) one must assume that these laws are none other than the conditions of truth themselves. "Hence it is quite superfluous to speak of laws of thought when we could just as well deal with the conditions of truth itself." Moreover, the only way in which we can discern that a given law of thought holds for all beings whatever is by discovering that it is a law that holds for truths in themselves.

Even those logicians who wish to restrict themselves to laws of human thought cannot maintain their position without difficulty, for they would wish to hold "that the circumstance that these laws truly hold for humans should be objectively true, and not merely appear to be so. It is astonishing that some laws of thought, including complicated empirical laws of association of ideas, are said to hold of men objectively while objections are made to the view that propositions such as 'what is is, and what is not is not' are something more than mere necessities of human thought."

While the so-called doctrine of elements does not generally treat representations, propositions and truths with sufficient abstraction, the opposite objection can usually be made to the chapters on methodology. Since handbooks are meant for people, not angels, methodologies must contain sufficiently detailed and concrete instructions. They should, for example, contain a good deal on definitions, though "definitions are obviously not required for every thinking being, but only where beings are addressed who, like men, have confused ideas and use symbols to aid their thought".

In a note, B. criticizes Hegel, the first part of whose logic deals with being and the second with thought. This is an extreme reaction to the customary one-sidedness: instead of advancing from a discussion of thoughts to a discussion of truths and propositions in themselves, which would be the "next higher thing", he had the extravagant idea to advance to the laws of being in general. "This confusion (for that is what I take it to be) was greatly facilitated because the concept of a truth in itself has been almost completely forgotten, and is often confused with the concept of being as such."

BOOK ONE

THEORY OF FUNDAMENTALS

§ 17. *Purpose, Content, and Divisions of this Book*

B. points out that it is very often the case that we reverse our judgment on some matter, and that there is hardly any assertion that was not doubted by someone. This may lead us to wonder whether there is truth in any of our judgments. He takes such a general scepticism to be "an unfortunate mood" and an "aberration", and suggests that a work in logic is a good place to dispel such doubts. To do this, he will first define the concepts of proposition and truth in itself. Then he will show that there are a good many, and not just à few, of them. Finally, he will demonstrate that at least some of them are known, i.e. "that not all judgments are false".

§ 18. *Refutation of some Objections*

A sceptic who doubts his own and everyone else's existence is not very likely to take up the study of this book, and even if he does, it will not force him to accept its reasons. B. replies that he cannot force a sceptic to accept his reasons, but that there may very well be many a sceptic who does not like his affliction. If such a one were to take up this book "in just about the same way in which he uses food and drink, even though he doubts their actual existence" he could be benefited by the encounter.

But if anything is proven to a sceptic, the proof must not presuppose anybody's existence. But how could one talk to him without making this presupposition? B. replies that "it is true that a perfect sceptic believes neither in his own nor in anybody else's existence. We will therefore never be able to convince him of a specific truth, if we *presuppose* in its proof his or somebody else's existence. By 'presuppose' I here mean that the assertion of his existence is introduced as one of the *premises* of our proof. But, as a matter of fact, by merely speaking to a man or trying to agree upon the meaning of certain words, this is not done. For there is a difference between *acting* according to a certain presupposition while presenting a proof, and using this presupposition as a premise in the proof. A mathematician who uses certain drawings and drafting equipment when he presents a proof acts on a number of presuppositions which are taken from mechanics, optics, and other sciences; but should one therefore say that the theorems on which he here acts appear as premises in his proof? and that everybody who is to find his proof convincing is forced to admit them as true? Surely this cannot be maintained and neither can one say that a doubter must first faithfully presuppose his own and our existence if the subsequent truths are to appear convincing to him, merely because we appear as addressing him when we demonstrate them to him."

PART I

Of the Existence of Truths in Themselves

§ 19. *What the Author Means by a Proposition in Itself*

I wish to show as clearly as possible what I mean by a *proposition in itself*. In order to accomplish this, I want to define first what I mean by a *spoken* proposition or a *proposition which is expressed in words*. With this name I wish to designate any *speech act*, if through it anything is asserted or expressed; that is to say, whenever it is one of the two, either true or false in the usual sense of these words; or, as one can also say, if it is either correct or incorrect. (Such speech acts will usually consist of several words, but occasionally only of one.) Thus, for instance, I call the following sequence of words a spoken proposition: 'God is omnipresent'. For through these words something is asserted; in this case something true. But I also call the following sequence of words a proposition: 'Squares are round'. For through this form of words something is also stated or asserted, although something false and incorrect. On the other hand, I do not call the following expressions propositions: 'The omnipresent God', 'A round square'. For through these expressions something is indeed *represented* but nothing is stated or asserted. Consequently one can, strictly speaking, neither say that there is anything true, nor that there is anything false in them. Given that it is understood what I mean by a spoken proposition, I should like to note that there are also propositions which are not presented in words but which somebody merely thinks, and these I call *mental propositions*. Obviously, in the expression 'spoken proposition', I differentiate the proposition itself from its articulation. In the same way I differentiate a proposition from the thought of it in the expression 'mental proposition'. A *proposition in itself* I call that particular entity which one must necessarily associate with the word 'proposition' if he wants to follow me in the above distinction. It is that very entity which one thinks of as being a proposition when one asks whether or not somebody has articulated it, or whether or not somebody has thought it. The same entity I mean by the word 'proposition' if, for brevity's sake, I use it without the additional phrase 'in itself'. In other words, by *proposition in itself* I mean any assertion that something is or is not the case, regardless whether or not somebody has put it into words, and regardless even whether or not it has been thought. In the following example the word 'proposition' occurs in the sense I have here laid down for it: 'God, being omniscient, is cognizant not only of all true but also all false propositions; not only of those which are held true by created beings or those of which such a being merely had a notion, but also of those propositions which nobody holds to be true or even conceives of or will ever conceive of'. Many similar examples can be found. I hope that this makes intelligible the concept of a *proposition in itself*, but in order to give the reader a firmer grasp of the concept and to persuade him that he has understood me correctly, I wish to make the following remarks.

a. If one wants to associate with 'proposition in itself' the meaning that I have here stipulated, then he must not think of the original sense of the expression. Thus he must not think of something actually proposed, which would presuppose the existence of a being that does the proposing. We should always abstract from such figurative associations which adhere to the original meaning of technical terms, even in many other sciences. Thus in mathematics, when we consider the concept of a square root, we should not think of a root as it is known to the botanist, nor should we think of a geometrical square.

b. Just as one should not think of a proposition as something which is proposed by somebody, he should not confound it with the *idea* which is present in the consciousness of a thinking being, nor with a *belief* or a *judgment*. It is indeed true that every proposition, if by no other being, is at least thought and represented by God and, if it is a true proposition, God will also acknowledge it as true. Thus it occurs in the divine understanding either as a mere idea, or as a judgment. Nevertheless, a proposition is still different from an idea and a judgment.

c. For this reason one must not ascribe being (existence or reality) to *propositions in themselves*. Only the mental or asserted proposition, i.e. the *thought* of a proposition, likewise the *judgment* which contains a given proposition, has existence in the mind of the being that thinks the thought or makes the judgment. But the proposition in itself which constitutes the content of this thought or judgment does not have existence. It is therefore as absurd to say that a proposition has eternal being as it is to say that it originated at a certain moment and ceased at another.

d. Finally, it is obvious that a proposition in itself can be about thoughts and judgments, although it is itself neither a thought nor a judgment. In other words, it may contain the concept of a thought or judgment as one of its parts. This is exemplified by the proposition that I have given above as an example of a proposition in itself.

NOTE: After what has been said, it may be quite clear what should and what should not be understood by a proposition in itself. Still, the following question may cause some embarrassment. In Savonarola's *Compendium aureum totius logicae* (Leipzig n.d., bk. 10, no. 18), the following example occurs under the heading: "A proposed insoluble (i.e. a proposition which destroys itself) should be neither conceded nor denied: 'This is false' (given that the subject refers to that very proposition" that is 'This (what I just now say) is false.'

The question is whether this form of words deserves the name of a *proposition*, and moreover, whether this proposition is true or false. Savonarola says about such collections of words that one should neither affirm nor deny them: "And if it is said that every proposition is either true or false, then it should also be said that these are not propositions. For the definition of a proposition, namely that it is a true or false announcement, does not truly apply to them. However, it has the *appearance* of a proposition. Just as a *dead man* has the figure and appearance of a man without being a man, so these are called *self destroying propositions* or *insolubles*, but not propositions simpliciter." What does the reader think? One could think that Savonarola is right, especially as the *subject* of a proposition can never be that proposition itself, just as a part cannot constitute the whole. Still, I dare profess the contrary opinion, and I believe that common sense, too, will decide in my favour. Which teacher of languages will hesitate to call the words 'What I now assert is false' a proposition with a complete meaning? However, there still is the objection that this proposition is its own subject, which seems as absurd as to assert that a part of a whole constitutes the whole itself. This objection, it seems to me, disappears if a distinction is made between the *proposition as such* and the *mere idea* of it. Not the proposition itself, as a proposition, but rather the idea of it is the subject term of that proposition. That this distinction is well founded can be seen from the fact that not only here, but in all cases the thing itself must be distinguished from the concept of it, if one does not want to become entangled in gross absurdities. However, if I call the above speech act a complete proposition, I must also be prepared to call it either true or false. As one might suppose, I do the latter and say that the proposition 'What I now assert is false' is itself a false proposition, for it is equivalent to the following 'What I now assert I *declare* for false and *do not assert it*'. And this is, of course, not true. However, from this it does not follow at all that I have to assert the following proposition, 'What I now assert is true'. Savonarola seems to have believed this, and since the latter proposition seemed as absurd to him as the former, he preferred to seek refuge in the assertion that none of the two word sequences are proper propositions. I say, on the other hand, that the proposition 'What I now assert is false' merely has the peculiarity that its contradictory cannot be construed in the same way in which it is done with many other propositions (whose subject term has only one object). That is to say that we cannot find the contradictory of this proposition by merely prefacing its predicate 'false' with the negation 'not'. This cannot be done in our case since a change in the predicate causes at the same time a change in the subject. For this subject, i.e. the concept that is expressed by the words 'what I now assert' is different when I say 'What I now assert is false' as when I say 'What I now assert is not false'. A similar phenomenon is encountered with all propositions whose subjects and predicates contain a reference to the propositions themselves or to any one of their parts. Thus, for example, the following two propositions, which to judge from their wording seem to be contradictory, are both true: 'In the present speech act, the third word from the end is the copula' and 'In the present speech act, the third word from the end is *not* the copula'. On the other hand, the following sentences are both false: 'The number of words in the present sentence is eleven' and 'The number of words in the present sentence is not eleven', for the last sentence does in fact consist of eleven words, since it has one more (the word 'not') than the first, etc. Therefore, the contradictory of the sentence 'What I now assert is—or I declare it for—false' is not the sentence 'What I now assert is true', but 'What I now assert, I assert'. But enough of such hairsplitting.

§ 20. *Justification of this Concept and its Designation*

Since it is customary to understand by proposition [*Satz*] a proposition put in words, or an articulated proposition, it is reasonable to expect an explanation why the concept of a proposition in itself should be introduced into logic.

1. Once it is admitted that it is sometimes useful to speak of truths in themselves, the introduction of the concept is justified. In addition, many known theorems which are said to hold of judgments also hold of propositions in themselves, and "it is a rule of good exposition not to describe something as holding under certain conditions if it is unconditionally valid."

2. The word 'proposition' was preferred to 'judgment' [*Urteil*], 'state-

ment' [*Aussage*] and 'Assertion' [*Behauptung*] since the others carry stronger overtones of agency.

§ 21. *That Others have already Used this Concept*

If the concept of a proposition in itself is as important as it is here made out to be, then it must be expected that others have thought of it before.

1. B. surmises that the ancient Greek philosophers already had the concept of a truth in itself, which presupposes that of a proposition in itself. He promises to examine this point in the sequel.

2. Logicians usually give a negative answer to the question whether there can be two exactly similar ideas on the grounds that in this case one and the same idea was thought twice. "But if someone distinguishes an idea from a thought of that idea, then he must also distinguish a proposition in itself from its appearance in the mind." A further point is this: most logicians maintain that there is no essential difference between two syllogisms which differ only in the order of their premises. "Fundamentally, every syllogism is only one proposition, namely a proposition of the form 'From the truths *A* and *B* follows the truth *C*', where the two truths *A* and *B* are mere members of a sum, i.e. no order of rank holds between them. Thus it is true that no essential change occurs if they are exchanged, provided that we take the syllogism as a whole, as well as the propositions *A*, *B*, and *C* to be propositions in themselves. By contrast, mental occurrences of propositions are ordered in time."

3. Some logicians seem to have had a quite distinct apprehension of the concept of a proposition in itself. "Thus Leibniz uses as equivalent the expressions *proposition* and *cogitatio possibilis* (*Dial. de Connexione inter Verba et Res* (C. I. Gerhardt, ed. *Philos. Schriften*, vol. VII, p. 190). This obviously presupposes that by propositions he meant propositions in themselves."

§ 22. *Diverging Views of the Concept of a Proposition*

1. By propositions most authors mean verbally articulated judgments. This is objectionable, since a judgment is commonly taken to be a proposition which is held to be true by some thinking being, hence a much narrower concept than the one here required. Also, it leads to a predicament in connection with hypothetical and disjunctive judgments. Their constituents can neither be propositions in the present sense, since they need not be articulated, nor can they be judgments for, "if someone makes the hypothetical judgment 'If Caius is wicked, then he is unhappy', he neither judges that Caius is wicked, nor that he is unhappy."

2. Some logicians use the word 'judgment' to designate any mental proposition, whether or not it is held to be true. This concept is still too narrow, as has been shown, "and we no longer have a specific symbol for the logically important concept generally designated by the word 'judgment' ".

3. Most logicians do not want to call anything that expresses a wish, a command, an order, etc. a proposition (cf. Aristotle, *De Interpr.*, 17ª4). "In my view, questions, wishes, entreaties, etc. and even mere exclamations, in the sense which they get through their context, must be regarded as genuine,

though often obscurely stated, propositions. Take the case of a question, e.g. 'What is the ratio between the diameter and the circumference of a circle?' Of course, this question asserts nothing about that which is in question, but it nevertheless states something, namely our desire to be instructed about the subject of the question. Hence it may be either true or false. It is false if it indicates the desire incorrectly. Thus, for instance, if the author of the above question did not really want to know the ratio itself, but merely whether it is rational or irrational, then the question would be false, and should have been phrased differently. It is true, of course, that we occasionally say 'I do not assert this, I merely ask'. It would seem that in such a statement we contrast question with assertion or judgment. Upon closer inspection, however, it turns out that the meaning of such a statement is actually this: 'I do not *assert* that this or that is the case, but I merely *ask*, i.e. I assert that I desire to know, whether or not it is the case.' Hence this case confirms that a question is a complete proposition. Just as something is asserted in every question, so also in every wish, command, etc. Hence they should all be called propositions. It is much more doubtful if exclamations, e.g. 'Woe is me' should be considered propositions. For, even if we take into consideration all the circumstances under which they were uttered, it is often an open question what the author had in mind or wanted to indicate. However, if we suppose that the author *intended* to indicate something through his exclamation, then this exclamation is clearly a proposition; all we can criticize is the ineptness of the verbal expression. The assertion of Reusch, that wishes, questions, etc. do not determine a relation between subject and predicate is clearly a mistake. The proposition 'I wish that Caius were an honest man' does indeed not determine the relation between Caius and honesty, but then, Caius and honesty are not subject and predicate of this proposition. The real subject of this proposition is the author, and the predicate is his wish that Caius be an honest man. In making the concept of a proposition in itself wide enough so that questions, wishes, etc. fall under it, I do not want to deny that they should be distinguished. They ought to be distinguished, and we owe a debt of gratitude to logicians who have written special monographs, e.g. "On Questions", etc., and thus have made us aware of the various kinds of propositions and their peculiarities."

4. Kant wanted to reserve the term 'proposition' for assertoric judgments, while calling 'problematic proposition' a contradiction in terms, "since a proposition [*Satz*] must posit [*setzen*] something" (cf. Kant, *Logik*, A 170). B. objects that problematic judgments, too, posit (assert) something. "For, if by a problematic judgment one means a judgment of possibility, e.g. 'man is corruptible', then the possibility of a thing is posited, in this case corruption. If one has in mind a judgment which expresses the indecision of its author, such as 'perhaps Caius is already dead', then it is this indecision which is asserted or posited."

Similarly, hypothetical and disjunctive judgments should be called propositions. "It is, of course, true that in the hypothetical judgment 'If *A*, then *B*' neither *A* nor *B* are posited, but something else is, namely that the supposition of *A* entails that of B." Similarly, in a disjunctive judgment '*A* or *B* or *C* . . .' what is posited or asserted is that one of them is true.

Finally, the view that a judgment becomes a proposition when it is actually formed is to be rejected, "since every judgment is a formed one."

§ 23. *Examination of several Definitions of this Concept*

In § 19 I attempted to make clear the concept of a proposition in itself. At that point, some readers may already have asked themselves why I did not give a brief *definition* of this concept instead of being so diffuse. In my view, defining a concept, i.e. indicating its components, is not always the easiest or even safest method to come to an understanding about it. Moreover, I have not been able to use this method in connection with propositions in themselves because I am not at all certain how this concept should be defined, and various attempted definitions given by others appear to be deficient. They either contain the concept to be defined unanalysed in some word or another, or have some other obvious defect. In order to prove this point, I now want to examine the most important previous definitions. This is rather difficult because never, or at least very rarely, was an attempt made to define the concept of a proposition in itself; the definitions usually aim at some related concept, notably that of a judgment or of a thought or verbally articulated proposition. For this reason we should not expect that any one of these definitions completely fit our concept, but we should be satisfied if we could find one which would meet our purpose after suitable adaptation.

1. An altogether correct definition we can hope to find only with scholars who had a distinct notion of our concept. However, one soon realizes that they either did not have the intention of giving a precise definition of this concept, or that they rested with whatever was given by their predecessors. The concept which is designated by Leibniz' expression *cogitatio possibilis* (*loc. cit.*) is not composed of the two concepts *cogitatio* and *possibilis* in the same way in which many other concepts are generated by connecting two concepts, for example the concept of a golden candlestick. This is composed of two concepts designated, respectively, by a noun and an adjective, namely candlestick and something golden. The golden candlestick is a kind of candlestick, but the possible thought is not a kind of thought, but merely a kind of possibility. Therefore, if we wish to make the above expression more precise, we will have to say that, according to Leibniz, a proposition is "the possibility of a thought", or, even more clearly, "it is something that *can* be thought or *can* constitute the content of a thought." There is indeed no doubt that this thinkability is a property of any proposition, but it is also obvious that it does not form part of the concept of a proposition. We can think the concept of a proposition in itself without reminding our-

selves that it has the property of being thinkable. This makes it sufficiently clear that the indication of this property does not belong in the definition of this concept. An additional factor is that this definition cannot be reversed. For, if we take the word *thought* in its widest sense, then mere ideas can be the content of a thought, and it would therefore be completely wrong to say that every possible thought is a proposition. If we want to correct this mistake, we shall have to narrow down the concept of a thought. But how? I see no other way but to declare a thought to be a mental proposition. But then we can obviously not define the concept of a proposition via that of a thought.

B. then rejects the definition that propositions are combinations of concepts, as too wide, since this definition also applies to complex concepts.

2. In the sequel, I shall examine several other definitions; I shall not consider them in historical order, but shall arrange them so as to facilitate the examination. Still, the oldest definition can be considered first.

The Greeks already asserted that "A proposition is what is either true or false" (cf. Aristotle, *De Interpr.*, 17^{a}3, and Sextus Empiricus, *Adv. Log.*, II, 12). The ancients, it seems, just used this definition in order to come to an understanding about the sense of the expression. It is, of course, admirably suited for that purpose; and since I do not know of any better method, I have used it myself in § 19. However, it cannot be accepted as a definition of the kind that is supposed to give us the components of the concept to be defined, if only for the reason that it contains a classification. For, in the concept of a proposition, such a classification does certainly not occur.

3. The same fallacy of classification is found in a second explication of Aristotle. (It is, of course, fallacious only if we want to use it as a definition.) It is found in *Anal. Prior.* (24^{a}16): "A premise is a sentence affirming or denying one thing of another." This sentence, taken as a definition, mentions components which certainly are not present: neither of the concepts of affirming or denying occurs as a component of the concept of a proposition. To affirm something simply is to assert that something is true; to deny something is to assert that something is not true. Thus, both concepts contain the concept of truth and therefore that of a proposition. Finally, to *assert* something clearly means the same as to judge. The concept of judging, however, is certainly no component of the concept that we mean by proposition in itself.

4. Similar criticisms are advanced against Wolf's definition "Judgment is

an act of the mind by which something is attributed to something else, or removed from it."

5. No satisfactory definition results if the concepts of affirmation and negation are replaced by the concepts of connection and separation.

6. According to Mehmel, the type of combination into which representations enter when they form a proposition is *immediate connection*. B. finds this unhelpful. The connection between negation and a concept *A* in 'non-*A*' appears to him as immediate as that between *A* and *B* in 'All *A* are *B*'.

7. Locke attempted to give a definition when he said (*Essay* IV, I, 2) "Knowledge then seems to me to be nothing but the perception of the connection and agreement, or disagreement and repugnancy of any of our ideas". . . . However, Leibniz (*Nouv. Ess.* IV, V, 1) already said: "Properly speaking, a proposition does not express agreement or disagreement. Two eggs agree and two enemies disagree. The question here concerns a very special mode of agreement or disagreement. Thus I believe that this definition fails to explain the point with which we are here concerned." I agree, and I believe that everybody will feel the same as soon as he tries to make clear to himself what the word agreement must mean, so that one can say that in the proposition 'Caius beats Titus' or 'The sun heats the stone' there is an agreement between the ideas Caius, Titus, beating; Sun, heating, stone. But in whatever way the words agreement and disagreement are taken: it obviously is not the agreement or disagreement of a pair of ideas as such, but the statement of this agreement or disagreement that constitutes the proposition. This can chiefly be seen from the fact that every proposition must be either true or false. But agreement or disagreement in and by themselves cannot be called either true or false, only the statement that among certain ideas such an agreement or disagreement is present. Hence if this definition is not to be altogether pointless, the concept of asserting, and thus of a proposition, must be included in it. But even then it is not true that a proposition asserts the agreement or disagreement among *ideas*. This is at most true of propositions that are about ideas, for example of the sentence 'The idea of a triangle contains the idea of a figure'. Of other propositions it should be said that they assert the agreement or disagreement of certain *objects*. Finally, whatever expression one chooses, this definition obviously applies only to true propositions but not to false ones. For how could one say of a false proposition that it asserts the agreement that holds between two objects or ideas?

8. The fallacy of classification which all the above definitions have in common, at least in the way in which they are usually stated, is

too obvious as that it should not have been avoided by those who wanted to give their definitions an air of logical rigour.

They sought a concept which could serve as the common genus of the concepts of connecting and separating or of agreeing and disagreeing. The concept of a relation seemed to meet the purpose. Consequently, a proposition was defined as *a relation* or the *determination of a relation* among ideas. Leibniz himself seems to have preferred the word 'relation' to the word 'connection' when he writes (*Nouv. Ess.* IV, I, 7) "I believe that one can say that [this] connection (the connection that is found between the subject and the predicate of a sentence) is nothing but rapport or relation taken generally."

The following objection must be made against this definition: If we take the word relation in the wide sense that is customary in scientific discourse, then there exists a relation between any, even the most remote, objects. Consequently it is indeed true that between two or more ideas which occur as constituents of propositions there always holds a certain relation. But the converse is not true, namely that two or more ideas which stand in a relation to one another form a proposition. It is even less defensible to say that this relation itself is a proposition. (The latter has actually not been asserted, and merely the determination of this relation has been called a proposition (or judgment).) However, it is not clear what the word 'determination' means in this connection. To my knowledge, in scientific writings this word occurs mainly in two senses: to determine some object either means to contain its ground or cause; or it means to state properties of that object. In this connection, only the latter can be meant. For if I say that in the proposition 'Caius is a scholar' the idea, or rather object, Caius, is determined by the property of being a scholar, I obviously want to do nothing except assert the property of being a scholar of the object Caius. (Nobody will seriously believe that my proposition contains the *ground* for the relation that holds between Caius and learning.) Thus, this definition already contains the concept of *asserting* and thereby that of a proposition. Also, instead of the concept of an idea, that of an *object in general* should here be used. For, in the proposition 'Caius is a scholar' no assertion is made about the relation between the ideas which we have of Caius and the property of being a scholar. Rather, the proposition makes an assertion about the things which these ideas represent. Klein (*Anschauungs und Denklehre*, Bamberg 1818, § 40) realized that one can do without the concept of a relation and says therefore that a judgment is the determination of an object through a concept. One can, of course, talk in this fashion, provided only that a determination is meant to be an *assertion.* But if this is so, then the concept to be defined is completely contained in the word determination, and for this reason one can be even shorter and can simply say that a proposition is nothing but a *determination.*

9. B. criticizes Maaß and Hofbauer, who define a proposition as the idea of a relation that holds between objects. But while a proposition always contains ideas, it is not itself an idea: proposition and idea are mutually exclusive concepts. This section anticipates the discussion of ideas in themselves in §§ 48 ff.

10. Malebranche and others replace 'idea' in the definition under (9) by 'recognition', 'awareness', or 'insight'. This lays them open to the additional criticism of having confused propositions in themselves with judgments.

11–19. B. rejects a number of definitions mainly because they employ the notion of consciousness and are therefore not suitable to define the concept of a proposition in itself.

20. Hobbes, Condillac, Bardili and others speak about the act of judging as if they envisaged all propositions as mere *equations.* In many other books, too, propositions are called equations. Thus the proposition 'Caius is a man' is written as 'Caius = Man', and the principle on which all syllogisms are said to rest is expressed by Condillac as "Two things equal to a third are equal to each other." Herder had the same opinion (*Metakritik*, Leipzig 1799, *Gespräche über Spinoza*). Recently, Beneke has again taken up this notion; he sees the essence of judging in the fact that activities of the human mind which are totally or partially equal can be stated to be totally or partially equal (*Erkenntnislehre*, Jena 1820, p. 20). Judging is for him the activity of asserting the equality of equals in the given (p. 40). It is impossible for me to agree with this. Without doubt, all equations are propositions, but not all propositions are therefore equations. The proposition 'Caius is a man' does not assert any equality between Caius and man. It merely says that the property of being a man belongs to the object Caius. It is even harder to mistake other judgments for equations, e.g. 'You should always tell the truth'. But even if it were true that propositions are merely equations, the above would not provide a definition of that concept, for the words 'state' and 'assert' contain it in unanalysed form.

21. Thanner's definition that the essence of the act of judging is the unity which is asserted or denied in the opposition of two concepts is criticized as unclear and figurative.

22. In Hegel's *Encyclopedie der philos. Wissenschaften* (Heidelberg

1817, § 115) it is said: "Judgment is the concept in its separateness, as differentiating relation of its moments, which are posited as at the same time isolated and identical with themselves, and thus oppose one another as singular and general. Judging is original dividing of the concept." I do not flatter myself to understand completely what Hegel meant by all the words in this most curious definition. It is obvious that he does not here define the proposition in itself, rather, he declares judgment to be an action of the mind. It is furthermore clear that he envisages a judgment as some sort of concept which is divided into *parts*. I have already said that I hold this to be false. He moreover says of the moments (i.e. constituents) of which a judgment is composed (subject and predicate) that they are posited as at the same time *isolated* and *identical with themselves*. This must be taken as even more figurative than the expressions *opposition* and *unity* in Thanner's definition above. Finally, the assertion that subject and predicate oppose each other as *singular* and *general* is true only of certain propositions in which the predicate is a wider concept than the subject, but it is hardly true of all propositions.

§ 24. *Several Senses of the Words 'True' and 'Truth'*

B. first points out that ambiguities, though reprehensible, are rarely misleading in ordinary usage. Confusion is said to arise from senses "which have been suggested by scholars, but which have fortunately not been generally accepted." He then goes on to enumerate the senses of 'truth' and 'true' that are "familiar in common usage".

1. Truth as a property of propositions in themselves. Example: "Of the three propositions 'There are winged snakes', 'They have died out', 'They never existed' one must necessarily be true, though I do not know which." If propositions have this property, then they state something as it *is*. B. calls it the abstract objective sense of 'truth'.

2. Sometimes a true proposition is called a truth. B. calls this the concrete objective sense.

3. If a true proposition is asserted we sometimes call the assertion a true judgment. B. calls this the subjective sense of 'true'. He prefers the expression 'correct judgment'.

4. Sometimes a collection of truths in senses 2 and 3 is called truth. "Thus Jesus said he came into the world to bear witness to the truth" (John 18, 37).

5. Sometimes, 'true' is used as synonymous with 'real', and 'genuine', hence as opposed to 'false', 'unreal', 'imaginary', as in 'This is the true God'. This use, paradoxically, allows one to say "that something is 'a true lie', but this simply means that a statement not only seems to be a lie, but really is one." B. calls this a transferred sense. A proposition using 'true' in this sense is always short for one using the word in one of the earlier senses.

§ 25. *What the Author Means by Truth in Itself*

By the expressions *truth in itself* or *objective truth* I shall mean what is understood by the word *truth* if it is taken in the second of the above senses (i.e. the concrete objective), which is probably the most common anyway.

Thus, to state it again, I shall mean by a truth in itself any proposition which states something as it is, where I leave it undetermined whether or not this proposition has in fact been thought or spoken by anybody. In either case I shall give the name of a truth in itself to the proposition, whenever that which it asserts is as it asserts it. In other words, I shall give it the name of a truth in itself whenever the object with which it deals really has the properties that it ascribes to it. Thus, for example, the number of blossoms that were on a certain tree last spring is a statable, if unknown, figure. Thus, the proposition which states this figure I call an objective truth, even if nobody knows it. Since the present concept is very important, I want to make sure that I am not misunderstood. Hence I shall state the following fairly obvious theorems about truths in themselves.

a. Truths in themselves are a kind of proposition in itself.

b. They do not have actual existence, i.e. they are not something that exists in some location, or at a some time, or as some other kind of real thing. *Recognized* or *thought* truths have indeed real existence at a definite time in the mind of the being that recognizes or thinks them. They exist as thoughts of some kind that begin at one moment and end at another. But to truths themselves, i.e. the contents of these thoughts, no existence can be ascribed. Nonetheless, some truths, for instance the truths of religion, morality, mathematics, or metaphysics, are sometimes called eternal, as when it is said that 'It will be eternally true that vice makes unhappy' or that 'The straight line is the shortest line between two points', etc. But by this is meant merely that these are propositions which express a persistently (eternally) enduring relation. Other propositions, for instance 'The bushel of grain costs three *Thaler*' or 'It snows', etc. state a merely passing relation (one that holds at a certain time and place only.) Hence, in order to be true, such propositions require the addition of a time determination (and oftentimes also of a location): 'Today, in this place, it snows'.

c. It follows indeed from God's omniscience that each truth is known to him and is continually represented in his understanding, even if no other being is acquainted with it or thinks it. Consequently, there actually is no truth which is recognized by nobody at all. This, however, should not keep us from speaking of truths in themselves, since their concept does not presuppose that they must be thought by someone. The fact that they are thought is not contained in the concept of such truths, but it can nevertheless follow from some other circumstance (in this case the omniscience of God) that they must be recognized by God himself, if by no one else. In this respect the concept of a truth in itself is similar to very many other (actually all) concepts. We must distinguish their content, i.e. what we must think to have thought them, from that which belongs to their object as a mere property (and which we do not have to think in order to have thought them). Thus, the thought of a line which is the shortest between two points is certainly different from the thought of a line of which all parts are similar to each other. There clearly is a difference between these two thoughts, and similarly between the concepts in themselves that we think when we think these thoughts. Hence, the concept of a line which is the shortest between two points is altogether different from the concept of a line of which all parts are similar to one another. Nevertheless, there can be no doubt that a line which falls under the first concept, i.e. a line which is the shortest between two points, also has the property that is described by the second concept, i.e. all its parts are similar to one another, and vice versa. We can see from this example that we must not deny that two concepts are different from one another, merely because they are equivalent [*Wechselbegriffe*]: Thus, although all truths in themselves are known (namely to God), their concept must be distinguished from the concept of a recognized truth or cognition. Thus, to give another example, the logician must have the same right to speak of truths in themselves as the geometrician who speaks of spaces in themselves (i.e. of mere possibilities of certain locations) without thinking of them as filled with matter, although it is perhaps possible to give metaphysical reasons why there is no, and cannot be any, empty space.

d. I said above that a truth in itself is "a proposition which states something as it really is". These words are not to be taken in their original nor in their ordinary sense, but in a certain higher, more abstract sense. This sense follows, as I think, from the proviso "that I want to leave it undetermined whether or not such a proposition has in fact been thought or stated by anybody". Most readers will anticipate the remarks that I now want to make about the sense of each of the above words.

The word *proposition*, because of its origin from the verb "to posit' does indeed remind one of an action, of something that has been posited by someone. But we must abstract from this in the case of truths in themselves. They are not posited by anyone, not even by the divine understanding. It is not the case that something is true because God recognizes it as such; on the contrary, God recognizes something as being of a certain nature because it is so. Thus, for example, God does not exist because he thinks he does, rather, because there is a God, this God thinks of himself as existing. In the same way, God is not omnipotent, wise, holy, etc. because he thinks that he is. The opposite is true. He thinks himself as omnipotent, etc. because he really is so, etc.

Similarly, it should be clear that the verb 'to state' is also to be taken in a figurative sense; for no truth can literally state (say) anything.

The following is also easily overlooked: If we say that a truth states *something as it really is* this phrase must also be taken in a figurative sense. This is so, because not all truths state something that really is (i.e. has existence). This holds especially of those truths that deal with objects that do not themselves have reality, e.g. other truths or their constituents, i.e. ideas in themselves. Thus, the proposition 'A truth is not something that exists' does certainly not state something that exists, and yet it is a truth.

§ 26. *Some Related Concepts*

The following must be carefully distinguished:

1. Truths in themselves and recognized truths
2. Truth (of propositions) and reliability (of judgments)
3. Truth and reality: "Though there are truths that refer to something real . . . the truth is never itself that real thing, rather, . . . not a single truth, as such, has reality or existence."
4. Truth, comprehensibility, and recognizability. Comprehensibility is a wider concept than truth: "if something is true, then it must be comprehensible, but not everything that is comprehensible has to be true." On the other hand, 'recognizability' [*Erkennbarkeit*], though different from 'truth' is neither wider nor narrower than it. It seems to contain the concept of truth in it: "If my observation of the usage is correct, the word 'recognize' is always used of true propositions, never of false ones." 'True cognition' [true knowledge] as well as 'recognition of truth' are redundant expressions. The *genus proximum* of cognition (as well as error) is judgment or opinion. A cognition is a true judgment. "Recognizability of an object is the possibility of pronouncing a true judgment upon it." Hence the concept of truth seems to be a constituent of the concept of recognizability. That truth and recognizability are different concepts can also be seen from the fact that truth does not admit of degrees, while cognitions have degrees of vividness as well as reliability.

§ 27. *The Concept of a Truth in Itself has been Recognized before*

B. remarks that the concept of a truth in itself is frequently employed by everybody, as when it is said that a certain truth is known or unknown to someone. But we are not often conscious that we use this concept. Only occasionally have philosophers become aware of it, e.g. Sextus Empiricus (*Adv. Log.*, II, 9 and 88) gives the Epicurean and Stoic definitions of truth. In both definitions, truth is opposed to falsity and not to error. B. takes this as an indication that truths in themselves were intended.

"It is well enough known that among the Schoolmen the realists at any rate declared the true to be the *existing*. The error that lies in this expression seems to have contributed much to the disagreement between them and the nominalists, but it proves clearly enough that they did not envisage truth as a kind of cognition. In the logic of Thomas Aquinas as in many other writings of the time it is stated that *Verum et ens convertuntur*. Apparently, they wanted to say by this no more than that a truth is a proposition which states something as it *really* is." Leibniz, too, is said to have expounded the same concept in *De Connexione inter Res et Verba* (C.I. Gerhardt, ed., *Philos. Schriften*, VII, p. 190). Hence he dismissed Locke's distinction of mental and verbal truths as of no more interest than a classification into parchment and paper truths.

§ 28. *Probable Constituents of this Concept*

B. attempts to state a definition of 'truth'. The Genus is proposition, "if by it is understood not a string of words, but merely the sense [*Sinn*] which a certain string of words can express." He notes that "for every proposition, especially if it is to be a true one, there must be a certain object with which it deals (subject) and also a certain something which is stated about this object (the prédicate). . . . Could we then not say that a truth is a proposition that states something about its subject that really belongs to this subject?" This is quite true as an assertion, but it will do as a definition only (a) if it is reversible, (b) if it does not contain the definiendum in the definiens, and (c) if it states exactly what we think "whenever we think the concept of truth."

B. notes that (a) is indeed fulfilled. Concerning condition (b) he states that the constituents of the concept of truth are the concepts of a *proposition*, an *object*, *stating*, *belonging*, and *something*, none of which contains the concept of truth; but, in addition, there is also the concept designated by the word 'really', which seems to be none other than the concept of truth. (Instead of 'really' we might say 'in truth'.) Hence, the definition seems circular. "However, this objection is groundless, for we can altogether omit this word, or rather the concept indicated by it, without causing any essential difference. For at bottom, there is no difference between saying that *P really* belongs to *S*, and saying that *P belongs* to *S*." Hence, the correct definition is this: "A proposition is true, if it states what belongs to its subject." Close attention discloses that this is indeed what we think by truth. Hence condition (c) is also fulfilled.

B. then notes that a number of philosophers agree with him in this definition, most notably Aristotle, when he uses the expressions 'This belongs to that' and 'This can in truth be stated of that' interchangeably (*Anal. Prior.*, 49ᵃ6).

§ 29. *Other Explanations of the Concept of Truth*

Definitions of 'truth' given by other logicians are discussed.

1. B. maintains that a considerable number of logicians agreed with him in their definitions, notably Aristotle, Epicurus, Chrysippus.

2. Other philosophers restricted the concept of truth to thought or recognized truths. Bolzano cites Carneades, Locke, Leibniz, Kant and others, and then criticizes at length Tetens, who had asserted that truth is the correspondence of thoughts with things. If the relations between our ideas are the same as the relations between the things of which they are ideas, then, he maintained, these ideas are true (*Philosophische Versuche*, Leipzig 1777, I, 533) B. replies:

"Can one really say that the ideas of a house and of a garden stand in the same relation to one another as these objects, i.e. house and garden themselves? The ideas 'house' and 'garden' have always the same immutable relation to one another, while the objects, houses and gardens, can stand in a variety of relations, depending on their characteristics. The objects *God* and *world* are related to one another as cause to effect. God is the creator of the world. But who would want to maintain that the *concepts* of God and world stand in the same relation; that the concept 'God' is the creator of the concept 'world', etc.? Tetens himself admits that he does not mean that the concepts are related to one another in all respects as their objects are, but that they are so related only in *certain* respects. I ask, in which respects?"

B. then claims that the definition that truth is the agreement of our ideas with their objects applies only to judgments of the form 'This is *A*', e.g. 'This is a tree', etc., and that a better way of stating this definition is that "truth consists in this: that our ideas of the objects are compounded from ideas of qualities which these objects really have." He criticizes this definition as being merely subjective and in any case too narrow, since it applies only to the propositions of the form 'this is *A*'.

3. The above definitions are more or less close to common usage. Now B. turns to a discussion of philosophers who define truth in an entirely different way.

a. Truth is the universally accepted (Aenesidemus, cf. Sextus Emp., *Adv. Log.*, II, 8). It is indeed the case that a judgment held true by all thinking beings, including God, is true, but the converse does not hold.

"It is perhaps thought that this difficulty can be overcome, if a judgment is declared to be true whenever every thinking being must accept it 'if it is in a suitable situation and uses its faculty of judgment properly'. But how can we prove that there are not some beings who even under the most favourable circumstances and careful observation of the rules of judgment still judge the same object in diverging ways? If we say that beings are to be called thinking or intelligent only if they must judge uniformly under equal conditions then, I should think, the whole definition dissolves into an empty tautology."

b. Others believe that they can find the essence of truth in a *conformity to the rules of thought or cognition.* Thus Gerlach (*Grundriß der Logik*, Halle 1817, § 219 f.) states "Truth is the property of our ideas of being formed in conformity with the rules of representing. Since all our cognizing is human, truth can only be a merely *human* truth, therefore not *absolute*, but only *relative*." B. objects to this that "if we understand by the rules of thought the rules which must lead us to the truth if we follow them, then we obviously commit a circle in our definition. . . . Moreover, in this case, *human truth* (understood as truth that man can discover) will not stand in opposition to absolute truth, but will be part of it." On the other hand, if we mean by laws of thought rules that merely increase the certainty that we have found the truth, then, "aside from the fact that we still encounter the same circularity, it does not follow that a judgment is invariably true if it is formed in conformity with the laws of thought."

B. then considers two other conceptions of what laws of thought are, namely first, that the laws of thought are laws which our thinking necessarily follows; but these cannot be the rules of logic, since we often transgress the latter. Secondly, it might be maintained that the laws of thought are laws which are in conformity with the nature of our thinking. Here the difficulty is hidden in the "obscure word 'conformity' ". Moreover, they can conform to the nature of our thinking either absolutely or in order that our thinking meet a certain purpose. If the former, the preceding criticism applies. If the latter then this purpose must be indicated. If it be perfection of our thinking, it must be perfection in meeting a given end, and what could this end be but the discovery of the truth, in which case the definition would again be circular.

c. Some have claimed that a judgment is true if it is immutable, always agreed to, etc. B. remarks against this that the converse does not hold, and that long duration of an opinion, just as its wide dispersion, is a fairly reliable *index* of truth, but not the same thing, just as cheerfulness may be a sign of innocence.

4. The definitions under (3) were offered as improvements upon common usage, because their authors held that truth in the common acceptation of that term is altogether unattainable; in the following, B. discusses deviations, i.e. expansions of common usage that seem to him to be without any justification.

a. Some authors distinguished objective from subjective or apparent truth. For lack of a definition, it is impossible to guess what concept of truth-in-general these writers had, since they subsume under it such disparate things as objective and subjective truth.

Perhaps it was not their intention to envisage *subjective* or *relative* truth as a kind of truth-in-general; rather, they may have taken these expressions in analogy to the phrases 'apparent truth' and 'pictured fish'. "For by apparent truths we do not understand a kind of truth-in-general, but a kind of proposition, and a pictured fish is not a kind of fish, but of picture. If this is the case, then I have no misgivings about this concept. I merely wish to remark that I find its designation cumbrous and misleading."

b. B. now considers an "even more forced deviation from common usage." Many authors have divided truth into formal, logical, or analytic on the one hand, and real, material, metaphysical, transcendental or synthetic on the other, such that the latter group denotes truth in the ordinary sense. Of the former, many definitions are given, all very obscure. B. cites a number of them, and then attempts a comprehensible version of the way in which the expression 'formal truth' has so far been used. He takes it to be the case that a proposition was called formally true if it is free of contradiction. "Contradiction, however, can only occur when one proposition is compared with another. Therefore it is important to determine the class of propositions which must be examined in order to determine if a given proposition is formally true. The wider we make this class of propositions, the smaller becomes that of the formal truths." E.g. if this class embraces all material truths (i.e. truths in the ordinary sense), then formal and material truth coincide. "For if a proposition is not materially true, then it must contradict at least one material truth, namely its negation." But the class in question can be arbitrarily narrowed down. Hence 'formal truth' becomes a relative term: "The same proposition *X* could be called formally true with respect to a certain class of propositions and formally false with respect to another. In short, formal truth would then be what is normally called consistency of a proposition with certain others, and it would have been reasonable to retain this expression (cf. § 95)." However, the expression was probably not intended in this way. But if formal truth is not to be arbitrarily attributable to any proposition, then the class of propositions with which it must be consistent must either be given through that proposition itself, or must be fixed once and for all. One way of doing the former is to call a proposition formally true if all propositions that can be formally deduced from it are consistent with each other. But then only such propositions as 'An *A* which is *B* is not *B*' would be formally false, since 'An *A* which is *B* is *B*' can be formally deduced from it. ("A proposition is formally deducible from a given proposition *X* if it is deducible when all those parts of *X* that the logicians do not include in its *form* are taken as variable ".) On the other hand, if the class of propositions with which the proposition in question is to be consistent is to be fixed once and for all, one might take this class to be constituted by all *a priori* propositions. But in this case "even the coarsest lies and absurdities" would have to be called formally true.

In a note, B. comments on the usage of 'logical' and 'metaphysical truth' in precritical philosophy. He continues in a second note: "Hobbes (*De corpore*, I, 3, no. 7) asserts that the predicates *true* and *false* can be assigned only to *verbal expressions*. Truth, he maintains, does not reside in *things*, but only in *words*. For this reason only beings that have a language are said to be capable of truth. Furthermore, all first truths are said to be arbitrary and to owe their existence only to the pleasure of those who have given these and no other *names* to things. Perhaps it was only delight in extravagant formulations that caused him to make these assertions. (This assumption seems unlikely, because others have said similar things, e.g. Locke, and Hollmann, *Logik* (Göttingen 1746, § 126, note). If it was not this, then I have no explanation but that he confounded words with ideas. Hobbes only acknowledged things and signs of things, but not ideas of them."

§ 30. *Sense of the Assertion that there are Truths in Themselves*

To avoid misunderstanding, the expression *to be*, which I use in the assertion that there are truths in themselves, must be specifically defined. In its proper and most common sense, e.g. in the proposition 'There are angels', the word is to indicate the existence or actuality of a thing. In the present context, it is not to be taken in this way, since, as I have frequently said, truths in themselves have no existence What is meant, then, when we assert that there are such truths? Nothing, I answer, but that certain propositions have the character of truths in themselves.

About the *number* of these propositions, whether there are several or only one, I do not wish to say anything at this juncture, since we may consider our assertion established if we merely show that there is one single truth, or, what amounts to the same, if we show that the assertion that there is no truth is false.

If we combine this remark with the preceding one, then it turns out that the sense of the assertion which we want to prove could be most clearly expressed in the following way: 'The proposition that no proposition has truth, does not itself have truth', or, shorter, 'That no proposition is true is not itself true'.

§ 31. *Proof that there is at least One Truth in Itself*

The precise expression to which we have just reduced the assertion that there is at least one truth in itself (§ 30) cannot fail to prove the point even to the most myopic. That no proposition has truth disproves itself because it is itself a proposition and we should have to call it false in order to call it true. For, if all propositions were false, then this proposition itself, namely that all propositions are false, would be false. Thus, not all propositions are false, but there are also true propositions. There are truths, at least one.

NOTE: B. then points out that the proof which he has just presented has been known from antiquity. He mentions Aristotle (*Met.*, 1012ᵃ30), and quotes Sextus Empiricus (*Adv. Log.*, II, 55). He continues: "This proof could be conducted in several other ways. Thus, one need not use the proposition that everything is false, but could choose any other proposition '*A* is *B*' and could point out that, if this proposition is false, then the assertion that it is false is a true assertion. We can also proceed in the following way: if someone does not hold anything to be true, we could draw his attention to the fact that there are propositions, and if he doubted this, we could remind him that the words 'There are no propositions' do themselves contain a proposition, etc. I do not believe, however, that these proofs are any more persuasive."

§ 32. *Proof that there are several, even Infinitely Many, Truths*

1. From the preceding it follows that there is at least one objective truth, since the opposite assertion is self-contradictory. But perhaps there is only one objective truth, namely that there is a truth? To disprove this contention, I now want to show that there are several, even infinitely many, truths.

2. Given that someone asserts that there is only one truth; let us represent it, whatever it may be, by '*A* is *B*'. I will now show that in addition to this truth there must be a second one. For, to claim the opposite, one should have to assert the proposition 'Aside from the truth *A* is *B*, there is no other truth.' Obviously this assertion is different from the assertion '*A* is *B*', since it consists of entirely different parts. Hence, if this assertion were true, it would be a second truth. Hence, it is not true that there is only one truth, but there are at least two.

3. In the same way it can be shown that there are more than two truths for, whatever they are, it is clear that they are altogether different from the assertion 'Nothing is true but the two propositions *A* is *B* and *C* is *D*.' Thus, if this proposition is true, then it is a third truth, and we have mistakenly assumed that there are only two.

4. It is evident that this type of inference can be continued on and on. It follows that there are infinitely many truths; since the assumption of any finite set of truths involves a contradiction. Suppose that somebody wants to acknowledge only *n* truths; then these truths, whatever they may be, can be represented by the following *n* formulae: *A* is *B*, *C* is *D*, . . . *Y* is *Z*. By claiming that only these *n* propositions should be acknowledged as true, he asserts something which could be stated in the following form: 'Aside from the propositions *A* is *B*, *C* is *D*, . . . Y is *Z*, no other proposition is true'. This formulation makes it evident that this proposition has entirely different parts, and therefore is different from any of the *n* propositions '*A* is *B*', '*C* is *D*', . . . '*Y* is *Z*', taken by themselves. Since our critic nevertheless holds this proposition to be true, he vitiates the assertion that there are only *n* true propositions, since it is the $n+1$st.

NOTE: To my knowledge this simple way of proving that there are several or even infinitely many truths has never been used. Indeed it seems that it was considered satisfactory to force the sceptic to admit that there is at least *one* truth. The above method of proof can be altered in various ways. If one wants to avoid the apagogical form, he may proceed in the following way: If the proposition '*A* is *B*' is true, then the assertion 'The proposition that *A* is *B* is true' must undoubtedly also be a true proposition, and since the latter has different parts, it is a different proposition from '*A* is *B*' and thus a second and different truth. Similarly, from every true proposition of the form '*A* is *B*' the proposition 'therefore some *B* are *A*' can be deduced. The latter is a truth that is different from the given one, and so forth.

§ 33. *Discussion of several Objections*

B. discusses several objections against the assertion that there is at least one truth. He thinks that once this is seen, there will remain few doubts that there are infinitely many truths. The objections are (a) In order to be convinced by this proof, the doubter must ascribe to himself the ability to recognize truths. Thus he "presupposes what is to be demonstrated, namely that there are truths in themselves." (b) To be persuaded by this proof, the doubter must "presuppose as true the principle of inference that is used in it." (c) The proof presupposes that "the set of concepts expressed in the words 'all propositions are false' truly forms a proposition." Hence several truths are presupposed, and the sceptic might argue that he cannot make these assumptions without contradicting himself.

Against (a), B. claims that the ability to recognize truths is indeed necessary for the proof to be persuasive, but that this ability is not postulated as a premise of the proof. Hence there is neither circle, nor contradiction in the proof.

"(b). Thus, in order to be persuaded by a proof it is not necessary to presuppose that there are truths in themselves and that we can recognize some of them; nor is it necessary to acknowledge antecedently the validity of the forms of inference that occur in such a proof. It is indeed required that these forms of inference be valid, and that we do not doubt this validity *at the moment when we use them*, i.e. that we do not at that moment form the judgment that these forms of inference are perhaps invalid. But it is not at all required that we should have antecedently formed the judgment that these forms are valid. This should be clear even to a sceptic, if he would only observe himself during such an objection. For he draws inferences himself, which he could not do if each inference had to be preceded by a judgment acknowledging the validity of the underlying principle." The only way in which a sceptic could remain unconvinced would be to doubt the form of inference itself. But, this form is so natural and well known, that only imbecile and mentally disturbed persons could doubt its validity. "Everybody can pretend to doubt, but nobody with a developed faculty of judgment can doubt the validity of that form of inference for a moment."

(c). ' 'Every sentence is false' is a proposition' was indeed asserted as a premise, but B. claims that nobody can doubt its truth. "It is true that if someone admits that this proposition is true, then it is no longer necessary to persuade him that there are truths. But the fact is that we do not always realize how absurd is the assertion that there are no truths. The only merit of the above proof is that it makes this absurdity evident."

In a note, B. discusses Sextus' contention (*Adv. Log.*, II, 15) that the existence of truths cannot be proved, for the truth of the proof would have to be proved in another proof, etc. B. claims that the truth (or rather validity) of a proof does not have to be proved. "For if it is valid, then the reader is in the end more firmly convinced, the more often he thinks it through. He concludes from this effect, not from a new proof, that the proof is valid."

PART II

Of the Recognizability of Truth

§ 34. *What the Author Means by a Judgment*

Since I want to show in this part that we can have certain cognitions, I shall have to define the concept which I designate with the word 'cognition'. Since this concept, in my opinion, includes the concept of a judgment, I want first of all to reach an understanding concerning the latter.

1. I wish to take the word *judgment* in the same sense in which it is commonly used, as, for example, in the following proposition: 'God's judgments, unlike the judgments of men, are infallible', etc.

2. If this should not suffice, let me add the following. There is a certain common constituent in the concepts which are designated by the words 'to assert', 'to decide', 'to opine', 'to believe', 'to take for true' and similar words. In each of these concepts this common constituent is combined with a different additional concept. If we omit these additional concepts and merely think of what the designata of these words have in common, then we have in mind what I call *judging*.

3. A consideration of the following propositions is a third means of ascertaining exactly what I mean by the word 'judgment'. (a) Every judgment contains a proposition which is either true or false. In the first case the judgment is called correct, in the second incorrect. (b) Every judgment has existence. (c) A judgment does not have its existence by itself but only in the mind of some being which forms the judgment. (d) There is an essential difference between a judgment and the mere thinking or representing of a proposition. Thus, for example, at this very moment I am thinking the proposition that there are pygmies; but I merely think it and I do not assert it; that is, I do not judge it to be the case. (e) In God's infinite understanding every true proposition is present as an actual judgment. False propositions are also present in God's understanding, not, however as judgments, but merely as ideas of objects about which he judges. (f) Judgments of which we are conscious are activities of our mind which follow, and depend upon, a stage of mere observing and representing. Nonetheless, the act of judging is only indirectly dependent upon our will, namely only in so far as we have a certain

arbitrary influence upon what ideas we wish to entertain. (g) Each of our judgments, depending on the considerations that precede it, is carried out with greater or less force. I call this the confidence of the judgment. (h) Just as it is not a matter of our arbitrary decision whether a given judgment turns out affirmative or negative, so the degree of confidence with which we judge is not subject to our immediate control. (i) If a proposition appears just as probable as its contradictory then we can judge neither that it is true, nor that it is false, rather, we doubt. Thus to doubt a proposition means to have an idea of this proposition; but for lack of a sufficient reason neither this proposition nor its contradictory is asserted.* Just as in some cases we judge, whether or not we want to, so in other cases we doubt. (k). Nonetheless we sometimes say: 'You should not doubt, you can believe this, you may—or must believe this, you may have full confidence in it', etc. But these are mere manners of speaking which can only mean that if one were to pay proper attention to certain ideas then he would judge with such and such a degree of confidence.

These propositions will probably suffice to make clear the concept I connect with the word 'judgment'.

In a note, B. considers the etymology of the German word '*Urteil*'. In a second note, B. acknowledges that most logic books use the word 'judgment' in the same sense in which he uses it, except that it is sometimes not carefully enough distinguished from a merely entertained proposition. "What some logicians call a problematic judgment is actually no judgment at all but merely an entertained proposition, or to be more precise, the mere idea of a proposition, to which the subject may not assent at all. This confusion of a proposition with a judgment may have led certain logicians to claim that an assertion of the form 'if *A*, then *B*' contains two judgments, namely *A* and *B*. In my opinion, if somebody made this assertion, then he neither judges that *A*, nor that *B*, rather, he judges that the proposition *B* is a consequence of the truth of the proposition *A*."

§ 35. *Examination of other Definitions of this Concept*

B. objects that most other definitions are either circular or too wide.

Some logicians have tried to explain the concept by either stating the faculty which produces judgments, or the purpose for which they are made. But this does not explain the nature of judgments "it only touches upon an external property of judgments, a relation between them and our faculties and purposes." Moreover, if it is true "that the understanding has judgments as its effects, then the understanding must be defined as the faculty of

* If somebody doubts, then, in so far as he doubts, he does not judge. But if he pronounces the proposition 'I doubt', then he really does form a judgment.

judging, and the circle becomes evident." Others have tried to define 'judgment' in terms of such concepts as perception, awareness, consciousness, comprehension, cognition, etc., but all of these concepts must be explicated in such a way "that they already contain the concept of judgment as a part, each with the additional conception that the judgment is true or correct. One cannot say that somebody perceives or becomes aware of something, or that he is conscious of it, cognizes or comprehends it if he does not make a judgment. Similarly, we do not consider the forms of words 'to perceive or become aware of a thing, to be conscious of a thing' as synonymous with the expression 'to judge about that thing'. The reason for this is not that perceiving, becoming aware or conscious of something is not in itself already a form of judging. Rather, the reason is that those judgments which we call perceptions have a different content from judgments about the perceived object. For the former are judgments which have the perceiver as their subject, while the subject of the latter is the perceived object. Thus, if someone says that he has perceived the rose in front of him, he must have made the judgment 'I see a rose', even if he has not put it into words. But since this judgment does not have the rose but himself as its subject, we say quite rightly that he has not yet formed a judgment about the rose, but has so far only perceived it. (In my opinion, it holds quite generally that each perception is a judgment, so that we must attribute the ability to judge even to animals which have perceptual powers (though they may not have the ability to remember that they have made these judgments; hence they may not be able to judge that they have made a judgment, and thus not have any clear consciousness.) It is even more obvious that the expressions 'to comprehend' and 'to cognize' already contain the concept of judgment. But all of these words in their proper use contain the additional concept of correctness in them. If somebody merely fancies that between objects *A* and *B* such and such a relation holds, we do not say that he perceives or notices it, or is conscious of it, or cognizes or comprehends it. Hence it follows not only that these definitions are guilty of a circle, but also that they are too narrow, since they only apply to true judgments."

Other theories hold that judgments are ideas. But if ideas are taken to be constituents of judgments, then the definition is in error, and if 'idea' is taken to encompass judgments as well, then the definition becomes too wide: even though it now covers all judgments, it covers mere ideas as well.

Sometimes a definition of the concept of judgment is attempted in terms of the concept of thinking. But either 'to think' is taken as synonymous with 'to judge', and no explication is achieved, or it is taken as encompassing mere ideas, and the definition is again too wide. "Someone who merely *thinks* the proposition 'lightning rods are harmful', without believing that it is really the case, i.e. someone who does not judge can still be said to *think* about the relation between ideas and their objects." B. then points out that judgments are also not a kind of sensation: he distinguishes sensations from their ideas and again from judgments about them.

§ 36. *What the Author Means by a Cognition*

To convey what concept I connect with the word 'cognition' [*Erkenntnis*, knowledge], let me note, to begin with, that I take the forms of words 'to cognize something, to have knowledge or cognition of something, to know it, perceive it, comprehend it, etc.', all in one and the same sense, by abstracting from the various accompanying ideas which each of them carries with it, and by concentrating upon what they have in common. Thus by the word 'cognition' [knowledge] I mean any judgment which contains a true proposition, or, what comes to the same thing, which is true or correct. Thus I take each cognition to be a judgment; but the converse does not hold since there are incorrect judgments, which I call errors, as opposed to cognitions. Every cognition presupposes a being who judges and among whose judgments are some correct ones.

§ 37. *Justification of this Concept*

B. argues that his definition of cognition [*Erkenntnis*] conforms to ordinary usage. He points out that the phrase 'true cognition' [true knowledge] is a pleonasm. Also, "when we say that somebody has come to a recognition of his errors, we mean that he has realized that his opinions were wrong". We do not mean that he knew or cognized propositions that are false.

§ 38. *Other Definitions of this Concept*

B. begins by criticizing those logicians who divide cognitions [knowledge] into true and false, because they violate common usage. He then takes exception to definitions, such as that of Kant, which maintain that to cognize is to apply an idea to an object. He points out that the expression 'An idea is applied to an object' is ambiguous. "We can say of an idea that it applies to an object, if it is the idea of this object. If we take the expression in this sense, then all ideas which have an object, for example 'triangle', 'sun', 'man', etc. must be called cognitions; and only very few ideas, for example 'nothing', 'round square', etc., are not cognitions in this sense because they have no object at all." In another sense, an idea can be said to be applied to an object, if the object is the subject of an affirmative judgment, where the idea is the predicate. In this sense, every judgment, even if it is false, must be called a cognition. "For even in a false (affirmative) judgment, the idea which is its predicate is applied to the object which is represented in the subject. Thus if I form the false judgment 'the earth is a cube' I have the conscious idea 'cube' and I apply it to the object earth. Nonetheless no one will say that anything was cognized in this mistaken application."

Finally B. takes exception to attempts at defining 'cognition' in terms of intuition since there are many objects which we cannot intuit, but of which we can nevertheless gain true cognitions, for example God and spiritual beings.

§ 39. *Sense of the Assertion that Man can Know some Truths*

There could be some doubt, B. maintains, whether children and idiots know any truths. It is not necessary, however, to remove these doubts. What is to be attempted here is "to convince any reader, who has any doubts, that he, in his own person, has the knowledge of some truths".

§ 40. *How it can be Shown that we Know at least One Truth*

1. A person who has doubts not just about a particular class of propositions, but who doubts that we can know anything at all is to be called a complete or total sceptic.

2. While a person is such a complete sceptic, he cannot form any judgment, no matter what its content. "For a judgment is nothing but a proposition which the subject, with more or less confidence, takes to be true. If anybody forms a judgment, he gives us to understand that there is at least one proposition which he takes to be true (with more or less confidence) at the moment when he judges; he believes (with more or less confidence) that there is at least one truth; hence he is not to be called a complete sceptic. Consequently, in order to heal a sceptic, all we have to do is to get him to make a judgment whose truth is so irresistible to him that he tries in vain to doubt it, when, a moment later, we call his attention to the fact that in this judgment he has recognized at least one truth."

3. It might seem that the sceptic could be confounded if we simply could get him to answer any question whatsoever, even if he answers it with a 'I do not know'. Hence it seems that the most convenient method would be to get the sceptic to admit that he doubts everything, "but if we show him his error by pointing out that he was mistaken when he thought he doubted everything while in fact he did not doubt this one proposition, we would only give him another proof of the uncertainty of his own judgments and thus confirm his suspicions against all judgments."

4. It would therefore seem better to get the sceptic to admit his doubt in some particular proposition. But the proposition would have to be of such a kind that the sceptic continued to doubt it even after his confidence in his opinion grows. Thus we could ask him whether there are men on the moon, whether they have houses, etc. But even in such a case the sceptic could retract his initial answer 'I do not know' by claiming later on that he did not know whether he did not know, and so forth. "But could there not be a truth which is even more irresistible than the ignorance of a certain matter? A truth of which the sceptic must realize that he confirms it the moment he doubts or denies it? Now I think that there is such a truth and it is this: that he has ideas. For anybody who admits that he doubts proves that he has ideas; and even if he retracts this admission again and makes a change to the effect that it only seems to him that he doubts etc. etc.: the truth that he has ideas does not lose one bit of its previous certainty. Thus it seems to me that the best method for curing a complete sceptic is to ask him, either suddenly or after some conversation along the previously sketched lines, whether he does not at least have ideas? Whether or not he gives us the correct answer to this question, he will at least inwardly feel that it is true that he has certain ideas, and among others, even ideas of whole propositions, since otherwise he could not possibly doubt whether or not these proposi-

tions are true. If he feels this, we have won. For once he ceases to doubt that it is true that he has ideas he will no longer doubt that he has recognized this truth and that there is at least one truth which he knows."

§ 41. *Proof that we can Know Infinitely Many Truths*

If somebody admits that he has recognized truth *A*, he will then also admit that it is true that he has recognized the truth of *A*. Since the proposition 'I recognized the truth of *A*' is different from *A* itself, the recognition of a second truth has been admitted. This argument can be continued indefinitely.

§ 42. *Reply to several Objections*

At best, the arguments contained in the preceding two sections will suffice to liberate the sceptic from his doubts for a few moments only. He will sink back down into his doubts as soon as he recollects the ideas which caused him to be a sceptic in the first place. It would be easy to demonstrate to the sceptic that he contradicts himself as soon as he says that he doubts everything. But it has been pointed out already that the actual purpose, namely to heal the sceptic, is not advanced by this procedure. It is much more reasonable to discuss and refute the acknowledged or hidden reasons for his scepticism. In the sequel this is done in the form of a dialogue where *A* represents the sceptic, *B* someone who wishes to instruct him.

A: By agreeing that I have ideas, I admitted more than I should have. For by admitting this, I already presupposed my own existence, although I am not certain of it.

B: It is quite possible that there are moments at which you forget that you exist, i.e. it is possible that you do not think this truth while your attention is directed toward other objects. However, it is unlikely that you should have been confronted with the question whether you exist, that you should have become conscious of this question and that you should then have doubted whether you should answer it in the affirmative and that you should have said to yourself 'I do not know whether I exist'. It is easy to have an idea of, or put into words, the proposition 'I do not know whether I exist', but it is impossible to mean what it says. . . . It is also not necessary that your existence must be presupposed when you form the judgment that you have ideas, if by presuppose we mean that you must form the former judgment before you can get to the latter. For we do not infer that we have ideas from the fact that we exist, but rather from the fact that we have ideas, sensations, etc. we infer that we have existence. We infer: whatever has ideas must exist. Since we have ideas, we have existence. Therefore you have no reason to retract your judgment that you have ideas, and thus to lose the conviction that there are a few truths which you can know.

A: I am very much inclined to admit the truth of all this, but I begin to waver again when I consider the possibility that it could be the peculiar character of my reasoning faculty which makes it appear to me that these propositions are probable, while they are in fact false. For since there are beings who are wrong in some of their judgments, could there not be beings who are wrong in all of theirs, i.e. who do not know a single truth? How can I be sure that I am not a being of this kind?

B: I admit that there could be beings whose reasoning faculty is so deficient that, at least during a period of time, none of their judgments is true. However, it is certain that among the judgments which such a being forms during this period of time we will not find the judgments that it judges nor that it has ideas. For this judgment is not false, and the being would thus have formed at least one true judgment. Hence from the fact that you can form this judgment you can conclude that you do not belong to the class of beings who cannot recognize a single truth.

A objects that it is presupposed that he can recognize truth, for otherwise he could not recognize the truth of the proposition that he has ideas. *B* replies that it is true that he could not form this judgment unless he were able to recognize the truth, but the latter is *not* a presupposition of the former "it is not necessary that you must be convinced of the possession of this faculty before you can believe that that judgment is true. For we do not have to assume that we have the faculty to do certain things, before we can recognize that we can indeed do them. Rather, the opposite is the case: we realize that we can perform certain acts and we conclude that we have a faculty for these performances. Thus we do not believe that we can see or hear because we presuppose that we have organs of sight and hearing, but we conclude that we have these organs and faculties because we see and hear."

A: But what is gained by the recognition of a sterile truth like 'I have ideas'? I am much more concerned with making true and certain decisions about God, immortality, etc.

B: Very well, but if you have first recognized one truth, you can conclude that all those objections against your cognitive faculty which would prove that you cannot recognize a single truth are mistaken.

A: If I must, for the sake of consistency, attribute to myself the ability to form true judgments, then I can perhaps form a system of judgments which is internally consistent; but it does not follow that these judgments are objectively true. Because, to have objective truth, my ideas must agree with the objects to which they refer. In order to persuade myself of this agreement I must be able to compare my ideas with the objects in themselves. But this is impossible since I have knowledge of the objects only through my ideas of them. I can never observe the objects themselves to see whether they agree with the images in which they appear to me.

B: If it is impossible, then it is unreasonable to demand it. This misguided demand comes about only because of a misleading expression which is sometimes used in the explication of the concept of truth. You have probably been told that truth (i.e. transcendental truth) consists in a certain agreement between our ideas and the objects to which they refer. But this expression is not quite correct. We should rather say that our judgments are true if our idea of a certain object is connected with the idea of an attribute which this object really has. Hence if you want to persuade yourself of the truth of your judgments it is not necessary to compare your ideas with their objects: this is superfluous anyhow since you know from the outset that your ideas of the object agree with them. For if one of your ideas does not agree with a certain object then it is, for that very reason, not an idea of that object but of some other object, namely of the object with which it agrees. I must, however, remind you that this agreement between an idea and one or several objects to which it refers or whose idea it is should not be thought to be a kind of resemblance. The idea 'something', and the things to which it refers (i.e. all things that there are) have no similarity whatsoever with each other. For it is not by virtue of similarity that an idea represents this or that or no object at all. Thus in order to persuade yourself of the truth of a given proposition you need altogether different methods than a completely impossible comparison of ideas with objects to which they refer or which they represent. The complete enumeration of these methods can only occur in the sequel, but I will say enough about them at this point to show you that they are possible, though it is necessary that you should understand the meanings of several technical terms which I shall employ. If the proposition in question consists mainly of concepts, as for example the proposition that virtue deserves respect, or that the sum of any two sides of a triangle is longer than the third side, etc. then its truth and falsity will depend upon the composition of these concepts themselves; hence there are many cases in which you can persuade yourself of the truth of a proposition simply by paying careful attention to the concepts of which it is composed. Thus it is possible to recognize the truth that virtue deserves respect because you are familiar with the concepts of virtue, of respect and of deserving. You could not claim that you are familiar with, or in the possession of, a concept *A*, if you could not distinguish it from another, *B*, i.e. if you did not know that *A*, together with certain other concepts, forms a true proposition, while *B* does not. You can recognize truths of this kind (i.e. purely conceptual truths) because you know the concepts out of which they are composed. It is different with judgments which contain intuitions [*Anschauungen*] of external objects. These may be of one of the two following kinds: 'this (what I see here now) is red', where the subject idea of the proposition is a simple intuition ('this') and the predicate idea a concept ('red'); or they are propositions of the following kind: 'the same object which produces within me experience *A* is also the cause of experience *B*'. The following judgment is of the latter kind: 'the pleasant fragrance which I experience just now is caused by the red object which I see before me (i.e. by the rose).' Judgments of the first kind cannot possibly be in error, and you form them immediately on the basis of the two ideas of which they consist. However the truth of judgments of the second kind (I call them proper judgments of experience) does not merely depend upon your ideas (i.e. the ideas 'red', 'fragrance of a rose', etc.) but also upon the characteristics of the external objects which are represented by them. Nonetheless, to persuade yourself of the truth of propositions of this kind it is not necessary to know these objects in themselves in addition to the ideas which they produce in you; that is to say it is not necessary to find out what effects, aside from their effects upon you, these objects might have. For since you do not assert anything except that the object which has produced the idea *A* in you is the same object that has produced your idea *B*, you are simply speaking about the effects which certain objects have upon you, but not about effects that they might have upon other beings. Hence to ascertain the truth of such a proposition, if not with complete certainty, then at least with sufficient probability, nothing is required but that you should have perceived on a number of occasions that experiences *A* and *B* occur simultaneously. In the sequel you will see that all your judgments fall under the kinds of propositions which I have enumerated and that it is therefore never necessary to go beyond your ideas in order to ascertain the truth of your judgments.

A: No matter how willing I am to have confidence in my judgments, everything becomes doubtful again when I call to mind how often even our most confident judgments contradict each other and it becomes certain that an error has been committed. I have often taken something to be an indubitable truth only to discover later on that it was an error.

B: It seems that you do not mistrust your experience of repeatedly making erroneous judgments. But please consider where these errors could have originated. An error can intrude into our judgments only in one of three ways. Either our mind is so arranged that even among the immediate judgments, i.e. judgments that are not derived from others, some are false. Or some of the forms of inference which we use to derive other judgments from immediate judgments are fallacious. Or, finally, one of the propositions which we derive by a merely probabilistic argument, hence which is not certain but only probable, is false. But if you consider the first two cases possible, i.e. if you think that one of your immediately formed judgments is false or that one of the immediate forms of inference is invalid, then you must distrust indeed not only some, but all your immediate judgments and forms of inference. For all of them have only one and the same warrant, namely your immediate consciousness. You would then have to distrust all judgments which you deduce from others and you should therefore not form any judgments at all. But since you cannot do this, since you are certain that you know at least some truths, you must conclude that at least those judgments which you form immediately, likewise those forms of inference which you do not derive from others, are not in error. On the other hand you should not deny that error can arise in the third way, namely if you accept propositions which merely have probability with a degree of confidence appropriate to this probability. It is rational to keep the possibility of error carefully in mind and therefore tone down your degree of confidence whenever doubt has smaller disadvantages than a confident judgment which is incorrect. Conversely, it is equally rational not to withhold your judgment where doubt causes greater damage than even error. In such a case you can save yourself a lot of inconvenience if you pay no attention at all to the fact that error is not altogether out of the question. This is rational, I say, because when we cannot avoid danger altogether it is rational to choose the least danger. Thus it may be a merely probabilistic inference which leads you to the judgment that a certain dish is not poisoned; nonetheless if you would rather not eat at all than be poisoned, it would be even more probable that you would lose your life; hence you act rationally if you pay no attention at all to this possibility and eat with good cheer. Generally speaking, the experience that you have frequently been in error, and the realization that you cannot avoid error in the future, should make you careful in your judgments, but not suspicious of all your judgments or, what comes to the same thing, prevent you from making judgments at all.

A objects that there are errors which cannot be avoided, for example errors that occur in dreams or in demented states. *B* remarks that the errors of dreamers and demented persons are not immediate judgments "for example when a feverish person complains about the great heat in his room, while it is actually quite cold there, he is not wrong in the judgment of immediate perception that he has a sensation of heat; he is wrong merely in the (only probabilistic) judgment concerning the origin of this sensation, which he seeks in the stove while it is in his own body". But we do not have to be afraid that we could be as frequently mistaken when we are healthy and awake as when ill or asleep. "For we can sufficiently explain how in these extraordinary cases erroneous judgments come to be made." In many cases there are changes in our body which usually take place only when we are confronted by certain objects. When we are asleep, the lack of fresh impressions, and the association of ideas, causes some ideas to reach a high degree of vividness, which otherwise arises only through the actual confrontation with an object. While we are awake, objects of our imagination can be recognized by contrast with actual impressions. This contrast is not present during our dreams, hence "the mere play of the imagination is sometimes mistaken for a picture of reality. But since we cannot similarly explain how we could be deceived even when we are wide awake and healthy and observe all rules of caution, there is no good reason why we should be as afraid of error in this case as in the other cases."

A objects that it might be possible that they are asleep or demented at this very moment since the dreamer always thinks that he is awake, and since the demented person is not aware of his condition. *B* replies that neither the sleeper nor the deranged person can become aware of their condition as long as it lasts "but those who are in fact awake and of sound mind are in a position to assure themselves satisfactorily about their own condition. As concerns being awake, are you seriously concerned that you might perhaps be asleep at this moment? Perhaps you merely are not clearly aware from which criteria you gather that you are awake; but you are in the possession of such criteria and you can discern your waking state with such assurance that you could not even persuade yourself that you are now asleep in the sense in which you slept last night." We know that we are awake whenever our ideas follow each other in a predictable way. In particular, we can judge "what impressions will follow when we behave in a certain way, especially when we carry out certain bodily motions." Still, it is possible, that our organs of sense have gradually changed or were originally inadequate, so that all judgments based on them are erroneous. But it is a mistake to think that we could not find out if we are in such a condition. We would only compare impressions with each other or find out whether other persons judge a given object in the same way we do. Finally, if we know that we do not have a passionate preference for a certain opinion, that we are aware of opposing reasons, that others, whose opinion on a certain matter diverges from ours, have not had a chance to carry out certain observations which are prerequisite to a sound judgment, if their diverging opinion can be explained from the fact that they have not paid proper attention to the matter, if all persons who become aware of our reasons agree with us, if certain other presumably unbiased persons testify, that we do not belong to the class of demented or overly imaginative people: then we have no reason to be afraid that we are not of sound mind.

A: Very well, but are we safe from deception even in the condition which we call healthy and awake? Could it not be the case that we are awake in a dream of a different kind, namely in a dream from which we awake only upon death, if ever?

B: If you want to say by this that the external objects might appear to us in an entirely different way after we are dead or in another condition, or that they could appear in a different way whenever we are in any way transmuted, then this is indeed true; but it is by no means a proof that we are presently wrong when we call sugar sweet, and bile bitter, etc., so long as we are aware that sweet, bitter, etc., are relations in which external objects stand to our body, and both stand to our mind. In time, these relations will, nay must, undergo a change. But if your comparison of our present condition with a dream is to mean that we are equally often mistaken whether we are awake and of healthy mind or asleep, even in the case of judgments which we form as carefully as possible, then I must contradict you and refer you to my earlier reasons. We are not free from error, but there is a whole class of judgments which cannot be in error, and others have such a high degree of reliability, and have such a content, that we act rationally if we don't disturb ourselves with the thought that they might be mistaken.

§ 43. *One of the most Reliable and Useful Criteria of Truth*

B. maintains that one of the safest and most useful criteria of truth is this: "if a judgment is confirmed, i.e. forces itself upon us, whenever we test it (i.e. whenever we turn our attention to all apparent counter arguments), then it deserves our confidence. That is to say we are not misguided if this circumstance makes us form this judgment with greater confidence. For, the more often we have undertaken such a test, the less likely it is that we should not have become aware of the error in our judgment. And if we did not trust a judgment under these circumstances, then we should trust others even less and thus should not trust any of our judgments."

§ 44. *Some Remarks about the Usual Treatment of this Subject*

1. B. remarks that Augustin already used the proposition 'I have ideas' to heal the sceptic; and that Descartes' *cogito* comes to the same thing.

2. B. claims that he has used the Cartesian *cogito* in spite of some objections that were advanced against it, for example by Beattie. One of the objections is that the argument can be rephrased as 'I am thinking, therefore I am', "where the word 'am' already comprises the concept of being. But this is a confusion between two very different concepts. For, as the copula in the proposition 'I am thinking' the word 'am' has an entirely different meaning than it has in the proposition 'I am'. Only in the latter proposition does it contain the concept of being. This can be seen from the fact that the copula 'is' is also used in propositions whose subject does not exist, as, for example, in the proposition 'the impossible is not real'." Mayer's claim that the concept of the self contains that of an existing being is somewhat more plausible. This would be true if we defined the I as a 'being' that is conscious of certain ideas. But it is not necessary to proceed in this fashion, especially since the concept formed in this way is redundant. We indicate the pure concept of the ego if we say that it is the something that is conscious of certain ideas. Now the truth that this something has existence or being does not lie in its concept as a merely analytic truth, but derives from it as a consequence.

4. Some philosophers, Euler for example, declare complete scepticism as irrefutable. B. points out that we can of course not force the sceptic to admit his error; we can at best make him feel inwardly that he is wrong. Moreover "he can artificially maintain his doubt only in the hours of idle contemplation. But as soon as an occasion arises where he would suffer damage through his doubts, he not only realizes this as well as anybody else, but even takes some trouble to get himself out of danger by appropriate evasive action. He then shows complete confidence in the correctness of his judgments, and we can hardly stop his efforts, even if we shout at him please do remember that everything is doubtful."

B. then objects to a remark of Euler's, namely that the sceptic could not pay any attention to our arguments without violating his system. B. points out that the sceptic could pay attention to these reasons without having specially decided to do so. Moreover "whether or not we have this confidence in a given truth does not always depend upon our free will, but depends upon the characteristics of the ideas which together form this truth. It also depends upon our way of considering it and upon the additional ideas which happen to be in our mind at the time."

5. B. here concerns himself with the distinction between Academics and Pyrrhonists. According to him, the former position is properly expressed in this way: "Only one proposition is certain, namely that no other proposition is certain." But it has been shown in § 41 that a great many further propositions can be derived from this. § 40 was dedicated to the refutation of the other party. It seems, however, that the self-contradiction in the Pyrrhonist position was not always clearly noted. B. quotes Sextus Empiricus (*Outlines of Pyrrhonism*, I, 206): "for in regard to all the sceptic expressions, we must grasp first the fact that we make no positive assertion respecting their absolute truth, since we say that they may possibly be confuted by themselves, seeing that they themselves are included in the things to which their doubt applies, just as aperient drugs do not merely eliminate the humours from the body, but also expel themselves along with the humours." "But this is a confession against his will that the sceptic ceases to be what he

claims he is as soon as he makes the claim. He is a complete sceptic only as long as he is silent and does not form a single judgment either in words or inwardly. And only as long as he remains in this condition can *others* truthfully claim of him that he does not recognize a single truth. But as soon as he says it himself or even means it, this condition ceases to exist and his judgment is false."

6. The method of Crousaz, to put the sceptic in a position where he must judge or act, is said to be effective only for the moment. To effect a basic cure the sceptic must be shown that he can trust his judgments without committing a vicious circle, or begging a question.

7. It is quite difficult to determine the relations between critical philosophy and scepticism. Kant, on a number of occasions, declared scepticism to be an error and a scandal for philosophy. Nonetheless his and his followers' doctrine about the objectivity of our judgments does not seem to make for a big distinction between critical idealism and scepticism. "For the *Critique* says in arid words that our synthetic judgments concern only appearances or phenomena, i.e. only objects of a real or possible experience, but not things in general or in themselves, or noumena. Only about the former can we say anything that does not already lie in their concept. It is reasonable to ask what is to be understood by the expressions 'thing in general', 'thing in itself', or 'noumenon'. Ordinarily, the adjunct 'in general' or 'in itself' indicates that the concept of the word connected with this adjunct is to be taken in its complete generality without any tacit limitation. And thus the concept of a thing in general would be the highest concept of all, the concept of an object, or a something. But then it would be absurd to say that we can know something about phenomena, i.e. about things of a special kind, but not about things in general. Because if you know certain kinds then it cannot be said of you that you are unfamiliar with the total species to which these kinds belong. Hence if you know several properties of plain figures it cannot be said that you know nothing about figures in general. Consequently critical philosophers must have understood by things in themselves or noumena something different from things in general. We must conclude from the distinction that they make between them and phenomena that only things that we cannot experience are to be called things in themselves. Hence they claim that we cannot make any synthetic judgment about an object which we cannot intuit, and that it is impossible to have any knowledge of a truth which concerns a super-sensible object, such as God, or our soul, no matter how important this object may be for us. Of course this is not total scepticism, still, they make us doubt exactly where it is most important for us not to doubt."

8. Even worse than the Kantian position is the view that all humanly attainable truth is only *relative* or *subjective*, i.e. "that even the most meticulous attention to the rules of thought will not allow us to reach anything better than judgments which must appear true to us as long as we are human, but of which it must remain open whether they are objectively true or false." (This is the position of Search, Lossius, Abicht, Krug and Gerlach.) "But it seems to me that this manner of protecting himself against the advances of scepticism differs from total surrender only in that the right to form judgments is maintained, though on this view this can hardly amount to more than the right to be systematically mistaken. This reservation, based

on the conviction that we do not have to fear a refutation of our errors, is nevertheless of practical importance since it removes the most pernicious consequence of scepticism. From a theoretical viewpoint, however, we must find it unsound and contradictory. For it holds of all assertions that to make them is to declare them for true (indeed objectively true). Hence we contradict ourselves when we declare that we are incapable of any knowledge of objective truth. For example, Gerlach says that a judgment is true for us, if it is formed according to the laws of our human thinking. He allows us to apply this definition and to maintain that certain judgments are subjectively true, because it *seems* to us that they conform to those laws which we take to be the laws of our thinking. Hence he permits us to suppose that we know two objective truths, namely that the apparent laws of our thinking are identical with the real ones, and that a judgment which appears to be in accordance with these laws really is in accordance with them."

9. "The adherents of identity philosophy take an entirely different road toward the refutation of scepticism. They start with the customary definition of truth, namely that it is the agreement of ideas with their objects or, to use their own language, the agreement of the subjective with the objective. But they claim to have discovered that no complete truth can occur if it were not the case that the subjective and the objective, i.e. the idea and its object, are in a certain respect one and the same. Thus Schelling says (*System der transcendentalen Idealphilosophie*, Tübingen 1800, p. 4), "How idea and object can agree is downright inexplicable if there is not, within knowledge itself, a point where both of them are originally one, or where there is to be found the most complete identity of being and representing." . . . I think that this aberration has been fostered by the customary definition of truth. To begin with, truth should not have been given out as an attribute of mere ideas, but only of complete propositions; and not as an attribute of mental propositions only, but one that applies to objective propositions as well. Finally if the definition had not been given in terms of the ambiguous word 'agreement' [*Übereinstimmung*] but if this definition had simply stated that the truth is a property of those propositions which attribute to certain objects an attribute that they really have, then it would have become not impossible, but much more difficult for our identity philosophers to apply their identifying techniques."

§ 45. *Some Remarks concerning the Basic Laws of Thought as they are Discussed in other Handbooks*

B. points out that in place of the investigations which he has conducted so far, most logic books concern themselves with a discussion of the so-called basic or most general laws of all thinking. He formulates two such laws in the following way: If an object has a certain property, then it has that property, and if an object has a certain property, then it does not lack it. He maintains that the qualification 'at the same time' is unnecessary because "the time at which we can truly attribute a certain property to an object belongs to the idea of that object and not to the copula of the proposition; hence it is attached to the subject idea of the proposition. The proposition 'Caius is *now* learned' is not to be construed as if the 'now' belonged to the copula; rather, such time determinations belong to the subject and the proposition is to be construed in the following way: 'Caius in his present state (or: the present Caius) is learned'. From this it follows that a pair of propositions such as 'Caius is now learned' and 'Caius was not learned ten years ago' have different subjects."

While such propositions as the laws of identity and contradiction, and the law of the excluded middle, are undoubtedly true, it is certainly not the case that they contain the ground of all conceptual truth or that they suffice to guarantee the truth and correctness of all our thoughts, as has been maintained. B. does not think that any worthwhile truths can be derived from them. It also seems improper to him to call them formal or negative criteria of truth. It is indeed the case that no truth may contradict them, but the same holds for any other true proposition. He further takes it to be improper to call these laws "laws of thought" because "this nomenclature may induce one to suppose that they are laws which are bound up merely with our (human) thinking", while in fact "these propositions express properties which belong to things as such". Hence they belong to ontology, where they were quite correctly placed by Wolff.

The balance of the section is an extended discussion of the treatment of the law of contradiction, etc. by various logicians and philosophers.

BOOK TWO

THEORY OF ELEMENTS

§ 46. *Purpose, Contents, and Sections of this Book*

Good order requires that we should first deal with the parts of propositions, namely ideas, then with propositions in general, then with true propositions, and finally with arguments [*Schlüsse*] which are a certain kind of proposition stating that relations of deducibility [*Ableitbarkeit*] or ground and consequent [*Abfolge*] hold between other propositions.

The name 'theory of elements' seems suitable for this book "for, any treatise is composed of ideas, propositions, especially true ones, and inferences."

PART I

Of Ideas in Themselves

§ 47. *Contents and Chapters of this Part*

In this part, internal and external attributes of ideas will be discussed. Internal attributes can be recognized in an idea taken by itself, while external attributes "belong to an idea only in relation to something else." Hence this part will be divided into four chapters dealing with the following matters: the concept of an idea in itself, internal attributes of ideas, relations between ideas and other ideas, relations between ideas and other objects.

CHAPTER 1

Of the Concept of an Idea in Itself

§ 48. *What the Author Means by an Idea in Itself, and an Idea which Someone has*

1. The word 'idea' has been used before, but generally only in such contexts where its sense could be readily surmised. Now a somewhat more detailed explication becomes necessary.

2. Anything that can be part of a proposition in itself, without being itself a proposition, I wish to call an *idea in itself*, or simply an *idea* or *objective idea*. This will be the quickest and easiest way of conveying my meaning to those who have understood what I mean by a proposition in itself. Thus, the combination of words 'Caius has wisdom' expresses a complete proposition. The word 'Caius' itself expresses something that can be part of a proposition, as we have seen, although it does not by itself form a proposition. This something I call an idea. Similarly, what is designated by the word 'has' and indicated by the word 'wisdom' I call ideas.

3. If a person does not know the concept of a proposition in itself, I would attempt to explain the concept of an idea in itself by deriving it from the concept that is designated by the word 'idea' in common usage. Everybody knows, or at least we can easily explain to him, what is usually called an idea: Whenever we see, hear, feel or recognize something through any outer or inner sense, when we imagine or think something without judging or asserting anything about it, we can be said to have an idea of something. 'Idea' in this sense is a general name for any phenomenon in our mind, whose diverse kinds we designate through the names 'seeing', 'hearing', 'feeling', 'recognizing', 'imagining', 'thinking', etc., as long as they are not judgments or assertions. Thus, what I see if someone holds a rose before me is an idea, namely the idea of a red colour. Also, what I smell as I approach this object is an idea, namely that of a special fragrance, commonly called rose-fragrance, etc. In this sense, every idea requires a living being as the subject in which it occurs. For this reason I call them *subjective* or *mental* ideas. Hence subjective ideas are something real. They have real existence at the time when they are present in a subject, just as they have certain effects. The same does not hold for the objective idea or idea in itself that is associated

with every subjective idea. By objective idea I mean the certain something which constitutes the immediate matter [*Stoff*] of a subjective idea, and which is not to be found in the realm of the real. An objective idea does not require a subject but subsists [*bestehen*], not indeed as something *existing*, but as a certain *something* even though no thinking being may have it; also, it is not multiplied when it is thought by one, two, three, or more beings, unlike the corresponding subjective idea, which is present many times. Hence the name 'objective'. For this reason, any word, unless it is ambiguous, designates only one objective idea, but there are innumerable subjective ideas which it causes, and their number grows with every moment it is in use. We usually call subjective ideas *equal* whenever they have the same objective idea as their matter, and whenever we disregard differences in vividness, etc. Thus the subjective ideas that occur in the minds of my readers when they see the following word: 'nothing' should be almost equal to one another, but they are nevertheless many. On the other hand, there is only one objective idea which is designated by this word. By contrast, the word 'port' designates two objective ideas which are distinguished in Latin by the words '*portus*' and '*porto*'. Finally, there may be objective ideas which are not present in the mind of any thinking being except God. The number of wine-berries which grew in Italy last summer is an idea in itself, although nobody knows this number, etc.

§ 49. *Distinction of the Concept of an Idea in Itself from some others that are Closely Related to it*

In order not to omit anything which would make it easier to grasp the truly difficult concept of an idea in itself, I must call attention to the difference between this concept and several others that are closely related to it.

1. In no. 3 of the preceding section I used the expression that an idea in itself is the *matter* of an idea in the ordinary or subjective sense. I did this for lack of a better word. This could be interpreted as though by idea in itself I meant nothing but the *object* to which the mental idea *refers*. This is not my intention. The object to which an idea refers or (as it may be called for short) the *object of an idea* I wish to distinguish not only from the mental idea, but also from the idea in itself on which the latter is based. I want to say that whenever a mental idea has one, none or several objects, the corresponding idea in itself must also have one, none, or several objects. They must, indeed, have the same objects. By object of an idea I mean that

something (sometimes existing and sometimes non-existing) of which we say that the idea [representation] *represents* it, or *of* which it is a representation. The object belonging to an idea is most easily recognized when it is a real (existing) object. Everybody will understand me when I say that Socrates, Plato and others are the objects to which the idea 'Greek philosopher' refers. This example also shows how important it is to distinguish an idea, subjective as well as objective, from its object. An objective idea, as I have pointed out, is never something *existing*. The object, on the other hand, to which an idea refers can have existence (as in the present example Plato, Socrates, etc.) There is the further difference that one and the same idea frequently refers to *several* objects, as with the idea 'Greek philosopher'. It is not so simple to distinguish between an objective idea and its object (if it has such an object) whenever this object does itself not exist. However, hardly anyone will deny that in the same sense in which the idea 'philosopher' refers to the objects Socrates, Plato, etc., the idea 'proposition' refers to things that are called the pythagorean theorem, the theorem of the parallelogram of forces, the law of the lever, etc. The only difference is that Socrates and Plato are existing things, while these propositions, as propositions in themselves, do not exist. Finally, there are also ideas that have no object at all, as the idea 'nothing', '$\sqrt{-1}$', etc. It is certain that what we think by the word 'nothing' is a subjective idea. Therefore, there must also be an objective idea that corresponds to it. But it is absurd to think of an object to which these ideas refer. Whenever an idea in itself has several, or none, or one, *existing* object, then the difference between it and its object can be easily enough recognized. The greatest temptation to regard an objective idea as one and the same thing with its object occurs when a subjective idea has only one object which does not exist, as for example the idea 'supreme moral law'. However, even here the difference is obvious as soon as we remember that the matter which is the basis of this subjective idea must be an idea, while the object to which this idea refers (the subjective as well as the objective) is a *proposition*.

2. To distinguish the *word* which is introduced to designate an idea from the idea itself is even more important than to differentiate between an idea and its object. A word is always a physical object. (It exists at a certain time and in a certain location.) It is always a combination of sounds or of written signs. But an idea in itself, as I have said, does not exist. Moreover, there are not only objective, but also subjective, ideas (thoughts), for which we have no words at all, and conversely, in many cases we have several words which

designate one and the same objective idea: more than enough differences not to confound words and ideas.

3. Finally it must be noted that the word 'idea' is taken in such a wide sense—not only in common usage but also in handbooks of logic—that sometimes complete propositions and judgments are meant by it. This is the case whenever one speaks of true or false ideas, for not ideas in themselves, but only sentences or judgments can be true or false. Sometimes such a wider use occurs even when a contrast between ideas and judgments seems to be intended, as, for example, in the following utterance: 'The harsh judgments which Caius makes about me are a consequence of the ideas of me which he has acquired.' By 'ideas of me which Caius has acquired' I mean nothing but certain judgments whose subject I am. The word 'idea' is here not to be taken in such a wide sense (as I have already pointed out in § 48, no. 2) thus we should never think of ideas in themselves as being propositions in themselves, but only as being actual or possible parts of such propositions. This does not mean that an idea cannot contain a whole proposition or even several of them as parts. For even complete propositions can be combined with certain other ideas in such a way that the whole which is thus formed does not assert anything unless further parts are added. Hence, such a whole will not be called a proposition, but a mere idea. Thus, e.g. the words 'God is almighty' expresses a complete proposition which recurs in the following combination of words 'Knowledge of the truth that God is almighty'; but this new combination of words no longer expresses a complete proposition. However, a new proposition can be generated through a further addition, e.g. when we say 'Knowledge of the truth that God is almighty can give us much consolation'. Consequently, what is expressed by the words 'Knowledge of the truth that God is almighty' alone, is a mere idea, although one that has a complete proposition as a component.

§ 50. *Justification of this Concept*

I can guess in advance that many philosophers will not readily accept my concept of an idea in itself. I foresee that I will be told how curious or even nonsensical it is to speak of ideas which nobody has. Nevertheless, I believe that I am justified in asserting that this concept has reference, and that it is necessary to introduce it into logic.

1. When I say of an idea that it has reference [*Gegenständlichkeit*] I mean nothing but that there are objects which are subsumed under it, where I want the 'are' to be interpreted in the same way as in the

phrase 'there are truths' (§ 30). My reason for defining the concept of an idea in itself in the way in which I have done derives from the concept that I have formed of propositions and truths in themselves. It seems indisputable to me that every, even the simplest, proposition is composed of certain parts, and it seems equally clear that these parts do not merely occur in the verbal expression as subject and predicate (as some seem to think), but that they are already contained in the proposition in itself. If this were not so, then they could not appear in the verbal expression of the proposition. It also seems obvious to me that the constituents of a proposition in itself, which is not thought, can themselves not be thought. They can thus not be mental ideas, but only what I have above described as objective ideas. Thus, anyone who is willing to admit that there are propositions in themselves which are grasped whenever a mental proposition is formed, must also allow for ideas in themselves which are grasped in the mind of a thinking being whenever a mental idea or thought occurs.

2. But if the concept of an idea in itself has reference, then it should be introduced into logic, for without a clear understanding of objective ideas and their distinction from mental ideas, no clear understanding of propositions and truths in themselves is possible.

3. B. here acknowledges that it is somewhat contrary to common usage to speak of ideas [*Vorstellungen*] which nobody has. But he is not aware that there is a more suitable expression, except perhaps 'concept' [*Begriff*], which he will reserve for ideas that are not intuitions [*Anschauungen*].

§ 51. *This Concept has been Used before*

B. here maintains that the concept of an idea in itself, though often not explicitly recognized, occurs in the writings of many others, often at the expense of consistency.

1. He is not certain whether he should claim that "we find a more or less clearly recognized concept of an idea in itself at the basis of Pythagoras' theory of numbers, Plato's doctrine of ideas, and Stilpo's and the nominalists' statements about universals. (The nominalists seem to have correctly noted that a concept in itself is not something existing, while the realists noticed that it is not a mere name.)" But most commonly the concept of an idea in itself occurs in contexts such as that of § 21, no. 2, where one and the same idea is said to occur several times. B. cites a number of examples.

2. Occasionally, the locution 'possible concept' is used to designate what B. would call a concept in itself.

3. B. cites among non-logical works particularly Locke's *Essay*, where an idea is defined as the *object* of the understanding when a man thinks.

§ 52. *Unsuccessful Attempts at Defining these Concepts*

B. points out that his own explications of 'idea' and 'idea in itself' are not definitions. Thus, the concept of a part of a proposition that is not itself a proposition is said to be coextensive with the concept of an idea, i.e. "neither narrower nor wider", but it is not the same concept, hence not a definition. Similarly, to call an idea in itself the *matter* of a mental idea does not provide a definition, nor does the locution that a subjective idea is the *occurrence* in the mind of an objective one. However, no other writer has been able to produce an appropriate definition of this concept either. One of the more notable failures is Locke, whose concept is at any rate too wide, since propositions, too, may be the object of the understanding, and since the word 'object' is here ambiguous.

B. then remarks about Hollmann's and Reusch's picture theory of ideas. He says that a criticism of this theory goes too far if it claims that ideas and their objects have no similarity whatsoever, since a number (though perhaps small) of common properties can be found for any two objects, no matter how dissimilar. But whether something is a picture depends not only on the degree of similarity. "It seems to me that one object is a picture of another, if they are sufficiently similar so that there are certain purposes and occasions for which it is suitable to examine the first, rather than the second. Since occasions and purposes differ, one and the same object can have several different pictures. . . . If this explains the concept of a picture correctly, then the idea of an object cannot, strictly speaking, be called its picture. It is not an object which we examine in place of another, rather, it is what arises in our mind when we examine this object itself." Finally, "such an explication does not fit all ideas, since some of them do not have any objects at all, and can therefore not be their pictures."

Others have claimed that ideas are signs of their objects. B. objects that 'sign' can mean either a conventional sign, or a symptom, and that in neither sense is an idea a sign.

§ 53. *How these Two Concepts have so far been Treated*

B. shows that the word 'idea' [*Vorstellung*] was used in different ways by others. The reason why he uses it in the indicated way is that the treatment of elementary logic "requires a word which comprises no more nor less than all those parts of a proposition that are themselves not propositions, no matter whether they are intuitions in the narrower sense, or concepts, whether the agency which causes their occurrence in the mind is intuition, imagination, memory, understanding or even reason."

He criticizes Born (*Versuch über die Grundlagen des menschlichen Denkens*, Leipzig 1791, p. 4) for claiming that "the concept of an idea is in itself clear and cannot be analysed since it is simple." He replies that the simplicity of this concept "indeed precludes an *analysis*, but it does not follow that a clarification is superfluous." Maaß maintained (*Grundriß der Logic*, Halle 1793, § 1) that "whatever consciousness shows us directly is of itself clear. Hence it is of itself clear what an idea is, and what it means to have an idea." B. replies that "consciousness tells us indeed what an idea is, but it does not tell us what we mean by the *word* 'idea' ".

Finally, B. rejects as most peculiar Kant's claim that the concept of an idea [*Vorstellung*] cannot be defined, because the definition itself would contain ideas (*Logik*, A 41 f.). "If this were so, then we could also not explain what a concept is."

CHAPTER 2

Internal Attributes of Ideas in Themselves

§ 54. *Ideas in Themselves do not have Existence*

B. states that in this chapter he will discuss attributes that belong to ideas in themselves regardless of any relations in which they may stand. First he will consider attributes that belong to all objective ideas, then attributes that belong only to some important kinds.

"One attribute which all ideas have in common is that they do not have real existence." B. reiterates the argument of § 48, which is said to have been mentioned there only incidentally, and must here be discussed again, since lack of existence is an attribute which belongs to all ideas in themselves.

§ 55. *Ideas in Themselves are neither True nor False*

"A second attribute which belongs to all ideas is that neither truth not falsity can be attributed to them. Only complete propositions are true or false. By ideas, however, we mean parts of propositions which are not themselves propositions. Hence, neither truth nor falsity can be attributed to them." If we still speak of true and false ideas, we use the expression elliptically in one of two senses. If we say, for example, that the idea of omnipotence is true of God, we mean that the proposition 'God is omnipotent' is true. Secondly, an idea is sometimes called true when it has an object and not merely appears to have an object. Thus 'body that is bounded by four equal sides' is sometimes called a true idea, while 'body that is bounded by five equal sides' is called false. This use is identified with the transferred use of § 24, no. 5.

In a lengthy note, B. cites various supporting opinions.

§ 56. *Parts and Content of Ideas in Themselves*

It is important to note that some, though not all, ideas in themselves are composed of parts. To begin with, we are conscious of certain parts in almost every mental idea. Let us use as an example the idea designated by the expression 'earthling' [*Erdengeschöpf*]. Obviously, we are to think by this expression the same as by the more extended expression 'a creature that lives on earth'.

The latter consists of several words, and this alone shows that the idea which requires all these words for its designation is also composed of several parts. Clearly, the idea 'earthling', contains the idea 'creature' and the thought that this creature lives on earth. However, if the *mental idea* is composed of several clearly distinguishable parts, then there can be no doubt that the *idea in itself*, which forms the matter of this mental idea, must also consist of parts. Hence, ideas in themselves are sometimes composed of parts. The sum of the parts of which a given idea consists is usually called its *content*. Consequently every complex idea must undeniably have a content.

By the content of an idea we mean the sum of the components of which this idea consists, but not the way in which these parts are connected. It follows that an idea is not completely determined by its content, but that two or more different ideas can sometimes be formed from a given content. Thus the two ideas 'a learned son of an ignorant father' and 'an ignorant son of a learned father' are quite different, though they have obviously the same content. The same holds of such ideas as '35' and '53', and others.

NOTE 1: I have asserted that every idea in itself is composed of at least as many parts as we distinguish in the mental idea whose matter it is. By this I mean that the former may well have more parts than we clearly distinguish in the latter. It is true that we think a certain idea in itself, i.e. have a corresponding mental idea, only if we think all the parts of which it consists, i.e. if we also have mental ideas of these parts. But it is not necessarily the case that we are always clearly conscious of, and able to disclose, what we think. Thus it may occur that we think a complex idea in itself and are conscious that we think it, without being conscious of the thinking of its individual parts or being able to indicate them. It follows that from the consciousness that a certain mental idea consists of several parts we may infer that the corresponding idea in itself also consists of several parts. But conversely, we cannot infer with certainty that an idea in itself does not consist of parts merely because we do not distinguish any parts in the corresponding mental idea.

NOTE 2: It is well known that the occurrence or lack of occurrence of certain thoughts in our mind does not altogether depend upon our own will, but takes place, for the most part, according to certain necessary laws. The most important of these laws is that if two thoughts have ever been present in our consciousness at the same time then either of them will tend to re-introduce the other. For this reason it is not sufficient to know which idea someone connects with a certain sign or word, in order to insure that this idea alone will come to mind as soon as we observe the sign. Nothing is more common than that an idea is produced in us by a sign whose intended meaning we well know to be more or less different from the idea which it actually produces. Sometimes we add certain parts which do not belong to the content of the designated idea, at other times we leave some of them out. Thus, for instance, we may be told that the words 'spheric surface' are meant to express nothing but the concept of 'that particular spatial entity which contains all the points that are equidistant from a given point' and that we are not to include in our concept the ideas that this thing 'is a closed and curved surface, and a surface which has equal curvature in each point, etc'. In spite of this information we will not immediately be able to keep the words 'spherical surface' from stimulating an idea in our mind which contains some of these thoughts. This shows that we must distinguish between the idea *in itself* which is meant by a certain sign, and the *mental* idea that is caused by the perception of this same sign. Hence we must also distinguish the constituents of the former from the constituents of the latter. As the preceding note shows, it can happen that we do not consciously distinguish certain parts of a subjective idea although we actually think them. But it can also happen that a subjective idea actually either lacks certain parts, or has more parts than the designated objective idea.

§ 57. *Of some Cases in which an Idea is only Apparently Complex*

If some ideas are indeed complex, then it will be one of the aims of logic to discuss the more important kinds of complexity in ideas. But first of all it will be appropriate to consider some cases in which it merely seems as if an idea is complex, while it is actually not. Every word in a language serves to designate an idea and some of them even complete propositions. Therefore, it is only natural to suppose that each idea is composed of at least as many parts as there are words in its expression. This supposition is well founded in the most common cases, but it has some exceptions, as the following examples will show.

1. Let the idea *A* be designated by the word $\mathscr{A}$ (e.g. the word 'animal'), and let it cover very divergent objects (e.g. horses, dogs, birds, fishes, infusoria, etc.). It is not very surprising that such a word is not always taken in its total extension. We often use it when we have before our minds an idea of only some of the objects that fall under it (e.g. only the quadrupeds or the domestic animals, etc.). We often do this without indicating through qualifications of the word $\mathscr{A}$ the particular secondary ideas that have been added to *A* (for example that we presently think only of domestic animals). In this

way our imagination connects with the word $\mathscr{A}$ certain additional determinations, especially those that have most often been thought together with A, so that they come to mind as soon as we hear the word. Now we no longer think by the word $\mathscr{A}$ *all* but only *some* of the objects that it may represent according to its definition (in the case of the word 'animal', e.g., we merely think of domestic animals, etc.). Thus we have foisted another, less extensive and more complex idea upon the same word. If we want to avoid this and think the idea A without any arbitrary additions, then it will be necessary to take care that hearing of the word $\mathscr{A}$ does not cause us to add any of the additional concepts that are frequently connected with it. This is commonly achieved by an addition to the word $\mathscr{A}$ such as *in general* or *as such*. This addition, therefore, does not have the purpose of adding something to the concept designated by the word $\mathscr{A}$; on the contrary, it is supposed to prevent us from arbitrarily adding something that is not to be thought by this word when taken in its proper sense. Thus the expressions $\mathscr{A}$ *as such* or $\mathscr{A}$ *in general* merely designate the idea A, and not an idea that is composed of A and an additional idea. For example, the expression 'quadrangle in general' is not supposed to indicate anything that the word 'quadrangle' cannot designate by itself, and actually designates when taken in its proper sense. But since we often take this word in a narrower sense by thinking of a quadrangle that is rectangular and equilateral, i.e. a square, it is sometimes necessary to add 'in general', to insure that we do not add anything to the idea 'quadrangle' than what belongs to it essentially. A similar case is the expression 'truth as such' (§ 25), where the sole purpose of the phrase 'as such' is to indicate that we should think the concept 'truth' in its pure form, as distinguished from recognized truths, etc.

2. There are other words whose purpose resembles that of the expressions 'in general' and 'as such', namely the words 'any' and 'every', the latter if it is not taken as collective but as distributive. In my opinion, these words differ from the additions 'as such' and 'in general' only because we usually use them when it is our purpose to clear a word from additional concepts generally associated with the word 'some' (more about this later). I take it that the expression 'any man' means no more than what we think by the expresssion 'man' alone, indeed what we must think by the word 'man' if we do not want to limit it arbitrarily to one or the other class of men; the only point of the addition 'any' is to prevent such a limitation. This is especially necessary if a word is frequently connected with additional ideas which are expressed by the word 'some' or other words, or are

tacitly added. . . . It is for this reason that words that are seldom or never used in a narrower sense and words that designate an idea of which we are not used to distinguish subordinate kinds, rarely occur with the words 'any' and 'every'. Thus, we say that 'in *any* triangle the sum of all internal angles equals two right angles' since several kinds of triangles are known, and someone might think merely of, say, equilateral triangles if the word 'any' were not added. On the other hand, if we speak of right triangles we do not say 'in any right . . .' but merely 'in a right triangle the square over the hypotenuse, etc'. The reason for this is that the addition 'any' is here found superfluous, since we do not usually distinguish several kinds of right triangles. Hence there is no reason to fear that someone may think only of a special kind of them (e.g. the isosceles right ones). There is a further reason to suppose that the expression 'any' does not change the essential content of an idea: we can add it to concepts that represent only one singular object. Thus one can say 'Any omnipotent being is also omniscient' although, as is well known, there is only a single being to which the concept of omnipotence refers.

3. Definite as well as indefinite articles, which are found in several languages, do not usually produce a truly complex idea with the concept with whose sign they are juxtaposed; their purpose is only to indicate more clearly what manner of concept is to be designated by the word.

This is the reason why propositions whose grammatical expressions differ widely can all have the same sense. A case in point are the following four sentences: 'man is mortal' [*Der Mensch ist sterblich*], 'men are mortal' [*Die Menschen sind sterblich*], 'every man is mortal', 'all men are mortal'. This shows clearly that the expressions 'man' [*der Mensch*], 'men' [*die Menschen*], 'every man', and 'all men', mean the same thing as 'man as such'.

In a note, B. draws attention to features of Latin and Greek to argue that '*A*' and 'any *A*' (*A quodlibet*) mean the same thing. In a second note he points out that there is a distinction to be made between a concept *A*, and the concept 'a certain *A*'. Sometimes 'a certain *A*' is expressed simply by the indefinite article, sometimes by the definite article. A further analysis of 'a certain *A*' will be given when sentences of the form 'a certain *A* has property *b*' will be discussed.

§ 58. *The most Important Kinds of Complexity in Ideas*

1. First of all we can, as I think, distinguish *immediate* and *remote* parts in any idea that consists of more than two parts. Actually, it is

possible to distinguish immediate and remote parts in the case of any whole that consists of more than two parts. For, if such a whole is divided into a number of parts smaller than the total number of parts, then several or perhaps all of these parts will themselves be composed of parts, and we can call the parts into which the whole is at first divided the *immediate* parts, and the parts of these the ***remote*** parts of the whole. However, if it is altogether arbitrary into how many and what parts a whole is at first divided, and if the thus formed immediate parts are not distinguished from the more remote ones for any important reason, then it would be pointless to differentiate them by such a terminology. Thus we must show that there is a point to distinguishing immediate and remote parts of ideas. This would be the case if we called those parts of an idea immediate for which language contains special words which allow us to express the originally given idea in an alternative way. In this sense the immediate parts of the idea 'earthling' are expressed by the individual words of the expression 'creature that lives on earth', while the parts into which the ideas 'creature', 'lives', etc. can be analyzed are more remote. . . .

2. B. takes up two points, namely, first, he repeats (see § 49) that complete propositions may be parts of ideas. Secondly, he points out that it was rarely doubted that ideas can have parts which are themselves ideas. "The objection that an idea which consists of several parts that are themselves ideas is not one, but a class of several ideas, can be easily answered: this does not follow any more than that a machine which has several machines as parts cannot itself be called a machine. Just as it is a single thing with respect to the effects which it has, and which its parts do not have, so an idea that is composed of several other ideas is a single thing with respect to the objects which it represents, and with respect to the position in a proposition that it can fill."

3. B. suggests a second sense in which immediate and remote parts of ideas could be distinguished: if an idea contains a proposition, then the ideas in that proposition could be called remote parts of the idea. All other parts could be called immediate.

4. B. maintains that ideas which are part of an idea must be somehow connected, since they form the parts of a whole. He distinguishes direct and indirect connection and suggests that in the concept 'nothing', the concepts 'not' and 'something' are directly connected. The same holds for 'having a duty', where the parts are 'having' and 'duty', while in the concept 'man who has integrity', the concepts 'man' and 'integrity' are connected indirectly, namely through the ideas of 'having' and 'who'.

5, 6. B. continues the discussion of connecting and connected ideas. In the idea 'man who has integrity', the idea of having connects two ideas into a proposition, or propositional clause, namely 'he has integrity', while the pronoun connects this propositional clause with another idea. Thus the idea expressed by 'who' has the double function of connecting the propositional clause with the idea 'man', and of forming the subject of the propositional clause.* B. then distinguishes the kind of connection generally expressed by a relative pronoun, and that expressed by the expression 'to have'. The latter will be discussed in the sequel. The connection via a relative pronoun always connects an idea with a proposition, where the pronoun also functions in the proposition itself, but not always in the same way. Compare 'a space which is empty', 'triangle whose sides are equal', 'a man whom nobody trusts', etc. Whenever an idea is in this way connected with a proposition, it could be called the main part of the resulting idea.

7. From the preceding it becomes clear that the constituents of an idea frequently occur in an order which is not arbitrary. If this order were changed, another idea would result, as I have said above (§ 56). It is evident that we must not think of this often essential order of the parts of an idea as a succession in *time*. For an idea in itself is nothing real, and hence we cannot say of its parts that they exist at the same time, nor that they follow one another at different times. It is different with *mental* ideas, which are something real. In the case of man, the parts of such mental ideas indeed follow one another in time. We generally think the part that is the first in an objective idea sooner than the one which is second in the objective idea, etc.

8. I claim that in a complex idea the constituents sometimes occur in a certain order and sequence; hence we may call one of the constituents the first, another the second, etc. But I do not claim that this is *always* the case. Could there not be parts whose order is arbitrary or, better, which occur in an idea in itself without any order and sequence but rather like members of a *sum*? If I am not mistaken, this is indeed the case. To give only one example: when we think the idea of an A that has the properties b, b', b'' . . ., then we are not usually capable of thinking the several properties b, b', b'' . . . all at the same time, due to the limitations of our nature. We think them one after the other, i.e. in a certain sequence which may be one or another. Still, we feel clearly that this temporal succession does not belong to the ideas in themselves, that they are not changed whether we think the properties b, b', b'' . . . in this sequence or in another, e.g. b'', b', b. . . . Thus, in the objective idea, the individual ideas of the attributes b, b', b'' . . . are constituents that do not occur in any special order so that, e.g., one occurs in the fourth place, another in the fifth, etc., but they all stand in only one place (namely the last).

* Perhaps B.'s meaning is best made clear by coining a locution like 'man such that he has integrity', where 'such that' has the connecting function, while 'he' serves as subject of the propositional clause. ED.

§ 59. *Interpretation of some Grammatical Forms, Especially the Form 'This* A'

1. The most common kind of complex ideas are those that conjoin a main idea with a proposition, e.g. 'man who is honest'. Although overtly different, ideas like the following actually are of the same kind: 'an error of consequence', 'fruitbearing trees', 'the memory of Julius Scaliger', since they can be appropriately transformed.
2. Although the main idea is usually the one preceding the propositional clause, there are important exceptions, e.g. 'a pictured fish', where not 'fish', but 'picture' is the main idea.

3. A specially important kind of idea is that which we usually express in the form 'this (or that) $\mathscr{A}$'. I believe that this expression is taken in two different senses. In one, which I call the more exact sense, 'this $\mathscr{A}$' means roughly the same as 'this, which is an $\mathscr{A}$'. Here, the idea which is designated by the word 'this' does not refer to any object other than the one that is represented by the complete idea: this $\mathscr{A}$, even if it is taken all by itself. The addition 'which is an *A*' expresses an attribute which already belongs to the object associated with 'this', and is employed merely for the sake of greater clarity. In this sense we take an expression such as 'This fragrance (which I just perceive) is sweet'. By the word 'this' here, we mean the particular perception we have at this moment. That it is a fragrance is an attribute which already belongs to the object represented by 'this'. Thus it is not the constituent '*A*' or 'which is an *A*' which restricts our idea to the particular object that it has. The case is different when we take the expression 'this *A*' in its *second* sense. In this case we want to have 'this *A*' refer to any *A*, as long as it does not lack the additional attribute that we designate by our 'this'. However, this additional attribute is usually that it is that particular *A* which we had thought immediately before. Thus, the expression 'these assertions' has only the following sense: 'all those assertions upon which we just now directed our attention' or 'of which we just spoke'. Hence, in the first sense of 'this *A*', the idea designated by 'this' is the main part, and the idea A (or rather the sentence 'which is an *A*') forms the secondary part, but with the peculiarity that this addition is not required to confine the extension of our idea to the one object that it has, since the 'this', taken all by itself, refers only to this one object. In the second case just the opposite holds. The idea *A* forms the main part, and the idea designated by 'this' occurs as the secondary part. It confines the extension of the total idea to the one or more objects which have the attribute indicated by 'this'. It must not be overlooked that ideas of the first as well as the second

kind often occur in propositions although the demonstrative pronoun 'this', or a similar one, does not explicitly appear. Languages that have definite articles frequently use just the article to express such a matter. Sometimes, no special sign is put at all, and one has to guess from the context that an idea of the form 'this $\mathscr{A}$' is present. Thus, in the proposition 'the table is round', not just any table is meant, but this table, and the meaning of the proposition is 'The table which we have now before us is round'.

§ 60. *Concrete and Abstract Ideas*

B. draws a distinction between abstract ideas (or *abstracta*) and their corresponding concrete ideas (or *concreta*). The abstract ideas are always attributive, while the corresponding concrete ideas have as their object things which have the attribute in question. Thus, if *b* is an abstract idea, 'something that has (the attribute) *b*' is the corresponding concrete idea (to be indicated by *B*). Given this definition, there are obviously many ideas that are neither concrete nor abstract, for example 'something', since it is neither attributive, nor has the form 'something that has (the attribute) *b*'. The same holds of the ideas 'nothing', 'this *A*', 'Socrates', and others. Abstract ideas are sometimes simple, while concrete ideas are always complex, since they consist of the ideas 'something', and of the clause 'which has *b*'. On the other hand, the linguistic expressions of concrete ideas are sometimes shorter than the expressions for the corresponding abstract idea; cf. 'animal' vs. 'animalhood'. "The reason for this is that during the invention of the languages *concreta* were signified by special words earlier than *abstracta*, and even now we have reason to speak of the former much more frequently than of the latter. Finally, the linguistic expression is sometimes misleading, since one and the same word may be used to stand for a *concretum* or for an *abstractum*, for example 'justice'."

In a note, B. criticizes several authors for distinguishing concrete and abstract ideas depending on how they arose in the mind. "Some ideas that I have called concrete, e.g. 'man' are called abstract if they have their origin in an awareness of those aspects which several objects have in common, where the differentiating features are disregarded. But it is obvious that every general concept (i.e. every concept that refers to several objects) must be called abstract in this sense."

He then criticizes Kant and others for defining 'abstract' and 'concrete' in terms of the use to which ideas are put, rather than by means of their inner characteristics.

In a second note, he points out that some logicians seem to have overlooked the difference between the extensions of a concrete and its corresponding abstract idea. "They have assigned the same objects to the former as well as to the latter. In my opinion, the objects that fall, e.g., under the abstract idea 'virtue' are quite different from those that are comprised under the corresponding concrete idea 'a virtuous person'. The abstractum is an attributive idea; hence it can comprise nothing but attributes. Thus 'virtue' covers the attributes veracity, charity, etc., while the concretum 'a virtuous person' has in its extension such persons as Socrates, Aristides, etc".

§ 61. *There Must also be Simple Ideas*

By a simple idea, as the word indicates, I mean an idea which has no parts whatever, whether mere ideas or complete propositions. I believe that I can show in the following that there are such simple ideas. Every object, even the most complex, must have parts that are not themselves complex, but altogether simple. If the number of parts of which a whole consists is finite, then the truth of this assertion is evident. For, in this case we must come to indivisible, i.e. simple, parts after a finite number of divisions, e.g. bisections. However, there may be wholes which contain an infinite number of parts, as we find, for example, in any spatial extension, any line, surface or body. In the case of such objects, no division, if it generates only a finite number of parts, like a bisection or trisection, etc., will yield simple parts, no matter how often we repeat it. This creates the illusion that such an object does not even consist of simple parts. I assert, nevertheless, that such a whole, too, must have parts which are simple. Complexity is an attribute which can obviously not exist without there being parts that produce it (i.e. parts that contain the reason or condition for it). If these parts are themselves complex, then they merely explain a complexity of a *certain kind* but not the *complexity as such* of the whole. In order to explain the latter sufficiently, there must be parts that are no longer complex, but simple. This is a condition for complexity that does not require any further condition.* Hence, e.g. lines, surfaces and bodies contain parts that are not further divisible, but simple, i.e. points, which, however, are not of the same kind as the wholes that they generate, since it takes an infinite number of them. Thus geometers, who take the word 'part' in the narrower sense of 'part of the same kind', do not usually call them parts. Therefore, any idea, however complex it may be, even if it contains infinitely many parts (if that is possible), must have parts that do not allow further division. These simple parts cannot be propositions, since every proposition, taken as such, is complex. Since the only parts of ideas are either propositions or other ideas, the simple parts must be ideas. Thus it follows that there are simple ideas. Since, according to § 56, there are also complex ideas, a legitimate classification can be founded upon this difference among ideas. I shall show in the sequel that this difference is of considerable importance.

* Hegel viewed the matter in the same way (*Wissenschaft der Logik*, Nuremberg 1812, vol. 1, p. 142).

§ 62. *There is no Idea with Greatest Content*

It should be noted that all ideas have a common property, namely that they can be transformed into different ideas by the addition of other ideas or complete propositions. The new ideas will have a greater content than the original one since they contain the given idea as a part. Consequently, one can say that no content of an idea is so great that it would be impossible to indicate another idea with a greater content. Hence the compossibility of ideas goes *ad infinitum*.

He gives several examples to prove the above point, one of which is that a new idea can be generated by conjoining two given ideas with 'and'.

§ 63. *Whether the Parts of an Idea are the same as the Ideas of the Parts of its Object*

It has often been said that an idea of an object, if it is correct, i.e. not merely assumed to be an idea of this object, has a certain agreement with it (cf. § 29). The obscurity of this expression caused some philosophers to think that this correspondence between an idea and its object is some kind of similarity in the composition of the two. Thus they assumed that the constituents of an idea must be ideas of the parts of its object. Thus Abicht says in his *Logik* (Fürth 1802, p. 362). "The concept of an object must allow us to distinguish as many parts of ideas in it as can be differentiated in the object of this concept." And p. 363: "the completeness of a complex concept is proved as soon as it is shown that its object has *such* and *only so many* parts." The assumption that a completely simple object can be represented only by a simple idea is a consequence of that opinion.

It seems to me that this view is erroneous. First of all, there are ideas that have no object at all, e.g. the idea 'nothing', 'round square' and others. It is evident that we cannot claim their parts to be ideas of the parts of their objects. Hence, the above assertion holds at most of ideas that have an object. However, if it is true that among the parts of ideas there are often complete propositions (§ 58, no. 2), then it will again not be possible to say that every part of an idea is an idea of a part contained in its object. It is often assumed that in such a case it is not the complete proposition, but only one of the ideas in it that designate a part of the object. This is indeed sometimes the case. Thus the idea of a right triangle, i.e. the idea of a triangle which has a right angle, contains the idea of a right angle in the proposition 'which has a right angle'. This idea does in fact indicate a part that occurs in a right triangle. The same holds for the ideas 'a mountainous land', 'a book of engravings', etc. But this is not always the case, i.e. not every proposition that occurs among the parts of an idea contains an idea that indicates a part of the object. Consider the following ideas 'A land without mountains', 'A book without engravings', etc. In these cases the constituent ideas 'mountains', 'engravings' obviously do not indicate parts that the object has, but those that it lacks. The same can be shown even more conclusively with ideas such as 'the eye of the man', 'the gable of the house', etc. Who could deny that in the first of them the idea 'man' and in the second the idea 'house' occur as constituents? Therefore, if the disputed opinion were correct, then the whole man would be part of his eye, and the whole house part of its gable, etc. Finally, there are objects that have no parts at all, while their idea is clearly complex. For example, any mental being is an altogether simple object, but its concept is composed of several parts. Thus, we shall have to give up the notion that every single part of an idea indicates a corresponding part of its object. But one could still think that, conversely, every part of the object must be indicated by one or more of the constituents of the idea. We can see at once that this cannot hold for those parts of the object which do not necessarily belong to it so as to make it an object falling under the given idea. No one will expect that in the idea 'flower', which comprises, among other things, this rose bush, there should occur constituents that show how many roses, buds and leaves this particular rosebush has. But it is perhaps different with parts that an object must have in order to be called the object of a given idea. If it is reasonable (and many seem to believe so) that the idea of every attribute that necessarily belongs to an object for it to be the object of a certain idea must occur in this idea as a constituent, then there can be no argument that for every part that necessarily belongs to an object so that it can fall under a certain idea, there must also be found a part in this idea which represents this part of the object (i.e. an idea that refers to this part of the object). For, that a thing consists of such and such parts belongs to its attributes. For reasons that I shall develop in the next section, I hold that opinion to be incorrect. Thus I have no reason to believe that the idea of an object must be composed of the ideas of all the parts that must necessarily belong to the object so that the object falls under the idea.

§ 64. *Whether the Parts of an Idea are the Ideas of the Attributes of the Object of that Idea*

B. here examines the theory that "Every idea of an object is the class of the ideas of all the necessary attributes of that object." He wants to show "that there are always several constituents of an idea which do not represent attributes of the corresponding objects, and, conversely, that every object has attributes which, though they belong to it necessarily, do not occur as part of its idea."

1. B. offers the following as a proof of the first part of his assertion, namely that there are sometimes constituents in an idea which do not represent any attribute of its object: if an object has the attributes b', b'', etc., its idea, fully analysed, would have the form 'something that has (the attributes) b', b'', etc.'. Thus it would also contain the ideas of something, of having, etc., hence more ideas than there are attributes in the object.

2. He gives several arguments for the second part of his thesis, namely that there are necessary attributes in an object which are not represented by any part of its idea.

a. There are many examples where a certain attribute is said to follow necessarily from the concept of a thing, although we are not conscious that we have thought this attribute in the concept.

b. Under the theory in question, coextensive ideas would have the same parts and thus be identical, which is clearly false.

c. We can see that such an object as $\sqrt{2}$ has infinitely many attributes, since it is the sum of infinitely many fractions: $1 + 4/10 + 1/100 + 4/1000 + 2/10000$. The idea of $\sqrt{2}$ determines that it is these fractions rather than others. Now the attributes of these individual parts (i.e. which numerator belongs to each denominator) can be envisaged as attributes of $\sqrt{2}$ itself, since any change in one of the fractions results in a change in the sum. Hence if it is the case "that any attribute of an object that follows necessarily from its idea must be considered part of this idea, then the idea $\sqrt{2}$ and similar ideas must be composed of infinitely many parts. Hence, in order to be able to say what we think by such an idea, we should have to think these ideas at least obscurely, hence think an infinity of ideas with our finite understanding."

d. B. notes that there are some fairly obvious counter-arguments against the above points, and it is for this reason that he does not want to rest his case with them. It will be said against (a) that an attribute that is deduced from a concept after extended consideration was already thought obscurely, and hence was present, in the concept itself. The counter to (b) is that coextensive ideas may differ from one another in that by one of them we think those parts clearly that are obscure in the other, and vice versa. Against (c) it might be said that the requirement to think an infinity of attributes with a finite understanding in a finite time does not entail an obvious contradiction as long as it is not required that these ideas are thought consciously.

These three arguments were not advanced in order to persuade the reader that the theory in question is mistaken. Rather the point was to show more fully what consequences follow from this theory. The following proofs, however, should be decisive. But before I state them, I wish to remind the reader that I have to prove my point only about ideas in themselves, not about mental ideas, since the latter are not under discussion at all at the moment. It is clear that an idea in itself does not comprise those other ideas which occur incidentally whenever we think it. Thus the objective idea 'triangle' does not include an idea of the letters *t* and *e*, though they could possibly be before my mind when I think this concept. Hence from the mere fact that a certain attribute comes to mind, that we have an idea of it, whenever we have an idea of a certain object, i.e. from the fact that our subjective idea of an object is always accompanied by an idea of that attribute, it does not follow that the *objective* idea of the object contains an idea of that attribute as a part. Even if it is the case that whenever we think the concept of an equilateral triangle the concept of equiangularity comes involuntarily to mind, it does not follow from this that the objective concept of an equilateral triangle includes the concept of equiangularity as a part. If all this is granted, it is not difficult to show that there are attributes which belong necessarily to the object of an idea without being represented by parts of that idea. The following consideration will show that this is the case with the concept 'equilateral triangle' and countless others.

All equilateral triangles have the attribute of having equal angles. But it is clear that the concept of having equal angles does not occur in the concept of an equilateral triangle, taken as such. For this concept is generated by connecting the concept 'triangle' with the proposition 'which is equilateral'. Now it is obvious that the concept of being equiangular does not occur either in the concept 'triangle' nor in the proposition 'which is equilateral'. Hence it does also not occur in the whole that is formed of nothing but these two parts. It may well be true that the idea of having equal angles does indeed appear whenever we *think* this concept, thus in the case of a *mental* idea of an equilateral triangle. But this does not concern the idea in itself, since it does not consist of any parts other than those given to it. Otherwise we would have to claim that it is impossible to combine the concept 'triangle' and the proposition 'which is equilateral' without adding a number of other parts, among them some that contain the concept of having equal angles. But this is certainly false, since the formation of an idea in itself is something altogether arbitrary, to the point that we can combine parts that indicate attributes which contradict one another. Thus even the combination of the following ideas gives us a concept, although one that does not have any object: 'A triangle that is and is not equilateral'. Possibly this proof becomes clearer to some readers, if we compare the concept of an equilateral triangle with that of an equilateral quadrangle. Just as the concept of an equilateral triangle is generated when we add the proposition 'which is equilateral' to the concept 'triangle', so the concept of an equilateral quadrangle is formed by adding the proposition 'which is equilateral' to the concept 'quadrangle'. Thus if the first concept contains the concept of having equal angles among its parts, then the same must be true of the second. But no one will say of the latter that it contains the concept of having equal angles among its parts, since it is not even true that every equilateral quadrangle is equiangular.

e. According to § 61, there must be simple ideas. If *a* is such a simple idea, then the proposition 'the idea *a* is simple' is true. Thus the concept 'simplicity' expresses an attribute which must belong to the object of the concept 'the idea *a*', namely to *a*. Hence, if the theorem under discussion were true, then the concept of simplicity would have to occur as a part in the just-mentioned concept. But this is not the case, for it neither occurs in the concept of an idea, nor in the concept of *a* itself.

f. Simple ideas, too, must, perhaps with a few exceptions, represent something. Let this something be of any character whatever, it must have the attribute 'of being something'. Hence, if every attribute of an object, which it must have in order to be the object of a certain idea, had to occur as a constituent of this idea, then all simple ideas would have to contain the idea 'something'. But in order to remain simple, they could contain no other ideas besides. Consequently, all simple ideas (with the exception of those that have no object at all) would have to be identical with one another, or, to put it better, there would be only one single simple idea, namely 'something'. Clearly, this is absurd.

g. However, if we admit (as we must) that there are several different simple ideas, then of any of them, *a*, we can assert that it is not another, *b*. But this assertion already states an attribute of the idea *a*. Hence it would have to be a constituent of the idea of this idea, i.e. a constituent of the idea 'the idea *a*', which is also absurd.

h. Finally, assume that there is an idea *A* which represents precisely one object, and that this object has the attribute *b*. Assume that it follows from this that *A* has a part that represents *b*. Then *A* would have to be of the form '*X* which has *b*'. Now the attribute *b* either belongs to all *X* or only to some. If to all, then we shall have to admit that *X* has itself the constituent *b*, and that it is of the form '*Y* which has *b*'. Hence, in this case, *A* would contain the part *b* doubly, or several, even infinitely many, times, depending on how far the argument is continued. In any case, there must be an idea which contains all the parts of *A* with the exception of *b* (even if the latter were repeated infinitely many times); let us call this idea *Z*.

What has just been asserted of *X* also holds for *Z*. Either the property *b* belongs to all or only to some of the objects that fall under *Z*. If to all, then *Z* is itself an example of an idea all of whose objects have an attribute that is not represented by any of the constituents of *Z*. In the second case, it follows that the idea '*Z* which has the attribute *b*' is not empty, i.e. that it has a referent. But this assertion states a necessary attribute of that idea, though it is not a part of the idea of that idea, i.e. of the idea 'the idea of a *Z* which has the attribute *b*'. For the thought that this idea has a referent lies neither in the idea of an idea in general (since there are also inconsistent ideas), nor would it matter if this thought were in fact contained in the ideas *Z* or *b*. For, from the fact that *Z* and *b*, taken severally, have referents, it does not follow that the idea of an object that has both attributes also has a referent. Thus we see that there are attributes that follow from the ideas of certain objects although they are not represented by any of their parts.

In a note, B. carefully considers and disputes Herbart's contention that the concept of a thing with several attributes is inconsistent.

§ 65. *Comparison of §§ 56–64 with the Received Theory*

B. considers his theory of the composition of ideas to be of great importance. A comparison with rival theories will make it more fully understandable.

All logicians, except Hegel, have agreed that there are complex ideas; but many have defined complexity in a different way. Thus Fries claims that the magnitude of the content of a concept is continuous. But for a magnitude to be continuous it must have an infinity of parts, like a line; and since there are simple ideas, not all contents can be of continuous magnitude. Also, the two ideas '*A* which has *b*' and '*A* which does not have *b*' differ from each other only in the 'not'. Hence their difference cannot be decreased, which contradicts the law of continuity.

B. claims that his theory that complete propositions can be parts of ideas is wholly new. At this juncture, he tries to anticipate some criticisms: "If it is objected that the part of an idea which I have called a proposition, is not actually a proposition since in such a context it does not *state* anything, then I would reply in the following way: It is true that the part which I have called a proposition does not assert anything about the object of the complete idea of which it is a part. Nonetheless, it states which determination must be added to the object represented by the main part of the idea, in order to arrive at the object represented by the total idea. Thus in the idea: 'a man who has integrity' the words 'who has integrity' state that the attribute *integrity* must be added to the object which is represented by the first part of this idea (man), if we want to get the object represented by the idea as a whole."

B. finds that, as a rule, logicians have thought that all complex ideas are compounded in the same way; some took them to be sums, others assumed that they were products. Also, opinions differ widely on the question whether there are simple ideas. But the most severely misleading error was the confusion between the part of an idea and the attributes of its object. "This confusion is apparent, for example, in Hollmann (*Logik*, Göttingen 1746, § 52) when he claims that it is an error to form a complex idea of a simple object or a simple idea of a complex object." But it was too obvious to be overlooked for long that it is often necessary to form a complex idea in order to represent a simple object. Conversely, however, it was generally taken to be impossible that an object, which is itself compound, should be represented by a simple idea or even by an idea which is not compounded of as many parts as the object. Kant seems to have been of this opinion when, in the *Critique of Pure Reason* (B 39 f.), he concludes from the fact that space is infinitely divisible that the idea of space cannot be a concept 'since no concept, as such, can be thought as if it contained an infinite number of ideas within it.' Hence his argument was this: "since space itself consists of infinitely many parts, therefore the concept of space must also consist of infinitely many parts. . . . In my opinion it is not necessary at all that a concept from which it follows that its object consists of a certain number of parts must consist of equally many parts (perhaps the ideas of these individual parts). The concept: 'the totality of truths' consists of a very modest number of parts, namely the parts into which the concepts of totality and of truth can be analysed. But the object which is represented by this concept, namely the totality of truths itself, demonstrably comprises an infinite number of parts."

B. then acknowledges his indebtedness to Kant, whose distinction between analytic and synthetic judgments clearly requires a sharp distinction between the attributes of an object, and the parts of the idea which represents it. "Kant taught us that there are truths (analytic truths) in which a component of the subject-concept is predicated of that subject; and that there are other kinds of truth (synthetic truths) where something is predicated of a subject which is not present among the components of the subject idea. He goes as far as to claim that the major concern of every science is the recognition of such synthetic truths, and that all the theorems of mathematics, physics, etc. are such synthetic truths. If this is acknowledged, then it is easy to see that there are innumerable attributes of an object which can be deduced with necessity from its concept, even though we do not think of them as parts of this concept."

Since Kant was so concerned with the distinction between analytic and synthetic judgments, one might have expected that he would have become aware of the distinction between the attributes of an object and the components of an idea that represents this object, "since the predicate of a synthetic truth can at best be an attribute of the subject but not a component of the subject idea." But in his logic he violates this distinction and reasserts the old doctrine "that content and extension of a concept vary inversely, since the more a concept has *under* it, the less it can have in it. I am going to give an explicit refutation of this proposition later on (§ 120). Here I only want to indicate that this doctrine simply rests on the false assumption that every idea *A* (e.g. the real), which stands under another idea *B* (e.g. the possible) must be composed of *B* and some other idea, and must therefore contain *B* as a component."

B. then speculates upon the reasons for the confusion between attributes and components. He thinks that the most common form of idea, namely '*A*, which is *B*' was taken to be the only form, and that the less important components 'which' and 'is' were neglected, so that only *A* and *B* were considered components of the idea. "This immediately led to the thought that the components of an idea indicate attributes of its object. From this belief came an inclination to assume that, conversely, every idea of an attribute (at any rate of a necessary attribute) must be a component in the idea of the object."

Finally B. remarks upon the importance of terminology. Once it was recognized that not every attribute of an object is thought in its idea, the concept of something that is thought in an idea was discovered. It would have been reasonable to give the name 'part' or 'component' of an idea to this concept. Instead, the attributes in question were called essential, original, or constitutive attributes. Though this terminology is not without merit, "it favoured the thought that a concept is nothing but a class of constitutive attributes, i.e. that there are no other parts in a concept (or an idea in general) than attributes". This then led to the confusion of the necessary attributes of an object with the components of its concept.

§ 66. *The Concept of the Extension of an Idea*

1. It has been noted before that for most, though not for all ideas, there exists something to which they refer. This something I have called their object or referent, using the word 'object' in its widest sense; in § 49 I have already tried to convey what concept I connect with it. In order to avoid any possible misunderstanding, it will be well to add the following: I always take the phrase "a certain idea applies to an object" to mean that this object is of such a nature that the given idea represents it, i.e. I use it in a sense which is different from the one employed when one says that in the proposition '*X* has (the attribute) *b*' the attribute *b* is applied to the object *X*. The expression 'is applied to' here simply means 'is asserted of', and the idea *b*, or rather the attribute that it represents, is *attributed* in this proposition to the object indicated by *X*. If this proposition is true (as the proposition 'God is omniscient'), then we can indeed say that the object represented by *X* (God) is at the same time *represented*—not by the abstractum *b*—but by the corresponding concretum, namely 'something that has the attribute *b*'. But in the case of a false proposition (e.g. 'man is omniscient'), it is not true that any object that falls under *X* (any man) is represented by the idea 'something that has *b*'. (Something that has omniscience.) This holds even though in this proposition the attribute *b* is asserted of, i.e. applied to, *X*.

2. Of ideas that have one or more objects I say that they have *reference*. On the other hand, of those that have no object corre-

sponding to them I say that they *lack reference* [*gegenstandslos*]. Once it is known of an idea that it represents certain objects, one can ask which and how many of them it represents. By indicating the objects to which a certain idea refers, we indicate the *range*, *extension*, or sphere of this idea. By these expressions I understand that particular characteristic of an idea by virtue of which it represents only those and no other objects. Hence only ideas that have one or more objects have an extension. This extension is indicated (by using other ideas, of course) when the one or several referents are individually enumerated. Thus the extension of the idea 'man' is indicated, if the beings that are represented by this idea are individually indicated. Of each one of these objects, also of each class of them, if it is not the total sum, we say that it is *part of* the extension of the given idea, or that it *belongs to* its range or *falls under* the idea, *is contained in* it, *is subordinated to* it, *can be subsumed under* it, or *is comprised or included in* it. Thus we say, e.g. that Julius Caesar belongs to the extension of the idea 'man', the duty of truthfulness to the extension of the idea 'duty in general', etc.

3. In order to determine completely the extension of an idea that has several objects, it is not sufficient (according to the definition just given) merely to determine the number of these objects. It must be indicated precisely which objects they are. The number of these objects, as any number, has a certain *magnitude*, which is called the *scope of the extension*. Many ideas have an infinite number of objects falling under them. Thus, for instance, the ideas 'line' and 'angle' include infinitely many objects to which they refer, since, as is well known, there are infinitely many lines and angles. Since the infinite in the respect in which it is infinite does not allow any determination, the scope of the extension of such ideas cannot in itself be completely determined, but only in relation to others, and in limited respects. For instance, it can be determined by saying that the scope of the idea A has the same size as that of another, B. Or the extension of A is part of the extension of B, or vice versa, etc. Since these methods of determination depend upon the relation of one idea to another, they will not be treated until the next chapter.

4. It can be seen from what has been said that the determination of the scope of an idea, or the answer to the question whether or not a given idea lacks reference and, if it does not, which and how many referents it has, cannot always be found by examining this idea alone. Rather, many other things must be taken into consideration, and often knowledge of the most fortuitous events is required. For example, a historical investigation tells us that the idea 'heirs to the

throne of Genghis Khan' has exactly four referents. Nevertheless, the extension of a given idea does not belong to its relations, but to its inner attributes. For the extension of an idea does not vary depending on whether we consider it in relation to one object or another. Rather, we must investigate which and how many objects of an idea are included in the class of all things. Since there is only one such class [of all things] there is only one extension for any given idea. Hence, this extension cannot be counted among the relations. Also, the circumstance that a given idea has an extension and which extension it has, is altogether invariable and cannot grow or diminish in time. Thus, e.g. the idea 'man', if this is what we call any rational sentient being that has lived on earth at any time or will live there, has its definite extension which never has, nor ever will, change. Also, the idea 'a presently living man' changes its objects only if we change the meaning of 'presently', i.e. if we change one of its constituents and thus the idea itself.

5. B. now discusses the imagery connected with such words as 'sphere', 'range', 'extension'. He maintains that the image does not have to be that of a body (sphere) or of an area (range); it could also be that of a line.

In a note, B. criticizes the view that the extension of an idea is the class of all objects in whose idea it is contained as a component. He maintains "that it is often, but not always, the case that an attribute which belongs to an object, or, rather, an idea into whose extension this object belongs, is a component of the idea of that object. Thus the attribute of being an equiangular figure belongs to each equilateral triangle. Hence we can say that an equilateral triangle belongs in the extension of the concept of an equiangular figure. But equiangularity is not a part of the concept of an equilateral triangle, which contains only the concept of equilaterality, from which the property of equiangularity merely follows."

In a further note, B. points out that the exact explanation of the *scope* of an extension meets with difficulties similar to the explanation of the related concept of the magnitude of a spatial extension. "It is because the concept of the extension of an idea is so closely related to the concept of a spatial extension (e.g. of a line, a surface, or a body) that we can use the latter as graphic representations of the former. . . . Now it is a mistake to define the length of a line or the content of an area simply as the quantity of points that are contained in it. In just the same way it is unsatisfactory to say that the scope of the extension of an idea is nothing but the quantity of all the objects that fall under it. The concept of the magnitude of an extended thing is properly defined in the following way: It is the magnitude assigned to that thing according to a rule which assigns magnitudes to the parts of that thing in such a way that the sum of the magnitudes assigned to the parts equals the magnitude assigned to the whole. Thus the length of the line *ad* is called a magnitude λ, if it is assigned to *ad* according to a rule such that if magnitudes α, β, γ are assigned to the pieces *ab*, *bc*, *cd* according to the same rule, then $\lambda = \alpha+\beta+\gamma$. Similarly I define the scope of an idea as a magnitude which will be assigned to the set of all the objects of this idea by a rule which makes it the sum of those magnitudes which this same rule assigns to the individual parts into which the total set might be divided. Now the magnitude of a spatial extension can be expressed variously, depending upon the choice of units; and the same holds for the scope of an idea. For example, if you take as your unit the scope of an idea which has only one object, then the scope of an idea which has ten objects is 10, and the scope of any idea equals the number of objects that fall under it. Ideas which have infinitely many objects will then have an infinitely large scope. But if we take as our unit the scope of an idea which itself has infinitely many objects, then it might be possible to find more or less exact numerical expressions for the scope of many other ideas that have infinitely many objects."

§ 67. *Ideas without Referents*

It is true that most ideas have some, or even infinitely many, referents. Still, there are also ideas that have no referent at all, and thus do not have an extension. The clearest case seems to be that of the concept designated by the word 'nothing'. It seems absurd to me to say that this concept, has an object too, i.e. a something that it represents. Somebody might in turn find it absurd that an idea or representation should have no object at all, and thus represent nothing, but the reason for this is in all likelihood that he means by ideas merely mental ideas, i.e. thoughts, and that he identifies the *content* of these mental ideas (i.e. the ideas in themselves) with their *objects*. It is reasonable to say that the thought 'nothing' has a content, namely the objective concept 'nothing' itself; but that the latter should also refer to a certain object, is an assertion that can hardly be justified. The same holds of the ideas 'a round square', 'green virtue', etc. We do and must think something by these expressions, but this is not the *object* of these ideas, but the ideas in *themselves*. Incidentally, it is evident from these ideas themselves that no object can correspond to them, since they would attribute contradictory properties to it. However, there are probably also ideas that lack reference, not because they attribute contradictory properties to their objects, but for some other reason. Thus the ideas 'golden mountain', 'a presently blooming vine' are perhaps without object, although they do not contain a contradiction.

NOTE. It might be objected that ideas of which I have said that they lack reference must nonetheless have extensions, since they are occasionally compared with respect to their extensions, and some of them called wider than others. For example we find it quite proper to say that a person who shows the impossibility of round polygons achieves *more* than one who merely shows the impossibility of round squares, since the latter follows from the former, but not conversely. This conclusion seems to be valid only if it is assumed that the concept of a round polygon is wider than that of a round square. I admit that the impossibility of round squares can be inferred from the impossibility of round polygons in general. I deny, however, that for this inference we need the minor premise 'round squares are a kind of round polygon', and that we thus have to attribute an extension to these two concepts in order to draw the above conclusion. It is after all the case that the assertion that no round polygons are possible follows from the proposition 'No polygon is round'. (Or every polygon is something that is not round.) Hence the conclusion that no square is round, and thus that round squares are impossible, follows from the minor premise that all squares (not only the round ones that do not exist) are also polygons. Moreover, in § 108, I shall discuss a sense in which the relation of subordination can also be applied to ideas that do not have reference.

§ 68. *Ideas with only a Finite Number of Objects, Singular Ideas*

To my knowledge, nobody has denied that there are ideas which refer to an infinite number of objects (§ 66, no. 3), but that there are also ideas that have only *one* object or a *finite* number of objects has not been generally accepted. Let me demonstrate both through obvious examples.

1. The ideas connected with the following expressions have obviously only one object: 'the philosopher Socrates', 'the City of Athens', 'The fixed star Sirius', likewise any idea of the form 'this *A*' (in the more exact sense of § 59). It might be objected that by each of these expressions we do indeed *think* of merely one object, but that our idea of this object actually fits several other objects as well. Or else it could be claimed that, though there may not be any objects in *reality* that have the attributes represented in our idea, such objects are at least in the realm of *possibility*. To this I reply: What we *think* by an expression is the same thing as the idea that we connect with it. Thus, even though it may be the case that the *words* 'The philosopher Socrates', etc., can be interpreted in such a way that they fit an entirely different object as well, the idea which we presently connect with this expression still has only one object, for the very reason that we *think* only one. Concerning the argument that there may be some objects that have all the properties that are mentally connected with our idea if not in reality, then at least in the realm of

possibility, the following must be said: our idea of the philosopher Socrates claims the reality of this object (about 2000 years ago). From this it follows that something that did not exist at that time cannot be an object of this idea. The same holds of all ideas that claim the reality of their object (either for a particular time or for all time). Of such ideas we cannot say that they include more objects than there are *real* things constituted in the way in which the ideas describe them, since merely possible things that are not real (at a certain time) are for this reason not among them. Other examples of ideas that have merely one object are 'highest moral law', 'pythagorean theorem', and the like, whose object is plainly a proposition in itself, hence something that does not have, nor can it attain, existence; hence it can be reckoned neither among the real nor the possible things. There is another whole class of ideas that obviously have and can have only one object, namely ideas of the form 'the class of all things that have (the attribute) *b*', e.g. 'the universe', 'the totality of mankind', 'the class of all truths', etc. It is the very form of these ideas that makes it impossible that they should represent more than one object (which is always composed of parts, and sometimes of infinitely many of them). (The only exception arises if there are no things at all that have the attribute *b*.) This is due to the fact that there can be only one class of *all* objects that have a certain property *b*. Some logicians have called ideas that have only one object "single ideas" [*einzelne Vorstellungen*]. Since every idea as such is only one it seems more correct to call them *singular* ideas. All ideas that represent several (at least two) objects, will be called, in distinction to these singular ideas, *common* or *general* ideas.

2. If singular ideas are allowed, then it cannot be doubted that among general ideas there will also be some that have only a *finite* number of objects. For, if *A*, *B*, *C* are singular ideas, then the idea 'one of the things *A*, *B*' will have only two objects, the idea 'one of the things *A*, *B*, *C*', obviously only three, and the idea 'one of the things *A*, *B*, *C*, . . .' will have as many objects as different things '*A*, *B*, *C*, . . .' are mentioned. The following idea provides another example of such an idea: 'A geometrical theorem known to Euclid'; this idea can certainly have only a finite number of objects, since no human mind comprehends an infinite number of truths. Similarly, the idea 'natural number between 1 and 10' has neither more nor less than eight objects, etc.

In a note, B. points out that it is not the task of logic to determine the actual sizes of extensions, and their relations to each other, but merely to define the concept of an extension.

§ 69. *Redundant Ideas*

1. B. chooses the name 'redundant' for ideas "which have more parts than is necessary to represent the objects which they represent." Hence these ideas contain parts which can be left out without changing their extension. Example 'a quadrangle whose opposing sides are parallel and of equal length'. Parts that can be omitted are called superfluous.

2. In the above example, either the part 'parallel' or the part 'of equal length' can be left out, so long as the other is retained. Such parts are called superfluous relative to each other. On the other hand, in the idea 'a triangle that has the *property* of being equilateral', the concept 'property' is the only superfluous part. Hence it is not superfluous relative to any other part and can be called *absolutely* superfluous.

3. B. here refers back to ideas of the form 'this *A*' (§§ 59, 68), which are also redundant.

4. "According to the definition given in no. 1, only ideas that have an object should be called redundant. Still, sometimes it is very reasonable to call an idea redundant even if it lacks reference, as for example the idea 'a round quadrangle that has corners'. To extend our concept so as to include ideas of this sort we may say that such an idea is redundant in parts α, β, γ, . . . under these conditions: It has certain other parts *i*, *j* . . . which we can consider variable; whenever *i*, *j* . . . are replaced by other parts in such a way that the resulting idea no longer lacks reference, then this resulting idea is redundant in part α, β, γ, . . . in the sense of no. 1. Thus the idea 'a round quadrangle with corners' is redundant with respect to the part 'corner' if we consider the part 'round' as variable: as soon as we replace the latter in such a way that the resulting idea has an object, it will be redundant in the sense of no. 1, i.e. it does not change its extension if the words 'with corners' are deleted".

In a note, B. examines likely objections. While it is possible to form redundant *verbal* expressions, can we also connect ideas if one of them merely repeats what another has already posited? For example, the expression 'an *A* which is an *A*' can easily be uttered; but the thought and the absolute idea expressed by these words is perhaps none other than the idea *A* alone. B. replies that, first, not all redundant ideas have the form 'an *A* which is an *A*' and, secondly, whenever we say the words '*A* which is a *B*' "so long as we replace *A* and *B* by meaningful words, our utterance indicates a certain compound idea; and whenever we attend to what we say, these words cause in our mind a corresponding mental idea. But since every mental idea has a corresponding idea in itself, there can be no doubt that there is an objective idea which is expressed by these words." Finally, the notion that the idea 'an *A* which is an *A*' is the same as the idea *A* is simply a mistake. They are not the same because they consist of different parts.

As a rule, a redundant idea is defined as one that gives more characteristics than are necessary for the determination of the object that is to be represented by it. "I abandoned this definition since the question whether a given idea *A* indicates a sufficient number of characteristics to designate the object it is to represent can only be asked and answered if the idea is not examined by itself, but as part of a proposition of the form '*X* is *A*'. This is the reason why this definition makes it appear as if the redundancy of an idea is a property that cannot be recognized in the idea alone, but only by a comparison with the object to which it is applied in a proposition. This is, of course, false. I can see from an idea alone whether or not it is redundant, as in the case of 'a sphere that has no corners', whatever object it may represent." B. then admits that his own definition, too, requires a comparison, namely of the idea in question with some other idea which has the same extension, but is not redundant. He points out that, nonetheless, redundancy is an inner and not a relative characteristic of ideas, since that second idea is generated from the first by deleting some parts. Finally, he remarks against the alternative definition that an idea can be redundant without even having a sufficient number of characteristics to determine the object to which it is applied. Thus the idea of a figure that has four sides and angles is redundant, although it would not suffice to determine its object, if we wanted to apply it to a square.

"In no. 3 I suggested that we should consider certain parts of an idea as variable, and examine what ideas result if we substitute any other ideas for these variable parts. In the sequel I shall frequently have to make use of this method. I hope that no one will object to the expression 'variable' and its cognates and that it will not be assumed that what I said contradicts the assertion that ideas are not something that exists. From the latter it would indeed follow that they cannot undergo variation. If I say that in a given idea, e.g. 'a wise man', a constituent e.g. 'man' is to be envisaged as variable and may be replaced by any other idea, this merely means that we are to survey all ideas that have the same constituents and order of constituents as 'a wise man', except that they have other ideas in the place of 'man'. Thus we are not actually concerned with a *variation* in the proper sense of the word."

§ 70. *Consistent and Inconsistent Ideas*

Just as an idea may contain parts which are consequences of other parts, so the opposite may happen, namely "that constituents are included in an idea which attribute properties to their object that contradict those that follow from the remainder." Since the properties that are indicated by such an idea contradict each other, they will not be found in combination; hence such ideas do not have objects. "Inconsistent ideas are of the kind 'an *A* that is at the same time *B* and *P*', where *P* and *B* are such that the two propositions 'every *B* is *M*' and 'every *P* is non-*M*' both hold"; they are best called *inconsistent* ideas, but the names 'empty', 'imaginary', and 'impossible' are also in use. All other ideas have been called '*consistent*', 'real', or 'possible'. The following misunderstandings must be guarded against:

a. "The expression 'empty' must not be connected with the content of these ideas. Thus one should not think of these ideas as without content, for all complex ideas have a certain content, namely certain constituents. They are called empty only with respect to their extension . . . it should be remembered that inconsistent ideas are not the only ones that are empty in this sense; there are many other ideas that do not have objects, although they are not inconsistent, for example the idea 'nothing'. The peculiarity of inconsistent ideas is that they lack objects because they attribute contradictory properties to any possible referent".

b. "It seems to me completely false to say of an inconsistent idea that the thought of it, i.e. its subjective idea, is impossible. We certainly have ideas of this kind as soon as we hear such combinations of words as 'a round square', 'a regular pentahedron' and similar ones. We should then be committed to the view that we think nothing at all by these combinations of words, or, at any rate no more than by such senseless words as 'abracadabra'. But we can prove that this is not the case since we can establish such truths as 'there can be no regular pentahedron' or 'there can be no negative square number'. In order to show that these propositions are true, we have to engage upon special investigations during which we have to reflect upon different objects in the two cases."

NOTES: If it is admitted that there are false propositions, then it must also be admitted that there are inconsistent ideas. "As often as we utter a proposition that asserts something false and impossible, for example the theorem of Hobbes (*Quadr. Circ.*), that the ratio between diameter and circumference of a circle is 5/16, and if we think this (or even imagine it to be true) and do not just say it with our lips, then we have before our mind's eye a certain inconsistent idea, namely the idea of an object which has the property attributed to it in this proposition. In the example, it is the idea of a circle in which the ratio between diameter and circumference is 5/16. Since every mental idea requires an idea in itself as its matter, there is no doubt that there is an idea in itself of such a circle and hence that there are inconsistent ideas." The objection that we can combine these words but not the corresponding thoughts may sometimes be well founded, but certainly not in all cases. Such mental ideas are impossible only in the sense in which we sometimes say of improper conduct that it is impossible.

It is also true that we cannot generally develop imagery that corresponds to an inconsistent idea, for example, for the idea 'a square that is round'. But the ability to develop imagery is not decisive for the distinction between inconsistent ideas and others, since the idea 'a body bounded by twenty-four equilateral triangles' does not have an image that is either more or less distinct than that of the idea 'a body that is bounded by twenty equilateral triangles', although the former is inconsistent and the latter is not.

Since it is often not obvious that a given idea is inconsistent, and since it is of great importance not to confuse consistent and inconsistent ideas, it is reasonable that the treatise of a given science should list certain truths containing inconsistent ideas. "Thus a treatise of geometry should state that one or even two straight lines cannot bound a plane and that therefore the idea of a plane bi-angle is inconsistent; it should state that a space cannot be enclosed by one, two or three planes, and that there are no four lines that can meet in a point at right angles to one another, that the idea of a regular body bounded by five, seven, nine, ten . . . nineteen planes contains a contradiction, and many other such truths which only assert that certain ideas are inconsistent. After all, analysis establishes a great number of important theorems that contain the inconsistent concept of $\sqrt{-1}$."

In a further note, B. comments that the distinction between real (consistent) and imaginary (inconsistent) ideas was not always drawn in the same way. Leibniz, e.g. criticized Locke (*Nouv. Ess.* II,XXX,5) for requiring sometimes that a real idea actually have an object, while at other times requiring merely that such an object should be possible. "Leibniz corrected this error by saying that an empty idea becomes imaginary if it presupposes the existence of its object. I agree completely with this view. For example, I call the idea of a regular chiliagon real, even if there is no actual chiliagon. For in this concept, as in all concepts of spaces, there occurs only the thought of possibility, not of actuality. On the other hand, the idea 'Alexander, the father of Philip', I call imaginary, since it has no real object, although it presupposes that there is such an object. It is indeed quite reasonable to distinguish two kinds of imaginary ideas, namely those where the assumption that they are non-empty violates a conceptual truth, and those where the contradiction arises in other ways." This will become clearer after the discussion of conceptual truths in § 133. A preliminary clarification is given through two examples: "an example of the first kind would be the concept of an equilateral right triangle, since the impossibility of such a triangle follows from mere concepts (from purely *a priori* truths). An example for the second kind would be the idea of 'Alexander, the father of King Philip' since the two contradictory properties which result from this idea for its object follow from it only after the addition of the empirical truth that Alexander was not the father, but the son of King Philip. This difference seems to have been in the minds of those who differentiated merely empty ideas from inconsistent ones and reckoned among the first group the idea of a man who lived to be 997 years old."

§ 71. *Two Consequences*

1. It follows from the definition of an inconsistent idea that is must be complex. Therefore, simple ideas are always consistent, and often have referents (cf. § 66).

2. Inconsistent ideas may be parts of consistent ideas, e.g. 'the mathematician who first used the concept $\sqrt{-1}$.'

§ 72. *What the Author Means by Intuitions*

Among the ideas so far discussed the most important are these: singular ideas as far as extension is concerned, and simple ideas as regards content. Ideas that combine these two properties, i.e. that are simple and yet have only one object, would be even more remarkable. The question is whether there are such things. The extension of an idea generally decreases only when its content is increased, i.e. if more determinations are included and the idea is thus made more complex; hence one feels tempted to doubt that any simple idea can have an extension so small that it has only one object. Thus the extension of the idea 'watch' is much decreased if we add to its content the determination that it should fit into a pocket, and thus form the idea 'pocket watch'. The extension of this idea becomes still smaller if we add a further part, namely that this watch is to have a golden case, and thus generate the idea 'golden pocket watch'. Thus it appears as though we have to include a great number of parts into the content in order to get an idea of smallest extension, i.e. one that represents only one object. It appears as if such ideas could never be simple. However, if I could prove that even among our subjective ideas there are many that are truly singular in spite of their simplicity, then it would follow that among the objective ideas some are also both simple and singular, since for each subjective idea there is a corresponding objective one. I think I can prove this in the following way: As soon as we direct our attention upon the change that is caused in our mind by an external body, e.g. a rose that is brought before our senses, the *next* and *immediate* result of this attention is that the *idea* of this change arises in us. Now, this idea has an object, namely the change that takes place in our mind at that very moment, and nothing else. Thus, it has only one object and we can say that it is a *singular* idea. On this occasion other ideas, some of them no longer simple, are also produced by the continued activity of our mind; similarly, complete judgments are made, especially about the change itself that has just taken place. We say, for instance, 'this (what I just see) is the sensation or idea 'red' '; 'this (what I just smell) is a pleasant fragrance'; 'this (what I just feel upon touching a thorn with the tips of my fingers) is a painful sensation', etc. It is true that in these judgments the ideas 'red', 'pleasant fragrance', 'pain', etc. have several objects. However, the ideas which occur in subject position and which we designate by the word 'this' are certainly genuine singular ideas (§ 68). For, by the 'this' we mean nothing but this individual change which takes place in us, and not a change that takes place elsewhere, no matter how similar it is to ours. Moreover, it is no less certain that all these ideas are also *simple*. For, if they were composed of parts, they would not be the *next* and *immediate* effect that results from the observation of the change that just now takes place in our mind; rather, several simple ideas, namely the parts of this complex idea, would have been generated earlier and more immediately. Merely from the fact that we employ several words to designate such a subject idea, e.g. 'this (which I now see)', 'this (which I now smell)', etc., we must not conclude that these ideas must also be complex. I have already pointed this out in §§ 59 and 69. Hence it is established that with each observation of a change that just now takes place in our mind, ideas emerge in us which are simple but have only one object, namely the just-observed change itself, to which they are related as the next and immediate effect is related to its cause. Thus the only issue left is to find a suitable name for this kind of idea. I believe that the word 'intuition' [*Anschauung*] is taken, if not in the same, then in a similar sense since Kant introduced it into the terminology of logic. Since I know no better or more suitable word, I should like to use the word 'intuition' not only for subjective ideas, but also for the corresponding objective ones. Thus, any simple singular idea will be called an intuition. I shall call it *subjective* whenever the idea itself is subjective, otherwise *objective*. . . . Hence ideas of any sense, even ideas that do not originate in any outer sense at all, will be called intuitions, so long as they are simple and have only one object.

NOTE: It will probably be difficult for many to understand the assertion that there can be ideas which have only one object although they are simple. All those who think that an idea must be composed of as many parts (constitutive characteristics) as its object has parts or attributes, will never admit that any object can be comprehended through an idea that is peculiar to it, unless this idea is composed of very many parts. It will not matter to them whether this object is external or whether it is a certain change that takes place in our mind. I believe that I have shown in §§ 63 and 64 how mistaken this opinion is. Nevertheless, even those who have abandoned it may still find it incomprehensible that one solitary simple idea could possess enough properties to represent only one, and no other, object, e.g. this individual change that just now takes place in our mind, and nothing else. They may argue along these lines: we cannot think so many simple and yet distinct ideas. Often we cannot perceive the slightest difference the two ideas 'this red' and 'that red', which we have when we look at this or that rose petal. Is this not proof that each of these two ideas has several objects? I reply that even simple things may be different from one another in infinitely many respects. According to Leibniz' well-known principle, they even *must* be different. Thus, from the fact that all intuitions are simple ideas it does not follow that there cannot be an infinite number of them, such that no two of them are exactly alike. Moreover, consider two ideas which are the *next* and *immediate* effect of our concentration upon two different changes that just now take place in our mind. Nothing is more easily understood than that these two ideas are as different from one another as their causes, and this in spite of their simplicity. For it is quite correct that different causes have different effects. It would be incomprehensible if we should expect different effects from equal causes.

We are not always in a position to indicate the actual difference between two such ideas, e.g. 'this red' and 'that red', but this is not particularly surprising, and does not prove that there is, in fact, no difference between these two ideas. For, in order to *indicate* their difference more than the mere presence of two ideas is required: We must form one or more ideas of these ideas and their attributes and then make judgments about them. Incidentally, that the two ideas 'this red' and 'that red' are truly singular ideas, and that one of them has only this, and the other only that object can be seen simply from the fact that through one of them we *represent* only this and through the other only that object. From this alone it follows that the first has only this, and the second only that object. For, the assertion that we *apply* the same idea first to this and then to that object, while it fits both, would only be correct if we were speaking of a *predicate* idea such as the idea 'red' in the two judgments 'this is red' and 'that is red'. But of a *subject* idea that occurs in a judgment, such as the ideas 'this' and 'that' in our example, one cannot say that it is merely applied to one of the several objects that it represents. Thus, an idea occurring as a subject idea in a judgment always occurs in its complete extension, and it is an error to believe that in the judgment '*this* man is a scholar' or in '*a* man was immaculate', or '*some* men are vile', the idea 'man' forms the true subject idea, as will be shown explicitly in the sequel. Let it suffice here to remark that if the idea 'this' in the judgment 'this is something red' were not a singular idea it would be altogether inexplicable how we know that we have two objects before us when we form the judgments 'this is red' and 'that is red'.

§ 73. *Concepts and Mixed Ideas*

1. Given that there are such ideas as I have just described under the name of intuitions, and that these ideas are quite important, it can hardly be doubted that there are ideas that are not intuitions and do not contain any intuition as constituents, and that they are important enough to deserve a special name. I call these ideas *concepts*, because I think that this word has been used with a very similar sense ever since intuitions and concepts were first contrasted. Thus the idea 'something' I wish to call a concept; it is not an intuition, since it has not one, but infinitely many, objects, and it does not contain any intuition as a constituent, for it is not complex at all. In the same way I call the idea 'God' a pure concept; first of all, this idea is no intuition, since it is not simple, though it has only one object. For, by 'God' I mean the particular being whose reality is uncaused. Secondly, this idea does not contain any intuition as a

constituent, since none of the ideas of which it consists is singular.

2. If a complex idea contains intuitions among its parts, I shall call it mixed, even if all its parts are intuitions (if this is possible). Thus the idea 'The rose, which disperses *this* fragrance' is mixed, since the constituent idea 'this fragrance' is an intuition.

3. Depending on whether the constituent that I take to be the most important (perhaps the main part, § 58) is an intuition or a concept, I shall call the whole idea either a mixed intuition or a mixed concept. Thus the redundant idea '*this*, which is a colour' I call a mixed intuition, since the main part of this idea, '*this*', is an intuition. On the other hand, the idea 'the truths contained in *this* book', I call a mixed concept, since the main part of this idea, 'truths', is a concept. To make the distinction more definite, I am going to call the rest pure intuitions and pure concepts, in contrast to mixed intuitions and concepts.

§ 74. *Clarification of the Preceding Definitions*

1. The intuitions that I have given as examples in § 72 all had objects which belong in the realm of *reality*, since they are *changes* that just now take place in our mind. I believe that this holds of all intuitions, of which *man* is capable, i.e. I think that the object of any humanly attainable (subjective) intuition must be a real thing. It seems to me that this follows from the concept of an intuition alone, as being a simple singular idea. For if an idea is to represent merely one object, although it is simple, then it must have something so peculiar (something that refers to only this object) that its origin in our mind can hardly be explained in any other way than that it is related to this object as effect is to cause. From this it follows at once that this object must be real, since it must be causally efficacious.

2. We cannot reverse this proposition and assert that we can attain an intuition of *every* real object. For in order to have an intuition of an object it is necessary that this object enter into that very special relation with us in which a simple idea is generated in our consciousness, an idea that refers to only a single object. It does hold, however, that we can easily form several mixed ideas of an object $\mathscr{A}$ (i.e. ideas that contain an intuition) once we possess an idea A that refers exclusively to $\mathscr{A}$ regardless whether or not $\mathscr{A}$ is real. For, once we have an idea A that represents only a certain object $\mathscr{A}$ we are in a position to use it in order to find certain relations that this object has to certain other objects, $\mathscr{X}$, $\mathscr{Y}$, . . ., that are *real and intuited by us*, and we will soon find a relation that is *peculiar* to $\mathscr{A}$. Such a relation

will then enable us to refer to $\mathscr{A}$ through a mixed idea which is partially intuitive, for we can now think of it as being the object which stands in such and such a relation to the intuited things $\mathscr{X}$, $\mathscr{Y}$. . . . Thus, if we have a flower before us, we do not have an intuition of the seed from which it sprang, but we can form a mixed idea that is combined with intuition and that fits only this seed, when we think of it as the 'seed of the flower which causes the present intuitions (of colours, scents, etc.)'. We often have in our mind a *subjective idea A* of an object, which we can make the object of a special intuition by directing our attention to it. Then we can use the latter idea instead of *A* (a simple instead of a complex idea) by thinking of the object as 'the object that is representable by the idea that is just now present in our mind'. Needless to add that this is always the case whenever we refer to an object of which we have just now formed a concept by means of the demonstrative pronoun 'this'.

3. By means of certain specific relations in which an object stands to other things that we have intuited we can easily form suitable ideas of a great many real things that surround us; these ideas, though they specifically fit their objects, are mixed. On the other hand the attempt to construct of nothing but pure concepts an idea that refers exclusively to some real thing (e.g. this house, this tree, this person), is bound to fail, though there are some exceptions. A pure and specific concept can indeed be given of God and certain powers and attributes of God, e.g. his omniscience, omnipotence, etc., and of the universe as a whole and of some other real objects. But the same does not hold of the rest of the innumerable real things, namely *finite* substances, whether taken singly or in combinations which comprise less than the totality of things.

It is indeed false that there are even as many as two real things that are completely equal to one another in all their (inner) attributes. One might hope that several of these inner attributes, each of which can be comprehended by a pure concept, could be used to form a concept that fits only this and no other object; but even if we suppose that for each object there is a finite number of inner attributes which no other object has in just this combination, all of which can be represented by pure concepts, it is clear, nonetheless, that we can never *know* whether the attributes that we have combined in our concept are really of such a nature. From the fact that *we* do not know a second object that has all these properties it does not follow that no such object exists in some unexplored region of the universe. Moreover, if the number of beings in the universe is infinite, then it can be supposed that not even an infinite number of inner attributes of an object will suffice for the formation of a concept that represents only this object and thus distinguishes it from all others. For, if the similarity of objects is a function of the number of independent attributes that are shared between them, then it could be the case that no similarity between two objects is the closest, i.e. so close that there is not a closer one. Hence no matter how many attributes we have already included in our concept, it could still be the case that there is a second object, or even infinitely many further objects, that have all these properties in common.

§ 75. *Some Remarks Concerning the Different Ways in which we Designate Intuitions and Concepts*

The difference between intuitions and concepts will become clearer if we consider the different ways in which they are respectively *designated* in language (*Wortsprache*).

1. It is impossible to generate for a second time an intuition that we once had, where the two subjective intuitions are to correspond to one and the same objective idea. The reason is that every subjective intuition has its own object, namely the external or internal change which is its cause, and every such cause exists only once, since every change that takes place at another time—even in the same subject—is already a second change. It follows that any two subjective intuitions have two different objects to which they refer. Therefore they must necessarily correspond to two different objective intuitions. It makes no difference how similar the just-perceived colour, scent, or pain is to a sensation that I had at some other time: they are still different, and the objective idea that refers exclusively to one can therefore not have the other for its object. However, if it is impossible that two subjective intuitions in the same individual belong to the same objective idea, then it is certainly also impossible to cause another person to have a subjective intuition which corresponds to the same objective idea as the intuition that is present in us. Thus, if by *communicating* an idea we mean arousing in another person an idea that belongs to the same objective idea as our own, then it follows that *it is altogether impossible to communicate intuitions.* The matter is different with *pure concepts*, which can be communicated in several ways, e.g. simply through words. Thus, all who know English will connect ideas corresponding to the same concepts in themselves with the words 'and', 'not', 'one', 'two', 'three', etc.

2. Still, we sometimes say that we communicate a certain intuition to someone else, but this merely means that we acquaint him with

various *attributes* of this intuition. In particular, if our intuitions are caused by the influence upon our senses of some external object, we may wish to indicate this object to others. If it is an enduring object, and if it recurs and is of sufficient importance, then a special sign is formed for it: a *proper name*. Thus, proper names always designate *mixed* ideas of the form 'the object that is the cause of my once having had such and such intuitions'. This holds not only of proper names designating external objects that influence our own senses, but also of proper names designating an object that has long since ceased to act upon our senses, e.g. Socrates. To the question what kind of intuitions our idea contains in such a case, I would answer that e.g. by 'Socrates' we mean a philosopher, 'who lived so and so many centuries ago in Greece and who was called Socrates'. If nowhere else, there are intuitions at least in the sounds of which the name 'Socrates' is composed. But proper names are not often used for the designation of objects. Most of them are designated by the description of a relation that holds only of them, a relation in which they stand to certain other objects that we already know. For this purpose, the relations of time and space are most useful. Even rough spatial and temporal determinations often suffice, such as those contained in the words 'now', 'recently', 'soon', 'here', 'there', etc., especially if we add the *kind* or *species* of thing to which the object we have in mind belongs. It is true that nobody knows what I mean when I point to the rose bush in front of me and say 'this thing'. You could not know whether I meant the whole rose bush, or only this rose, or this leaf, or what. I remedy this indeterminacy if I determine through a general noun the species of thing to which the intended object belongs, i.e. if I indicate that particular species of which there are not several at the indicated location and time. Thus, instead of 'this thing' I say 'this leaf', 'this colour', etc.

3. It is easy to see how this method, which is commonly used only for the designation of mixed ideas, can also be used, if scientific purposes require it, to designate pure intuitions. All we have to do is to proclaim that the determinations of time and place, and the determination of the species and kind of object are added for the sole purpose of making our idea *known* to the other, and that they should not be taken for a description of the constituents of this idea itself. The other person infers from this explanation not only that our idea is a simple singular idea, i.e. an intuition, he will also be able to judge which object has caused it in us and what other attributes it might have. That it is not possible to generate in someone else the *same* intuition that we have, has already been said in no. 1.

4. The following circumstance deserves special mention in this context. In all languages there are ambiguous words which sometimes designate pure concepts and sometimes mixed ideas. Even worse, we often jump from one meaning to the other without being clearly aware of it. This happens chiefly with the names we give to certain kinds of natural objects (especially the lowest species), such as 'man', 'lion', 'gold', etc. In one sense, we mean by these words simply things that have certain attributes, which we indicate or could indicate and which are representable through pure concepts. In this case we are willing to recognize any object as a thing of this kind so long as it has these attributes, no matter how much it differs in all other respects from the things that we have known under this name. We take the word 'man' in this sense when we decide to mean by it nothing but a being that combines a reasonable mind with an organic body. Hence we are prepared to give the name of man even to the inhabitants of the moon, if it can be shown that they are reasonable creatures that have an organic body, no matter how much they differ from us in mental power or formation of body. It is apparent that the idea which our word designates in this case is a *pure concept*. The idea 'man' would indeed remain a pure concept even if it were restricted to the point that only reasonable and sentient beings that are equal in all conceptually representable attributes to man as he is found on earth were called men. For there is only a determinate, though very large, number of attributes that are common to all reasonable and sentient beings on earth. Thus from the ideas of these attributes, i.e. from mere concepts, an idea can be composed which would represent what we call 'man', and does not contain any intuition. The matter is quite different if we decide that the name 'man' is to mean not (as before) *any* creatures that have certain attributes, but *precisely those* rational and sentient creatures that are found on earth, and no others, no matter how similar they may be. The idea designated by this name is no longer a pure, but a *mixed* concept that includes an intuition, because even after removing all parts of this idea that are pure concepts, the requirement remains that they must be beings that live on *earth*. Thus the intuition that is included in the name 'earth' remains. Hence 'man' is an example of a word that appears to be a pure concept, but is sometimes used so as to involve a real intuition. By contrast, there are also words that appear to designate a mixed idea but are sometimes used to express a pure concept. Such is the case with the words 'gold', 'silver', 'oxygen' and similar names of inorganic substances.* Our natural scientists are not

* Locke already recognized the double meaning of such words (*Essay*, II/V/8).

at all adverse to applying such names to any substance in the universe so long as it has the same inner attributes that this substance has on earth. However, we know only very few of the inner attributes of these substances other than through their influence upon further substances and finally upon our own senses and ourselves (i.e. our sensory and representative faculties). Thus we only know them through relations which they have to certain objects that are given through *intuition* alone. Consequently we tend to express these attributes in terms of relations and describe, e.g., gold as a body that causes the idea of a yellow colour in our organ of sight, thin layers of which transmit green light, that is 19 times heavier than water, etc. If we use this description as it stands in order to define the concept of gold, then our idea of this substance is indeed mixed with various intuitions and is an impure concept. But we can also look at the matter in a different way. We can take the effects of gold upon our sight and upon other objects of intuition merely as signs of certain attributes of gold that can be determined through pure concepts. For instance we can think of the words that 'it seems yellow to us' as being nothing but an expression for a certain inner attribute of gold which is the reason why gold causes the idea 'yellow' in an organ such as our eye, etc. If we understand matters in this way, then the intuitions which occur in the verbal expression of our concept of gold cannot belong to the content of this concept, but must belong merely to the means by which we *designate* those unknown inner attributes of gold that can only be determined through pure concepts, and of whose ideas the concept of gold is to be composed. Just as an *intuition* does not cease to be pure if we admit into its expression one or the other concept in order to designate it more clearly, so, on the other hand, a *concept* does not cease to be pure merely because we have to take recourse to certain intuitions in order to designate some of its constituents. If we look at the matter carefully, we find that this happens with *all* concepts whose constituents we communicate to others through *words*. For, in all these cases, we expect the listeners to think the concepts which arise in his mind when certain words are pronounced. With these words (the sounds) he becomes acquainted through intuition. At best we could make the following distinction: in the case of a word whose meaning is known to us we can mentally represent the concept itself which it designates. On the other hand, if someone specifies the constituents of the concept 'gold' merely by telling us that they are the concepts of those inner attributes of gold that cause the idea of yellow in our eyes, etc., then we still do not know what concepts they are. Thus we can say that the concept designated by the word 'gold' is a pure concept, but one that is not completely known to us, analogous to the value *X* in an unsolved equation.

§ 76. *Justification of these Definitions*

1. To show the usefulness of his definitions, B. advances the following considerations: (a) the concept of a *simple singular idea* is very close at hand, and that of a *concept* derives quite naturally from it; (b) there is the important distinction that no two subjective intuitions can correspond to the same objective intuition, which does not hold for concepts, and that concepts, but not intuitions, can be communicated; (c) a further important difference is that few real objects can be determined through pure concepts alone, but that, generally speaking, intuitions are required for that purpose. Still, since there are exceptions, "the question arises whether certain changes in the concept of an intuition (a suitable limitation) might not achieve a stronger contrast. It seems to me that this could be done through the following two definitions. All ideas that are neither themselves simple singular ideas nor contain any of them as parts are to be called concepts (same definition as above). On the other hand, simple singular ideas are to be called intuitions only if no pure concept can be found which represents precisely their object and no other. Here, the above exceptions would disappear, and under the new definition no object could be represented by an intuition and at the same time also by a pure concept. For this reason I am almost tempted to employ these definitions, if only their advantages were a little more pronounced. What difference does it make if (according to our first definition) we reckon God, the universe and some other objects among the *possible* objects of intuition which are also determinable through pure concepts? We put them among these objects because we do not know as yet whether they can be the object of a simple idea that refers to them alone. We do not assert that man has at present a subjective intuition of these objects, and neither do we contradict what the theologians quite correctly teach; namely that we can have no intuition of God in the present life, but that we shall achieve it in a future life. For, in the theological sense of the word, by intuition is meant something entirely different, namely simply an experience that has a very high degree of vividness and completeness."

2. Subsequently, B. tries to justify the employment of the two words 'intuition' (*Anschauung*) and 'concept' (*Begriff*).

§ 77. *Review of other Positions*

Although Kant was not the first to have drawn attention to the distinction between intuitions and concepts, he caused it to be generally recognized. However, his definitions are not satisfactory.

In his *Logik* (A 139) Kant maintained that an intuition is a singular idea (*repraesentatio singularis*) while a concept is general (*repraesentatio per notas communes*). "Concept is opposed to intuition [says Kant] since it is a general idea, or an idea of that which several objects have in common; i.e. it is an idea which can be contained in several others." On the other hand, in the *Critique of Pure Reason* (B 93) we read that an intuition is an idea in immediate relation to an object, while concepts relate to objects only indirectly (namely through the mediation of intuitions). . . .

According to the second definition, the difference between the two kinds of idea lies in the mediacy or immediacy of their relation to objects, where 'object' is obviously not meant to apply to every, even if non-existing, object, but only to really existing things. The form of words 'an idea is related to an object' presumably means that it is the idea *of* that object. Finally, the expression 'indirectly' is explained by Kant himself to mean that the concept is related to an object via an intuition that falls under it.

Aside from the fact that there are concepts which stand in no relation, either mediately or immediately, to any object, the following objection can be made: The concepts 'man', 'living being', and others represent something real just as immediately as any other idea, e.g. an intuition, such as 'Socrates'. Should we really say that the persons Socrates, Plato, etc. fall under the idea 'man' only mediately, by virtue of the fact that their intuitions fall under the idea 'man'? I think that just the opposite is the case. Only the persons (the beings themselves) but not the intuitions of them fall under the concept 'man'. In my opinion it is an improper mode of speech to say of a subordinate idea that it falls under a higher one: we should only say of its *objects* that they fall under the higher idea. It is not correct to say that the concept 'hydrochloric acid' falls under the concept 'acid in general', but all the objects of the former fall under the latter. Similarly, not the intuition of Socrates, but Socrates himself belongs under the concept 'man'. An idea under which the intuition of Socrates falls would have to be the concept of an intuition in general, or perhaps the concept 'intuition whose object is a human being', or a similar one, but never the concept 'man'. Perhaps there is a different interpretation of the phrase 'are related' in the locution 'intuitions are immediately, concepts only indirectly, related to objects'? Perhaps Kant's distinction between intuitions and concepts can best be gathered from a locution that he uses frequently, namely that objects are *given* through intuitions, but are merely *thought* through concepts. The locution that an object is given through an intuition seems to me to have the sense that if we have an intuition, we are entitled to conclude that there must be an object which produced it. The opposite is indicated if we say that through mere concepts we only *think* objects. There seems to be some truth in this.

We can indeed conclude from the presence of a subjective intuition that it was caused by an appropriate object which acted upon our understanding [*Vorstellungsvermögen*]. It is true that all other ideas which occur in our consciousness must also have causes; but in the case of a simple singular idea, we are entitled to infer that the cause is the real object itself which we are now representing. On the other hand, if we have a simple idea which does not represent a determinate real object, the question cannot arise whether there is a real object that corresponds to it, because the idea does not refer to any such object. Finally, if an idea is compound but represents a certain real object, then the immediate cause of its presence lies in our mind, namely in the activity through which its several parts were compounded. But whether and how an object of the kind which it represents exists at all, is a question which must be decided by investigating what caused this mental activity in the first place. No matter how well founded this distinction might be, it obviously holds only of subjective, but not of objective, intuitions that we can infer from their presence the presence of a corresponding object. Hence this entire discussion does not belong in the division of logic that is presently under consideration.

B. then represents Krug's position (*Fundamentalphilosophie*, Züllichau 1803, § 79, note 1) concerning the distinction between intuition and thought. In thought, Krug maintained, given ideas are further processed. The mind runs through the given ideas and retains what they have in common, and combines it into a new idea, "which is for that very reason called a concept (concept—since many characters are conceived in one representation). Hence the concept refers to its objects only mediately, namely through a mediation of the ideas from which it arose." B. objects that in this way the origin of such important concepts as 'something', 'nothing', 'and', 'obligation', 'impossible' can in no way be explained. It is further false (though Krug does not state it explicitly) that a compound concept which includes several attributes, consists merely of these attributes and of no other ideas. "The concept of an equilateral triangle consists of the ideas 'triangle', 'side', 'identity', 'have', etc. but nobody would wish to maintain that all of these ideas are attributes."

B. then criticizes Metz, who had maintained that even if a concept is applied to only a single individual, it does not lose its character of generality (*Handbuch der Logik*, Würzburg 1802, § 54). Against this B. calls attention to concepts that could not possibly be applied to several objects, e.g. 'universe'. Subsequently B. reviews several other positions.

§ 78. *Differences between Concepts with Respect to Content and Extension*

All intuitions are simple and singular, but among concepts we find every variation in content and extension.

"1. We need no special proof that there can be *complex* concepts, but we must show that there are also concepts that are *simple*. If one were to deny this, he would have to assert that all complex ideas, also all propositions, are nothing but an accumulation of mere intuitions, which is absurd. By thinking ideas of the form 'this', 'this', etc. in quick succession or simultaneously (where each of these demonstrative pronouns refers to some actual object), we do not unite them into a *single* idea. To bring about this unification, another idea is required, namely that denoted by the word 'and'. Furthermore, who could think even for one moment that the idea indicated by the word 'not' or the ideas 'have' and 'ought to' and a hundred similar ones could result from a mere connection of intuitions? It follows that there must be simple ideas which are not intuitions, i.e. there must be simple concepts."

2. It was shown in §§ 66–68 that there are ideas with infinitely many objects, a finite number of objects, one object and no object at all. In each of these types there are also pure concepts; the following examples are given in the above order of types: 'created substance', 'one of the four cardinal virtues', 'the society of beings with morally good will' (i.e. the so-called kingdom of God), 'a method of undoing the past'.

NOTE 1: It would doubtlessly be of merit if logic could establish a catalogue of simple concepts, especially if we could be persuaded that it is complete. But so far, all attempts at this have been conspicuous failures so that "I have no inclination whatever to contribute still another."

NOTE 2: B. considers the opinions of other logicians on the subject of simple concepts, and makes the following remarks:

"I maintain that in order to think a concept (whether it is simple or compound) we do not have to think any marks that distinguish it from other concepts, for these are not the concept itself, but attributes of it."

"We can reasonably suppose that a concept which we can in no way define (i.e. put together out of other ideas) is indeed not compounded. Actually, I do not maintain yet that we are capable of distinguishing simple and compound concepts, but only that *there are* simple concepts. Given the above definition of a concept, this follows from the fact that there are ideas and that they cannot be compounded from mere intuitions."

§ 79. *Whether the Ideas of Time and Space are Intuitions or Concepts*

The ideas of time and space form the subject matter of two special sciences, namely geometry and chronometry; we meet them everywhere in common life, and they are parts of most of our concepts and judgments. Still, there is no unanimity about the question whether these ideas have parts, and which, or even the question whether they are concepts or intuitions. Since I have just given my view of the difference between intuitions and concepts, I shall now consider what answer to the last question is entailed by these opinions. Actually this investigation is not part of the proper subject of a treatise on logic, but it will be excused in view of its importance.

1. There are quite a few ideas which we call ideas of time or space; with several of them it is not doubtful whether they must be reckoned among intuitions or concepts. It is clear that ideas related to time or space with more than one object are at least not *pure intuitions*, for they could then represent only one object. I have in mind here ideas such as 'moment in general', 'duration in general', 'points', 'distances', 'lines', 'planes', 'bodies in general'. They represent an infinity of objects, thus there can be no doubt that they are concepts; but perhaps they are not pure but contain an intuition as a constituent, like the concept of an inhabitant of earth. Actually, the dispute is only about ideas of time and space such as the following: 'all of infinite time', 'all of infinite space', 'this definite moment' (e.g. the present moment), 'this definite duration' (e.g. an hour), 'this definite point' (e.g. the centre of the earth), 'this definite distance' (e.g. between earth and sun), etc. All these ideas have only one object. If they were also simple, then I should have to admit that according to my definition they should be called *intuitions*.

2. However, remember that every (subjective) intuition that appears in our mind must have an *existing* object (§ 74). This gives us a method for proving that the aforementioned ideas are not intuitions. We do not have to make a decision on whether they are complex or simple, since the objects they represent are not actual (do not exist). This can be seen by anybody who *has* these ideas and can understand the above expressions; it is not necessary that he should also know whether they are simple or of which parts they are composed. I ask anyone who knows what the mathematician means by time and space whether he does not have to admit that only *objects* that are in time and space are real, but not time and space themselves. If time and space are something real [*wirklich*] then they must also have certain effects [*wirken*]. And what could they be? One says of time that it achieves this or that, e.g. that there is no pain that time does not ease, etc., but it is clear that this is just a figure of speech; it simply means that a certain change (e.g. the decrease of a presently violent pain) will not fail to occur in the course of time, since the causes that bring about such changes will certainly appear sooner or later. Not time itself is here held to be active, but the forces of things that bring about all changes and effects—always in a certain length of time. Sometimes we also say of space that it has effects; we note, e.g. that air expands as soon as it is given more space. But who would imagine that this giving of space is the actual cause for the

resulting expansion? Who does not assume a certain expanding force in the air which causes this expansion, after the opposing resistance has ceased? If time and space were something real, then their reality would have to be one of the two, either independent or dependent. In the first case they would be God and in the second created things that are subject to change. Now, nobody can really say either that space and time are God himself or that they are subject to change, since only things that are in time and space change but not time and space themselves. If time and space were something real then no two moments or durations, or two points or distances could be exactly equal since among existing things there are not two that equal one another exactly. But this is quite contrary to the concepts that mathematicians have about these objects. However, if two moments or two points are exactly equal (as all mathematicians continue to assert) and if time and space are something real, then the existence of a certain thing at this definite time and this definite place is something real that has no (sufficient) reason. For no reason could be given why this thing should be in this particular condition at this particular time and place and no other. It would not merely be humanly impossible to *give* such a reason, but there would *be* none, since the different times and places have precisely the same inner attributes. On the other hand, if the ideas 'this moment', 'this duration', etc. and the ideas 'this point', 'this distance', etc. are not ideas of something real (which must be admitted after the foregoing) then they are also not pure intuitions; I admit, as I have already said, that if they are complex, they may contain some intuition (in particular of something real that stands in these temporal and spatial relations).

3. If no individual moment and no individual point, taken by itself, is real, then neither is the class of all moments, i.e. *all* of infinite time, nor the class of all points, i.e. all of infinite space. Thus, neither of these two ideas can be called an intuition, although both of them have only one object (since there is only one infinite time and only one infinite space). But if the ideas of all of infinite time and all of infinite space are not intuitions, then they are pure *concepts*, for it is absurd to think that they contain an intuition as a constituent. What individual real object could it be that we unconsciously intuit in these ideas?

4. If we admit that the ideas of all of infinite time and of all of infinite space are pure concepts, then we can hardly refuse to admit that the ideas of a moment, a duration, a point, a distance and the rest (vid. no. 1) are all pure concepts. For, if there is no real object

that we intuit when we represent to ourselves all of infinite time or all of infinite space, then we also do not intuit such an object when we think of a duration, or a distance, or some other general idea connected with space or time.

5. To get a more precise notion what the constituents of the concepts of space and time are, consider the following: It is well known that we place everything that is real, with the possible exception of God, in a certain time; and if we want to assert that an attribute truly holds of some real thing, then we must always add a certain time at which this attribute belongs to that thing. This is of such universal validity that we may even say of the attributes of God that they belong to Him at a certain time, namely at *all* times. Thus we can assert that no proposition of the form 'the real thing *A* has (the attribute) *b*' expresses a complete truth, unless we include a time specification in the subject idea. Thus, for instance, the proposition 'I have a sensation of pain' or 'the earth is a planet' are not completely true unless we include a time specification in their subject idea, e.g. 'I, at the present moment' or 'The earth at the present span of time', etc. If we examine the matter more closely, it becomes apparent, as I believe, that by the word 'time' we mean nothing but *that particular determination of a real thing which is the condition for asserting that a certain attribute truly belongs to that thing*. From this concept all attributes of time can indeed be inferred, for example several contradictory attributes can be attributed to the same thing only if the times are different. This follows immediately from our definition, since propositions with contradictory predicate ideas can be true only if their subjects are different. Thus, if two contradictory attributes (e.g. ignorant and learned) are correctly asserted of one and the same substance (e.g. Caius), then it follows that different time specifications are present in this substance. For, if these time specifications were the same, then two propositions with the same subject idea and contradictory predicate ideas would both have to be true.

6. As concerns the concept of *space*, it will be admitted that by space we mean nothing but the class of all possible locations. Thus the only remaining question is what we mean by the locations of things. It is certain that every real thing has effects, and, if it is finite, that it suffers certain changes, and also causes certain changes in the other finite things that surround it. The nature of these changes depends obviously on nothing but two conditions: (a) on the forces which this thing and all the others have; and (b) on the locations at which it and the others are situated. We will then not be mistaken if

we say the following: *locations of (real) things are those determinations of these things that we must think in addition to their forces in order to comprehend the changes that they cause one in the other*; I not only believe that this proposition asserts a property that in fact belongs to space, but that it indicates the actual *concept* of space. Again, the reason is that all properties of space as they occur in geometry can be inferred from this simple definition.

NOTE: Kant treated this subject in a manner entirely different from mine, and since his doctrine has been accepted by almost all German philosophers, even those who differ from him in other points, it is worth while to subject Kant's reasons to careful scrutiny. Kant himself gave these reasons in the *Critique of Pure Reason* (*Transc. Aesth*, §§ 2 and 4, B 37 ff., B 46 ff.) in the following way: "Space is not a discursive or, as we say, general concept of relations of things in general, but a pure intuition. For, in the first place, we can represent to ourselves only one space; and if we speak of diverse spaces, we mean thereby only parts of one and the same unique space. Secondly, these parts cannot be considered as antecedent to the one and all-embracing space as being, as it were, constituents out of which it can be composed; on the contrary, they can be thought only as *in* it. Space is essentially one; its multiplicity, and therefore the general concept of spaces, arises entirely from the introduction of limitations. Hence it follows that an *a priori*, and not an empirical intuition underlies all concepts of space. For similar reasons, all geometrical principles, e.g. that in every triangle two sides together are greater than the third, can never be derived from the general concepts of line and triangle, but only from intuition, and this indeed *a priori*, with apodeictic certainty.

"Space is represented as an infinite given magnitude. Now every concept must be thought as a representation which is contained in an infinite number of different possible representations (as their common character), and which therefore contains these under itself; but no concept, as such, can be thought as containing an infinite number of representations within itself. It is in this latter way, however, that space is thought; for all the parts of space coexist *ad infinitum*. Consequently, the original representation of space is *a priori* intuition, and not a concept.

"Time is not a discursive, or what is called a general, concept, but a pure form of sensible intuition. Different times are but parts of one and the same time; and a representation which can be given only through a single object is an intuition. Moreover, the proposition that different times cannot be simultaneous cannot be derived from a general concept. The proposition is synthetic, and cannot have its origin in concepts alone. It is immediately contained in the intuition and representation of time.

"To say that time is infinite means no more than that every determinate quantity of time is possible only through limitations of one single time which underlies it. The original representation *time* must therefore be given as unlimited. But where the parts and every quantity of an object can be represented as determined by limitation only, the whole representation cannot be given through concepts (for these contain only partial representations), but it must be founded on immediate intuition."

This passage shows that (a) Kant did not declare all ideas of time and space to be pure intuitions, but only the ideas of total infinite space and of total infinite time; (b) there is some obscurity in the proofs of these assertions, for the sentence structure leaves it unclear if the opening passage, "For, in the first place, we can represent to ourselves only one space; and if we speak of many spaces, we mean parts only of one and the same space", is supposed to be a sufficient proof by itself that space is an intuition, or if parts of the following should be added to the proof. From what he says about time we may conclude the former: "a representation which can be given only through a single object is an intuition". I shall first take him in this sense and investigate the weight of this reason. Is it true that every idea that has, and can by its nature have, only one object is an intuition? It does not seem that way to me. For example, the idea 'God' can certainly have only one object. The same holds of the ideas 'universe', 'supreme moral law', and hundreds of others which no one will take to be intuitions. (c) Let us thus add the following: "These parts", he continues, "cannot be considered as antecedent to the one and all-embracing space as being, as it were, constituents out of which it can be composed; on the contrary, they can be thought only as in it as limitations". A contrast is here made between parts that *precede* their whole as its component parts (out of which it can be composed) and parts that can only be thought *in* the whole. What the nature of this antithesis is, is not clear to me. The meaning does not seem to be that parts of the first kind *precede* their whole in the actual sense of the word, i.e. in time. Otherwise it would have to be said of the other kind of part that they come into being at the same time or even later than their wholes. Since it is said of those parts that they can be thought of as existing within their whole only, the most probable interpretation seems to be that parts of one kind can be thought without thinking the whole, while ideas of the other kind of part contain the idea of a whole as a constituent. Such a difference between parts can in fact occur. One can, for instance, think of the parts of a watch without thinking the whole watch, while half a yard is obviously a part in whose idea the idea of its whole (of a yard) occurs as a constituent. Thus, the question is whether any limited space is a part of this second type, and whether every whole whose parts are of this nature must necessarily be an intuition. If we can show that the first part of this question must be answered in the negative, then we can leave the second open. Now I believe that even a person who does not think he knows all the constituents of the ideas of a point, a line, a plane, a triangle, and other spatial objects, can persuade himself that these ideas do not contain the idea of space as a constituent. Or can we seriously maintain that we have to represent all of space in order to think a single point? Admittedly, a point is sometimes defined as that which limits a line, a line as that which limits a plane, and finally a plane as that which limits a body. If these definitions are correct, it would follow that, in order to represent a point, we would first have to represent the line which it limits, and for the line we would have to represent a plane, and for the plane a body. But who does not feel that this is an error? Who has the slightest indication that his mind proceeds in this way?* But even if it were so, the idea of a point would require that of a body, but not that of infinite space. In my opinion the opposite holds. The idea of a point occurs as a part in the idea of a line, a plane and a body; and infinite space, as I think, is defined quite correctly as the class of all points. Thus I cannot agree with Kant when he maintains, as it seems, that the concepts of (limited) *spaces* (e.g. of triangles in general) are merely based upon *limitations* (namely of infinite space) in the same way and sense as the concept of half a yard is based upon that of a whole yard (i.e. includes it as a constituent). (d) He claims furthermore that geometrical principles are never deduced from general concepts but from intuitions. This may be true of past treatments of this science. I believe that it is not impossible to deduce each of the truths of geometry from mere concepts. (e) Against the assertion that space is represented as an infinite given magnitude, I have nothing to say except that I see no connection between it and its sequel. (f) I doubt that every concept can be thought as an idea that is contained in an infinity of others as their common character and that it thus contains all these ideas under itself. For not every

* There are lines with double curvature; it would be difficult for us to indicate the plane that they bound.

concept has to have several, or even infinitely many, objects; and if it does not have them, how can there be several, or infinitely many, ideas that are contained under it? (g) I, too, think that no concept can be said to contain an infinity of ideas *in it*, in the sense that no concept (which is thought by a finite being) can be composed of an infinity of constituent ideas, but I take this to hold not only of concepts, but also of all other ideas. No idea which is composed of an infinity of parts can be comprehended by a finite understanding [*Vorstellungskraft*]. It makes no difference whether this idea is a pure concept or mixed with intuitions. For this reason I do not believe that 'space' is an idea which is composed of an infinity of constituent ideas. The reason why this is supposed to be the case with space is 'that all parts of space coexist an infinitum'. It does not seem to me that this proves the point at all. For in order to form the idea of a 'whole that is composed of infinitely many parts' it is not necessary to represent these parts individually. I have thought that such a whole as soon as I have represented in suitable order the concepts that are expressed through the words 'whole', 'part thereof', 'set of these parts', 'infinite', etc. (The order is indicated through the expression 'a whole that is composed of infinitely many parts') (cf. § 63) (i) What is said of *time* is indeed true, namely that different times are only parts of the same (infinite) time. But it is inferred from this that time must be an intuition, since an idea which can only be given through one individual object is an intuition. I have already said why I cannot agree with the major premise. I do not deny that in the *pure* theory of time there are also synthetic sentences. But that they cannot be recognized through anything but intuition is an assertion that we will examine later (§ 315). However, the example that Kant gives on this occasion, namely "that different times *cannot be simultaneous*" is not a very fortunate choice. For, to be *simultaneous* means nothing but to be at the same time; things are called *different* when they are not the same. Thus it seems to be an identical sentence. (l) Again, I can only guess from the sequel how the following assertion is to be understood: "To say that time is infinite means no more than that every determinate quantity of time is possible only through limitations of one single time which underlies it". Afterwards, the claim is made that any object whose parts can be represented only through limitation (of the object) can never be represented through concepts. Thus if there is to be any connection between the two claims, then the meaning of the first assertion must be the following: The parts of time can be represented only through limitations of (the total infinite) time. And this is probably supposed to mean that any idea of a certain part of time contains as a constituent the idea of total infinite time. But I cannot admit this any more than the similar assertion about space. We can easily think what a moment is, or a duration, without thinking the idea of total infinite time. On the contrary, in order to think the latter, we must already have the idea of a moment, since infinite time is nothing but the class of all moments. Even the assertion that forms the major premise does not seem true, namely that an object whose parts can only be represented through limitations (of that object) can never be represented through concepts. I do not understand the reason that is given, namely "that concepts contain only partial representations". On the other hand, I think that there is a great number of objects whose parts can only be represented in relation to the whole (limitation in the same sense in which it holds of time) and which we nevertheless recognize through pure concepts. Which ideas are the major, minor and middle term in a syllogism is found out only through an examination of the whole; yet who would deny that the idea 'syllogism' is a pure concept? Now we have finished with Kant's reasons. Since this philosopher has himself admitted that he lacks the gift of precise exposition, let us see how one of his followers, a man of whom Kant said that he had understood and interpreted him very well, proved this important doctrine. In Schultz's examination of Kant's *Critique of Pure Reason* this subject is extensively treated (J. S. Schultz, *Prüfung der Kantischen Kritik der reinen Vernunft*, part I, 1789; II, 1792).

a. To begin with Schultz attempts (I, pp. 55 ff.) to show that all past attempts at defining geometrical objects, such as point, line, etc. were either mistaken or unable to give us an idea of the defined objects. He says (*ibid.*, p. 57): "Before the geometer can attempt a definition in his science, he must assume that everybody is immediately acquainted with the ideas of total infinite space and the different possible limits, directions, sides, areas in it. Then he can give names to these things, which are themselves already known." If the form of words 'to give an idea of an object' means the same as to generate for the first time an idea that fits this object (and only this object), then the question whether a certain definition is capable of giving us an idea of the defined object and thus acquainting us with it, depends upon the entirely accidental circumstance whether we hear this definition for the first time; if we heard it a second time, it could not give us that idea because we already have it. Hence no definition could be given of objects with which we became acquainted in early childhood, at a time when we were unable to grasp a definition. If this is what is meant, then I agree that nobody can give us an idea, by means of definitions, of a body, a plane, or several other geometrical objects. But I cannot see how it follows that every idea which we receive without the benefit of definitions is an intuition. We would have to reckon among intuitions a great number of ideas which are customarily counted among concepts, e.g. the ideas of an idea itself, of a judgment and a thousand similar ones; for it is plain that it is impossible to give us these ideas through definitions. But if it is true that every idea which cannot be given through a definition is an intuition, then it is clear that every idea is an intuition unless it is a simple concept, or can be represented through a combination of simple concepts. I fully agree that the ideas 'point', 'line', 'plane', 'body', etc. are not simple concepts; but it may well be possible that they can be represented as combinations of several simple concepts; from the fact that it has not been done, it does not follow that it cannot be done.

Secondly, one should not conclude that a definition is incorrect if it does not at once produce in the listener the entire image that is usually connected with the defined object. This image does not essentially belong to the idea and it is actually a shortcoming if the definition produces it. Do we not in fact have many assuredly correct definitions, even of geometrical objects, that do not create such an image? The definition of a dodecahedron as a body that is bounded by twelve equal sides is certainly very adequate. But do we at once perceive an image—a correct image—upon hearing this definition? Do we already know that the bounding planes are pentagons? It is no valid criticism of the definition of a line, for instance, if it does not help us in forming an image of a line, or, what comes to the same, if it does not at once tell us what a line must actually look like. It is sufficient if the definition does not hold for any other objects but lines, and holds for all lines, and if all properties of lines (and thus their looks) can be deduced from the definition. I believe that geometers could actually give such definitions without having to assume that their audience already have an idea of what infinite space is, what its different possible limits are, and what directions, sides, etc. are.

b. According to Schultz (*ibid.*, pp. 59 ff.), a second proof that space is an intuition lies in the fact "that total infinite space with all its parts and limits is *completely determined* not only with respect to quality and quantity but also with respect to location and situation. In the ideas of space and its properties nothing depends upon our choice, everything is inevitable in just the same way as—the *sensations* which we receive through our senses. Even through the most concentrated exercise of his imagination the geometer cannot think of a space which has more than three dimensions, etc. Not only the nature and magnitude of space but, what is even more remarkable, even the location and situation of each of its parts and limits is completely given and determined. A physical body can change its location in space, but space itself maintains its location. This puts it beyond doubt that the idea of space is intuition. If the idea of space depended on a concept, it would be impossible for the geometer to think two different points. For his concept of one point would be exactly the same as that of the other point. Hence the understanding does not have the least inner characteristic through which to differentiate one from the other; their difference consists merely in the fact that we represent them in two different locations of space".

Against this I should like to make the following remarks:

α. Assume that "Since total infinite space is completely determined with respect to quality and quantity because the understanding cannot make the slightest change in its idea" entails that space is an intuition: then the general proposition must hold that the idea of any object which is something single and individual, and in whose idea the understanding cannot make the slightest change, is an intuition. But where is the proof of this? Is not it the case that God is an individual and single object whose characteristics are all completely determined in such a way that the understanding cannot make the slightest difference in them? Schultz replies to this objection (*op. cit.* part II, p. 37): "The reason for this is that aside from the perfect being there are several independent beings of somewhat more limited perfection; hence we simply have to think away the limitations in order to form the concept of a being of infinite perfection. But as concerns space, there are no things of the kind which would allow us to form the concept of an infinite single space by abstracting from certain limits". The futility of this excuse is obvious, and it contains a tacit admission that the indicated proof, at least in the stated form, is invalid. More and more excuses of different kinds would have to be thought up in order to counter other examples of singular concepts. What could the objections be against such singular concepts as 'universe', 'supreme mortal law', 'smallest even number', etc.?

β. Schultz finds it remarkable that no part of space changes its location; this seems to indicate a somewhat confused concept of space. In my opinion every space is a possible location of things; and to say that a space does not change its location is the same as to say that no location is other than it is; hence it is a simple identical proposition.

γ. It will be recalled that, just like Leibniz, I find a proof that space is not real, and that its idea is therefore not an intuition, in the fact that we cannot determine a single point in space through mere concepts, since they are all completely like each other. It is most remarkable that Schultz draws the exact opposite conclusion from this circumstance. He says, "If the idea of space rested upon a concept, then it would be impossible for the geometer to think two congruent areas, for his concept of one of them would be exactly like his concept of the other". He presupposes that it is impossible to form a concept of two or more exactly like objects, and that such objects must be represented through intuitions. I take this to be an error which refutes itself as soon as it is asserted. For when it is said that it is impossible to represent through concepts several exactly like things, then several things are talked about, hence represented, and their ideas are certainly not intuitions but mere concepts. Hence the very act of asserting the preceding proposition contains a proof that several exactly like things can be represented through mere concepts. It may be asked by what characteristics the several exactly like things can be distinguished in our ideas; I reply that they are distinguished through the differences in their *relations* to each other. Schultz demands that those who think that the idea of space is a mere concept "ought to provide even a single concept through which one point in the periphery of a circle can be distinguished from another". I think that this can indeed be achieved. Every point in this periphery differs from every other point in its relations to it and other points: These relations can easily be represented through mere concepts.

c. Then Schultz (*ibid.*, p. 102) arrives at Kant's argument that "No limited space can be thought except as within the totality of co-existing space, i.e. that the idea of a limited space becomes possible only through the idea of the totality of space. For to think of space as limited by a surface that is an *absolute* limit, i.e. a complete cessation of space, is a complete denial of our concept of space." I admit that the latter is the case; but I cannot see how it follows from the preceding part. Why should it follow that if somebody thinks a certain *limited* space, for example a cone, and does not think the space outside this cone, then he denies that there is such an outside space? And only such a denial would lead to the indicated absurdity. If the form of argument which Schultz allows himself here were valid, then one would have to require of somebody who thinks a cone that he should also think all the infinitely many lines which are formed when this cone is cut by a plane, and in infinitely many other ways. For if somebody were to deny the possibility of even one of these sections, he would deny the entire conception of space.

d. Then an argument is produced (*ibid.*, p. 108) according to which it follows from the continuity and infinite divisibility of space that the idea of space is an intuition. "It follows from the infinite divisibility of space that space is a compound thing and yet does not contain simple parts; it follows from the continuity of space that a geometrical point as it produces a line must traverse all points on that line even though there is no point which is the *next* after the initial one. All these are representations which embarrass not only our *understanding* but even our *imagination*. These representations cause the geometer so much effort and concern that he would certainly recognize the continuity and infinite divisibility of space as the most absurd fantasy, if he were not immediately and apodeictically certain of them. Is it then reasonable to suppose that the understanding, which finds these concepts so incomprehensible, has formed them itself?" He adds in part II, p. 27, "If the idea of a line were a concept of the understanding, then every finite line, since it is infinitely divisible, would have to consist of infinitely many parts. But this is an obvious contradiction, for a quantity is called infinite if it can never be thought as completed; consequently a totality which consists of an infinite quantity of parts can never be complete, i.e. it cannot be a finite thing", etc. It may indeed be true that the answer to the indicated question will cost the understanding "much effort and concern" but how does it follow from this that the concept from which these problems arise cannot be produced by the understanding itself? Does it perhaps follow that *all* objects whose consideration causes difficulties for the understanding are not created by it but are given in intuition? According to critical philosophy, the doctrine of God is confronted with immense difficulties but it would be absurd to claim that the idea of God is given through intuition. It is further claimed that the continuity and infinite divisibility of space does not just lead to difficulties, but to obvious contradictions, if space is considered a concept. It is said to be contradictory that a finite thing like a straight line which is bounded on two sides should consist of an infinity of parts. I openly confess that I cannot see either how this consequence disappears if space is taken to be an intuition, since this does not change the appropriate theorems of geometry one bit; or on what basis such a consequence can be called a contradiction; though I am aware that even some great philosophers have subscribed to these objections. I think it incontrovertible that there are totalities which consist of an infinity of parts. The class of all truths is certainly one of them, and so is God's knowledge, which extends over all truths (to give an example of something real). It is indeed true that each such whole is infinite in a certain respect, namely in the respect in which it consists of infinitely many parts. But it does not follow from this that it is infinite in every respect, and that there is not some other respect in which the same object can be considered as something finite, for example as a finite magnitude. With respect to the number of points which it contains, the straight line which lies between points *a* and *b* is indeed infinite, but with respect to its length it is only a finite thing. Furthermore, the claim that an infinite set can never be completed is true only if we mean by it that we will never come to an end when we *count* it. But there is no good reason to suppose that such a set *cannot exist at all.* The two examples which I have cited suffice to demonstrate the opposite. Hence I conclude that the ingenious Schultz has not advanced any defensible argument for the intuitive nature of space and time. I am also not aware that since then different and stronger reasons have been discovered; cf. Krug's *Wörterbuch* (*Allgemeines Handwörterbuch der philosophischen Wissenschaften*, 5 vols., Leipzig 1827–29) and *Metaphysik* (*System der theoretischen Philosophie*, Pt. II, Königsberg 1808).

§ 80. *Ideas of Attributes and Relations*

No one will doubt that there are ideas of mere attributes. But since there are quite a few different kinds of attributes, one distinguishes several kinds of ideas that refer to them. Their elaboration in logic is all the more necessary as the largest and most important part of our knowledge concerns merely the attributes of things, or, rather, since every truth, and thus every cognition, can be envisaged as the indication of an attribute of certain objects. At present we can examine only distinctions that can be discerned in a given idea taken by itself, and not through comparison with something else. Hence we will only be able to arrive at two classifications that belong in this context. According to the first we classify attributive ideas into those that represent *actual*, *internal*, or *absolute* attributes, also called properties, and those that represent merely *unactual*, *external*, or *relative* attributes; the latter are also called relations. In order to explain this classification, I must first give a more precise account of the two related concepts of *attribute* and of *having*, although I have used these two concepts quite frequently.

1. By the word 'attribute' [*Beschaffenheit*] I mean exactly what is generally meant by it, if it is not used in its narrower sense, but in the

wider one in which every passing state of affairs, every fleeting change forms at least a temporary attribute of the given object. Anything that *belongs* to an object, whether always, for a very short time, or only for one moment, is an attribute of that object for the particular time. The word 'have' is also commonly used in two senses, a wider and a narrower one. According to the first we can say of any attribute which is found in an object that the object *has* it, as in the proposition 'the soul of man has immortality'. In the second sense we mean by it a mere *possessing* (i.e. the capacity for a kind of use) of a certain object. In this sense we say 'man has hands'. In the first sense what is had is always an attribute. In the second it can also be something that is not an attribute at all, e.g. hands, money, etc. I shall use the verb 'have' always only in the first and wider sense, hence for attributes alone. Thus I can say, *whatever is had* (*quodcunque habetur*) *must be an attribute.*

If I am asked what parts I take these intimately related concepts to have, I must admit that I am uncertain about their parts and about the exact relation between them. It seems to me that one of them is absolutely simple, and the other is composed of it and a few other parts. Perhaps it is the concept of having that is simple and the concept of an attribute is generated from it in such a way that an attribute is always something that is had. But it could also be the concept of an attribute that is simple, and the concept of having could have it as a part. I am not certain what the answer is, but the former alternative seems more likely.

2. By making reference to the parts of propositions (which I shall discuss in the sequel (§ 126)) the concept of an attribute could be defined somewhat more clearly and could be distinguished from a wider concept, namely that of *determination in general.* It will be shown later that every proposition has the form '*A* has *b*', where *A* and *b* stand for a pair of ideas of which the first is called the *subject* and the second the *predicate* idea. If the proposition is true, then the idea that occurs in the place of *b* (the predicate idea) must be the idea of a genuine attribute, and conversely, every idea of a genuine attribute must be usable as the predicate idea of a true proposition. We may say that any idea of an attribute which occurs as *b* (as predicate idea) in a proposition forms a *determination* of the object that is represented by the subject idea *A*. But the converse does not hold. Not all determinations of objects require a predicate idea in a proposition where this object is the subject. Rather, there are ideas that serve as *determinations* of objects without being *attributes* of them. These ideas have the peculiarity that they can never occur in the place of

the predicate idea (*b*) but only as parts of the subject idea (*A*) itself. Of this sort are especially the determinations of *time* and *space* of existing things, because the time in which an existing thing is located, and during which certain attributes can in truth be attributed to it, is not an attribute of this thing. For this reason, the idea of this time does not occur in the predicate, but in the subject idea of the proposition. This holds analogously also of the spatial determinations of things.

3. After these preliminary remarks about the concept of an attribute, I want to define that of a *relation*, from which the division of attributes into *internal* and *external* directly follows. It is easy to see that every object will have its own attributes. A whole which has several objects *A*, *B*, *C*, *D*, . . . as parts is, as such, a special object, which is essentially different from its parts. It is obvious that each whole will have certain attributes which its parts do not have. If I am not mistaken, these attributes are what we call *relations between those parts*. In particular, this holds when we think of the objects *A*, *B*, *C*, *D*, . . . on one hand and the attribute *x* of the whole on the other, as *variable*, i.e. if we think that other objects *A′*, *B′*, *C′*, *D′*, . . . which are of the same kind as *A*, *B*, *C*, *D*, . . . have an attribute that is, although not the same, yet of the same kind as *x*. Thus, for instance, if the line *A* is twice as long as *B*, this is not an attribute that belongs to any one of the two lines, *A* and *B*, taken by themselves, but to the whole that consists of both of them. Furthermore, if we replace these lines by others, then the new whole will not always have the same, but only a similar attribute, e.g. that one line is three times as long as the other. Thus, one line's being twice as long as another we call a relation that holds between these lines. For the same reason we call the circumstance due to which it can be said that Alexander the Great was a son of King Philip a relation that obtains between the former and the latter. The reason is that this circumstance is again an attribute that belongs neither to one nor the other alone, but only to both jointly, and that it would be changed if we were to put any other persons in the place of *A*. and *Ph*.

4. According to this definition, a relation *x* that holds between the objects *A*, *B*, *C*, *D*, . . . is an attribute that actually belongs only to the whole composed of *A*, *B*, *C*, *D*, . . . as such. In spite of this we can at least say of any individual part, e.g. *A*, that it has the attribute of 'forming a whole with *B*, *C*, *D*, . . ., which has the attribute *x*'. Such an attribute of *A* we call an *external* attribute. Hence, by an external attribute of an object we mean an attribute which consists merely in the fact that the object has a certain relation to another

object. Thus the attribute that a line has a length of two inches is an external attribute of this line, since this condition holds only if the indicated relation in fact holds between the line and an inch. Attributes which are not external, i.e. which are not a relation of the given object to some other object, we call *internal* attributes or *properties*. According to these definitions, wherever we find a relation (namely between the several objects A, B, C, . . .) there is also an inner attribute, though in a different respect (namely with respect to the whole $A+B+C+\ldots$). But the converse also holds: wherever there is an internal attribute b, even of a completely simple object A, there is also a relation, though in a different respect. Because, if b is an attribute of A, then b and A together form a whole which consists of an object and an attribute that it has; this fact is a relation that holds between A and b. . . . But that not all relations are of this kind has already been shown by the examples that I have given above. An idea that represents an internal attribute, I wish to call a *property idea*, and ideas that represent external attributes, I wish to call *ideas of relations*.

5. Let me add the incidental remark that we distinguish two kinds of relation: Sometimes the objects A, B, C, D, . . . all make the same contribution to the attribute that belongs to the whole which they form; at other times this is not the case. Relations of the first kind I wish to call relations of *equality*, also *mutual relations*; those of the second kind relations of *inequality*, also *unequal* or *unilateral* ones. The distance that two spatial points have from one another is a relation of equality, for both points contribute in the same way to this relation. On the other hand, the direction in which one of these points lies from the other is a relation of inequality: both points do not contribute in the same way to the determination of this direction. If a relation of equality holds between the points A, B, C, D, . . ., then it must be possible to represent this relation through an idea in which the ideas A, B, C, D, . . . all occur in the same way, i.e. with the same connections, etc. The opposite holds of relations of inequality. Hence if we want to speak correctly we should not speak of the distance of point *a from b*, but of the distance *between* points *a* and *b*. The first expression seems to indicate that point *a* makes a different contribution to the distance between the two than point *b* does.

NOTE 1: B. points out that according to the definition of no. 2, some relation holds between any objects, even if, as we say, they have no relation to each other. Secondly, every relation requires at least two objects. "Although we sometimes say that a thing stands in a relation to itself, it seems to me that in these cases there are actually two objects, but they are considered attached and connected and in a sense the same thing. Thus, when we say that man stands in a certain important relation to himself, namely that his conscience is his own judge, the two objects are the man and his conscience, hence not the same thing." Thirdly, more than two objects can be members of a relation.

B. is not so certain that the concept of a relation can be reduced to that of an attribute, but he claims that if this reduction is impossible, then there can be no reduction or analysis of it at all, and the concept is simple.

Not every characteristic of a set can be construed as relation between the members of that set. "Thus the attribute of the number 13 to be a prime number can hardly be considered one of its relations even though it is actually an attribute which does not belong to it alone, but to the totality which consists of it and all other numbers. On the other hand, if we are told that Caius knows Sempronius, then we must agree that 'knowing Sempronius' is an attribute of Caius, which can be called an inner attribute as well as any other (what could be more on the inside than our own ideas?). Nonetheless, once we notice that it is Caius who has this knowledge, while it could have been some other being, and that he has knowledge of Sempronius, while it could have been knowledge of somebody else, then the knowledge appears as a relation that holds between the two. Examples of this sort lead me to believe that we give the name of a relation to an attribute of a whole only if we think of the objects A, B, C . . . as well as of the attribute x as variable, i.e. if we think that there could be certain other objects A', B', C' . . . which are of the same kind as A, B, C . . . and which, as a set, have a property x' which is different from x. We do not call the attribute of the number 13 to be prime one of its relations, since the concept of this attribute carries with it the requirement that it must be compared with the totality of all other numbers. Since there is only one such totality, we cannot even raise the question whether the number 13 would still have this attribute if we chose some other totality instead. The opposite holds for the other example. I also think that this explains why we consider certain attributes of things, namely colour, odour, etc. to be inner attributes, even though careful consideration shows them to be mere relations between the things and our sensory organs. The reason is that we consider the nature of our sensory organs to be invariable."

B. then considers the definition of a relation as something that can be recognized only through a comparison of objects. He points out that this definition does not deal with relations between things, but with our recognition of certain attributes, which is a different matter. Also, some *properties* cannot be recognized except by making a comparison with some other thing. Thus we need a ruler to find out that a certain triangle has the property of having a relation 3 : 4 : 5 between its sides.

NOTE 2: B. considers the following contention of Tetens (*Versuch über die menschliche Natur*, Leipzig 1777, vol. I, p. 275): "Relations exist merely in the understanding; they are mere *entia rationis*. Properties, on the other hand, are something objective." B. replies: "In the same case in which we can and must say that a property exists, namely when the object to which it belongs has itself existence, we must also ascribe existence to relations, i.e. we must say that a relation exists whenever the objects between which it holds also exist. If a rose has existence, then not only its red colour exists, but also the relations which hold between this rose and other real objects, e.g. that it stands in a pot at my window, etc. This will be readily admitted, since an attribute which is a relation in one respect is a property in another. The fact that one leaf is smaller than another, that one is over another, etc. is a relation between the leaves, but with respect to the rose it is a property. In the second respect it would have to be called real, but in the first merely mental, if it is true that properties exist outside of our ideas, but relations only in them."

B. thinks that the reason for holding this theory is that properties are much more obvious than relations.

NOTE 3: Locke and others have distinguished real and ideal relations. Real relations cannot be changed except if something is changed in the thing itself, while with ideal relations the thing need not change. "I think that, in order to have a change in the relation between *A* and *B*, one of them must necessarily change; and if both of them are existing and finite things, then both must change; if we assume that initially there was a change only in *A*, a change in *B* must result because of the so-called *nexus cosmicus*. Hence the relations between existing and finite things are all real. But that does not mean that there are no ideal relations. For if an object *A* cannot change, e.g. God or some non-existent object like a truth in itself, then its relation to some other thing *B* can certainly change only if the latter changes. For example, God stands to man in the relation, sometimes of a judge, and sometimes of a compassionate redeemer, and this is not due to a change in Him, but in us. The same holds when the relation between our understanding and a certain objective truth changes, e.g. when it changes from an unknown to a known truth."

§ 81. *Ideas of Matter and Form*

1. In compound objects, two kinds of attribute can be distinguished; a reference to the first kind indicates out of what parts the object is composed, a reference to the second kind indicates the manner of composition. Ideas of the first sort are called ideas of matter, of the second ideas of form. 'Is made of brick' can be an idea of the matter of a house, 'is two stories high' an idea of its form.

2. Conceivably, the matter of an object can always be indicated by mentioning internal attributes only, though sometimes there seems to be an option where either internal or external attributes will serve. The form of an object can often be indicated only through relations.

In a note, B. indicates that the word form (*forma*) was first used by Cicero to refer to species. Aristotle used εἶδος to stand for both, the opposite of matter, and for species. "The word 'form' is also used in the sense of kind or species, when logic is called a formal science (§ 12) and when we speak of the forms of ideas, judgments, and arguments." Often 'form' and 'matter' are construed as contraries, although 'form' means no more than the attributes of an object which make it the kind of object it is.

In a second note, B. distinguishes the sense of 'form' in the present section from another sense: "When I speak of the ideas, propositions and arguments that fall under a certain form, then I mean by 'form' a certain concatenation of words or signs in general, which can represent a certain kind of idea, proposition, or argument. Thus, when the letter *A* represents any subject-idea, and the letter *b* any attributive idea, then the expression '*A* has *b*' is the general form of any proposition whatever, since all propositions can be represented by this concatenation of signs."

§ 82. *The Idea of a Class, in Particular the Idea of a Class of Individually Listed Objects*

1. To explain the meaning of 'class' [*Inbegriff*], B. gives a number of near synonyms: 'collection', 'combination', 'gathering', 'whole'. He is not certain about the parts of this idea, but he ventures the definition 'a something which has composition'. 'Composition' [*Zusammengesetztheit*] he takes to be simple.

2. He then considers the special case where the members of a class are individually listed, or given by name. What are the parts of the idea 'the class of *A*, *B*, *C*, *D*, . . .'? Clearly, *A*, *B*, *C*, *D* . . . are such parts, but there must be some further part, "for if we think the parts *A*, *B*, *C*, *D*, . . . one after the other or at the same time, and nothing else, we have not thought the *class* of them." One might think that the missing part is the idea indicated by the word 'and', repeated $n-1$ times for n class members. This notion must be rejected, for in '*A* and *B* and *C* and . . .', *A* is connected only with *B*, and the class of *A* and *B* only with *C*, etc.; yet, "in thinking the objects *A*, *B*, *C*, *D*, . . . as united into a class, we do not determine in what order they are to be connected, which are connected immediately, and which only through the mediation of others. I think that the idea of connection occurs only once, just as the linguistic expression 'the class of *A*, *B*, *C*, *D*, . . .' indicates, no matter how large the quantity of *A*, *B*, *C*, *D*, . . .". It is true that we can think and mention the members of a class only in a certain order but we often say explicitly that this order should be neglected. B. does not think that 'lack of order' is part of the concept, "since it is only the mode of our expression that necessitates the stipulation to abstract from the order. . . . And so it seems to me that the idea 'Class of *A*, *B*, *C*, *D*, . . .' has no other parts than the ideas of the individual objects *A*, *B*, *C*, *D*, . . . and the concept that is indicated by the word 'class'."

§ 83. *Further Ideas which have the Idea of a Class as a Part*

1. Sometimes we wish to refer indifferently to any part [member, *Teil*] of a class of enumerated objects. B. takes the idea 'any part of the class of objects *A*, *B*, *C*, *D*, . . .' to be composed of the idea of the class of enumerated objects, and the idea designated by the word 'part' and "by a part I mean each object out of which a class is composed".

2. The above definition could be criticized on the following grounds: Consider the class composed of Caius, Sempronius, and Titus. "We could call any limb on any of their bodies, or any class composed of two of them, e.g. Titus and Sempronius, a part of the class. On the other hand, when we say that each of these persons is an individual capable of administering a

certain office, we do not want to say that the arms of Caius, or the two persons Caius and Titus together, are suitable for this office. It seems therefore that the two expressions 'Each of the objects A, B, C, D, . . .' and 'Each part of the class A, B, C, D, . . .' do not have the same sense. This objection vanishes once we realize that according to the definition of 'part' that I gave in no. 1, only the objects of which we think as constituting a class are to be envisaged as parts of that class. We see furthermore that according to this definition things that are merely parts of these parts, likewise those that are themselves classes of such parts, have no claim to this name. It is true only of classes of a special type that 'parts of a part are parts of the whole', as they say. We shall presently deal with these."

3. B. wishes to call the ideas defined in no. 1 distributive or part-ideas, while a suitable name for the idea of a class itself would be 'collective idea'.

4. Sometimes we wish to talk about a class which contains certain enumerated objects, but want to leave it open whether or not it contains others besides. B. describes such a class in this way: 'a class which has each part of the class A, B, C, D, . . . as a part'. This definition covers the class A, B, C, D . . . itself as well as any larger class containing the parts A, B, C, D. . . .

In a note, B. considers the objection that the idea of a class of which it is undecided whether it contains any parts besides A, B, C, D . . . has a greater extension as the idea of a class of objects A, B, C, D . . ., although its content is more complex. B. gives another counter example to the contention that a wider concept must always be simpler than a narrower one: "Nobody will deny that the concept 'root of the equation x^3-3x^2-x+3' is wider than the concept '$+3$', since it comprises, in addition, the concepts '$+1$' and '-1'; nevertheless, the concept '$+3$' is contained in it as a part."

§ 84. *Concepts of Sets and Sums*

B. introduces the notion of a set as a class where the manner of connection between elements is not specified. Sets fall into two kinds: those where the parts of parts are themselves parts of the class, and those where this is not the case. He calls the former 'sums'. "It follows from the concept of a sum that it is not changed if the order of its parts is changed, and that it is not changed if one of its parts is replaced by the parts of that part."

In a note, he contends that "the meaning which the mathematicians connect with the word 'sum' when they claim to get a sum whenever they conjoin a couple of expressions with a '+' is the same as mine. It is perhaps different in the sense that they apply the concept only to *magnitudes*—as their science requires. However, they take the word in an entirely different sense when they undertake to find the sum of a set of given number-expressions, for instance $1+1/2+1/4+1/8+$*ad infinitum*. In this case they mean by 'sum' the simplest possible expression that is equivalent to the given set of number-expressions."

§ 85. *The Concept of a Sequence*

An important and frequently encountered sort of class is the *sequence*. I wish to call a class of objects . . . K, L, M, N, O, . . . a sequence, if for each object in this class, *M*, there is another object *N* in the same class, such that either *N* can be determined from *M*, or *M* from *N* in accordance with a law that applies *uniformly* to the total class; that is to say one of these objects can be determined merely through the relation in which it stands to the other. Hence I call the four propositions '*A* is *B*', *B* is *C*', '*C* is *D*', '*D* is *E*', a sequence because for each of them there is precisely one other proposition in this class, such that one of the two holds: either the predicate of the first is the subject of the second, or the subject of the first is the predicate of the second.

The individual objects . . . *K*, *L*, *M*, *N*, *O*, . . . whose class forms the sequence, I call the members [*Glieder*] of that sequence. The rule which holds for the entire class, such that for each member another one can be determined through the mere relation that it holds to it, I call the *formation law of this sequence* [*Bildungsgesetz*]. Two members *M* and *N* which stand to one another in the relation indicated in the formation law . . . I call a pair of *adjoining*, *neighbouring* or *immediately succeeding* members. One of them, perhaps the member that has been used in order to determine the other one from it (this is a matter of choice), I call the *earlier* or *preceding* and the other one the *later* or *succeeding* member. Hence in the above example the propositions '*B* is *C*' and '*C* is *D*' are a pair of immediately succeeding members, and if '*B* is *C*' is to be called the preceding member, then '*C* is *D*' must be called the succeeding member. A member of a sequence that is both a preceding and a succeeding member is called an *inner* member, one that is preceding, but not a succeeding is called *first* or *initial* member, one that is only succeeding is called an *end*, or *final*, member. Either of the last two are called *outer*, or *bounding* members. The idea of a sequence does not have to include the ideas of all its members. Out of the idea of one member, together with the idea of the formation law, and a few other concepts, we can construct an idea that refers exclusively to a given sequence and to nothing else. For example the expression 'a sequence whose first member is 1, and where each member is formed from its predecessor by doubling, and none of whose members is to be considered its last one' indicates precisely the infinite sequence 1, 2, 4, 8, 16 . . . *ad inf.*

In a note, B. considers the following set of objections: (a) 1, 3, 6, 10, 15, 21 is a sequence, but the relation between members is neither arithmetic nor geometric. Reply: The word 'relation' is to be taken in the wider sense of § 80. The formation law of this sequence, which determines the relation between two members, is $n = n\frac{n+1}{2}$. (b) In some sequences a member is determined not from one, but from several preceding members. Reply: by indicating which member precedes a given member, we also determine the predecessor of the predecessor, etc. Hence even in this case the succeeding member is determined by its predecessor, at least mediately. (c) In some sequences, some or even all members are alike. Reply: In these cases "not the objects themselves, but our *ideas* of them are envisaged as members of the sequence. These ideas can be different from one another . . . while the represented objects do not differ". As an example he gives 1^0, 1^1, 1^2, 1^3, . . . (d) Some sequences are not based on real objects. Reply: same as in (c). It is obvious how the ideas $\sqrt{-1}$, $2\sqrt{-1}$, $4\sqrt{-1}$ . . . can be envisaged as a sequence, although none of them has an object.

B. next points out that "the relation in which one of a pair of members stands to another suffices to determine a sequence only if a class of objects *A*, *B*, *C*, . . . *X*, *Y*, *Z* is given from which the members of the sequence are to be chosen. For example the propositions '*A* is *B*', '*B* is *C*', '*C* is *D*', '*D* is *E*', would not form a sequence if the formation law were simply that the subject of each member is to be the predicate of its predecessor. This relation does not suffice to determine the new member, since it indicates only its subject, not its predicate. The determination becomes complete, as soon as we indicate that the members of the sequence are all to be chosen from the class of propositions '*A* is *B*', '*B* is *C*', '*C* is *D*', '*D* is *E*'."

The indicated concept should be wide enough to cover not only mathematical sequences, but also sequences considered in the philosophical sciences; but it is not wide enough to include the so-called "continuous sequences", as when a span of time is called a continuous sequence of moments, and a line is called a continuous sequence of points. B. objects to this usage because, "as is well known, in time there are no two moments which are so close together that there is not one, nay infinitely many, between them, and the same holds for points on a line." Time could be called a class of moments, and a line a class of points, but neither should be called a sequence. Even in the case of continuous sets, some have talked of neighbouring elements, but this is a plain contradiction. "It is indeed true that there are contradictory concepts, such as $\sqrt{-1}$, which are of great usefulness. But if we want to employ such concepts, we must first show in the same manner in which mathematicians have done it for $\sqrt{-1}$ how they can be used safely and profitably."

§ 86. *Concepts of a Unit, of a Manifold, and of a Totality*

We must now consider several further concepts because they will be frequently applied in the sequel.

1. The first is the much-used concept of a *unit*. Each object that has a certain property *a*, or is subsumed under the idea 'something that has *a*', I want to call a *unit of kind A* in the concrete sense of the word 'unit', or a *concrete unit* of kind *A*, or, still shorter, *one A*. . . . The property of a thing by virtue of which it can be envisaged as a concrete unit of kind *A*, or can be subsumed under the idea of *A* as an object, I want to call the *abstract unit of kind A*. By 'unit in general'

in the *abstract* sense I accordingly mean nothing but the property which every thing has by virtue of which there exists an idea under which it can be subsumed as an object. Hence almost every object can be called a unit only in a *certain respect*, i.e. only in relation to a certain idea under which it can be envisaged as an object, while we allow that the very same object does not constitute a unit in other respects (i.e. in relation to other ideas).

2. A class [*Inbegriff*] whose parts are concrete units of a kind *A* I call a manifold, indeed a *concrete manifold of kind A*. . . . The property of a concrete manifold by virtue of which it is a manifold is called its manifold in the *abstract* sense or the *abstract manifold*. A manifold of *A* which contains nothing but one *A* and one *A* is called a *Two* of kind *A*. A manifold which consists of Two *A*, and, in addition, one *A*, and which does not contain anything else, is called a *Three* of kind *A* etc.

3. Finally, the class which contains all and only objects that are subsumed under the idea *A*, or, more briefly, in which every *A* occurs as a part and in which every part is an *A*, is called the *class of all A*, or the *all* or *total* of the *A* in the *concrete* sense. The property which makes a concrete all or total to be such an all or total is called the totality in the abstract sense or the *abstract totality*. The total of all *A* is often designated by the expression 'every *A*' which, according to § 57, no. 2, is also used to designate an entirely different idea, namely that of an *A* in general. In order to distinguish these two senses of one and the same expression it is customary to say that in § 57 it is understood as *distributive*, but in the present paragraph as *collective*.

4. B. here draws a distinction between the class of all and only *A*'s (as introduced in no. 3), and the class of all *A*'s, where it is undetermined whether there are elements that are not *A*. He maintains that the latter case occurs in such examples as "What all men together cannot do, cannot be done by an individual."

§ 87. *The Concept of Quantity, Finite as well as Infinite*

1. Even in logical investigations the concept of quantity occurs so often that we cannot leave it unmentioned. In my view, we say of an object that it is a quantity whenever we think of it as belonging to a kind of thing of which any two can exhibit only one of the following two relations to one another: they are either equal to one another or one of them occurs as a whole which contains in it a part that is equal to the other. Hence in the comparison of two quantities of the same

kind we always presuppose without proof that it must be the case either that these two quantities are equal or that one of them is larger, i.e. that it contains in it a part which is equal to the other.

2. This definition shows which manifolds or wholes, as well as units on which they are based can be regarded as quantities, and under what circumstances. If we attend in a given manifold only to that particular property that does not change when one of the units of this manifold is replaced by another unit of the same kind, and if we abstract from the mode of connection between the parts, then, I assert, we regard this multitude or whole as a quantity. Under this assumption a comparison of two manifolds will always have one of two results: either they are equal to one another or one of them contains in it a part that is equal to the other. The first will be the case whenever we are able to transform one of the manifolds into the other by merely exchanging each unit that is contained in it for a unit that is contained in the other. The second will be the case whenever there are units left over in one of the manifolds after all the units that are contained in the second have been exchanged for units contained in the first. Keeping in mind this way of envisaging manifolds of a certain kind, it is easy to understand how the unit and the totality that belong to the same kind can also be recognized as quantities of that kind.

3. Consider a sequence formed of manifolds of kind *A* by making the *Two A* the first member and by deriving each succeeding member from the preceding one by adding another *A* to it (or, rather, to a manifold that is equal to it). Any manifold of kind *A* that occurs in this sequence as a member is called a manifold of finite quantity or simply a finite manifold of kind *A*. On the other hand, a manifold of kind *A* which is of such a nature that every finite manifold of kind *A* occurs as a mere part of it, that is, which contains in it a part that is equal to any finite manifold of kind *A*, I wish to call a manifold of *infinite quantity* of kind *A*, or simply an *infinitely large* or *infinite manifold* of kind *A*.

4. Consider a sequence which is formed by taking a unit of arbitrary kind *A* as the first member and in which every further member is a sum which is derived by adding a new unit to a thing that is equal to the immediately preceding member. Every member of this sequence I wish to call a *number*, provided it is thought under an idea which indicates the way in which it was derived. One can easily see that any finite manifold can be represented by a number as far as its quantity is concerned, but that no number can be given for the *infinite* manifold, which is why we call it *uncountable*.

NOTE: I hope to be pardoned if I stay with the concept of *infinity* a little while longer, since it is of such importance among the concepts that have here been defined, and since there are so many mistaken opinions about it. Generally, a manifold is said to be finite if it can be *indicated* or *determined*, and infinite if it is greater than any finite manifold. But what could be meant here by the words 'indicate' and 'determine' [*angeben* and *bestimmen*]. To *indicate* an object may mean the same as to form an idea of it (an idea that holds precisely for this object).... In this sense, an infinite set can very well be indicated, since for any infinite set we can form an idea that applies precisely to it. If we could not do this, how could we speak of it? But even if we could not form such an idea, I do not think that the fact that we humans are, or are not, able to do a certain thing should be used as a standard of calling something finite or infinite. This last reason clarifies another point, namely that by 'indicate' we must not mean a picturing in imagination, and still less in reality. After all, it is quite possible to exhibit in reality an object which contains an infinity of parts (e.g. of simple substances). Every body consists (so I believe) of an infinite number of parts; hence if we display it we have before our eyes an infinite set. But perhaps the word 'determine' is better than the word 'indicate'. We say that a thing is determined or determinable if of any two contradictory attributes (*b* and non-*b*) only one belongs to it. According to the axiom of the *universal determinability of all things*, every object, of whatever kind, must be determined in this sense.... An infinite thing does not constitute an exception to this rule; it must be the case that of any two contradictory properties, such as *b* and non-*b*, one applies and the other one does not apply to it. Hence something else must have been meant when the infinite was called undetermined or undeterminable. Presumably what was intended is that the infinite is not determined and cannot be determined merely by indicating the relation in which it stands to a finite set or to a mere unit; and this is indeed true. For merely by saying that a set is larger than any finite set one does not determine which things belong to it and which do not. But does it follow that a thing is undetermined in itself, if that thing cannot be determined in some *special* way, e.g. through a relation to a given object? Certainly not. For example, we can in no way determine the length of a line through a relation that it has to a given point. Nevertheless, this length is anything but undeterminable. The same holds for infinite sets. They cannot be determined through their relation to the unit or to any finite set, but there are methods through which we can completely determine an infinite set. For example, the set of points that lie between two given points *a* and *b* is unquestionably infinite, but it is completely determined as soon as we indicate points *a* and *b*; this indication determines precisely what belongs to this set and what does not, and there is not a single point whose status remains doubtful. Perhaps none of this would have been overlooked if it were not for the fact that the concept of the infinite is a general concept, and if we had not adopted the custom of representing all infinite sets or quantities by one and the same sign, namely ∞. Since there are distinct infinite sets, we do not determine all of the attributes of such a set merely by declaring it to be infinite, just as we do not determine some other set by merely declaring it finite. Nonetheless, all infinite sets were to be represented by one and the same sign, and one could and had to accept the following equations: $\infty + 1 = \infty$, $2\infty = \infty$, $\infty/2 = \infty$. It was concluded from this that the infinite is undetermined. Somehow it was not realized that the following similar equations also hold, where the sign F merely means that a set is finite: $F+1 = F$, $2F = F$, $F/2 = F$.

Another definition of the infinite says that it is that which cannot be increased. I believe that this is quite false. Thus, although the set of all points in a line *ab* is indeed infinite, we can increase it when we add to it the set of points which lie in the continuation of *ab*, namely in the piece *bc*. The claim that the infinite set is larger than any *possible* set was even less correct. For a set that is larger than any possible set must, for this reason, be *impossible*, i.e. there is and can be no such set. . . .

Several famous philosophers, among them even Leibniz, have hesitated to assume an infinity of things that are real, e.g. an infinity of substances in the universe or even in a body that is large enough to be perceived. The reason usually given for this, namely that all real things must be thoroughly determined, while the infinite is undetermined, has been removed by what I have said above. I have already said . . . that at least *God* must be admitted to have a reality that is infinite. . . .

Gerdil (*de l'infini absolu*, in the *Mélanges de Turin*, vol. 2) has given as a further reason against the actual infinite that there can be no point of transition from the finite to the infinite. This reason carries no weight for me, as I hold the necessity of such a transition to be unprovable.

Some think that an infinite set of realities can at most be assumed for God or the substances in space; some even admit that an infinite number of succeeding future states has nothing contradictory about it, but they resist the assumption that an infinite number of succeeding states have *already gone by*. I assert, however, that an infinite number of succeeding states can be distinguished in any kind of change, however short it may be, even if all the states have already passed. (This can be done e.g. in any motion from one location to another.) Even the shortest duration includes an infinite number of moments, and if between any two of these moments the object has been continuously in the same state, then we must, I believe, deny that the change lasted through the whole duration. But then it is clear that an object which has changed through a given period of time has passed through an infinite number of different states, one after the other, since there is an infinity of moments in that duration that followed each other.

§ 88. *Exceptive Ideas*

B. here maintains that exceptive ideas, i.e. 'all A except a, α, . . .' are the same as the idea of mathematical subtraction. "To subtract N from M means to find a thing X whose addition to N gives a sum equivalent to M." Similarly, to think the idea 'all A except a, α, . . .', is the same as 'that which, together with a, α, . . ., constitutes the class of all and only A's'.

On the other hand, if the excepted elements are not enumerated, but indicated through a common property, the idea in question is not properly called exceptive. It is an idea of the form 'All A that do not have the property b'.

§ 89. *Affirmative and Negative Ideas*

1. The concept indicated by the word 'not' [non, *Nicht*] is simple. Concepts of which it forms a part are called negative in the wider sense.

2. Negative concepts in the narrower sense are those in which the negation occurs an odd number of times in immediate succession.

3. Even after the concept is so limited, there are still two kinds of negative ideas.

a. "Among one kind I count those that have the form 'non-A', and simply posit the negation of a certain idea A, without requiring that in the place of the negated idea some other idea, not even that of a something, should be thought. These ideas I call purely or absolutely negative, and I have no doubt that there are at least some of them." An example is the idea 'nothing'. To claim that in the idea 'nothing' that of something is tacitly contained, is to say that the idea 'nothing' is contradictory, namely 'something which is not something', which is absurd.

b. The second type of negative ideas are the so-called partially negative ideas. This group comprises all ideas of the form 'A which is not B'. An example is the concept of a curved line, which is a line that is not straight. Ideas of the form 'something which is not B' come closest to absolutely negative ideas, but posit something without indicating what it is except that it is not a B. The subject of the following proposition is of this sort: 'what has no corners is not a triangle'. "Since every proposition must be about something, and since there are no words which could designate an object other than the words 'what has no corners', the idea which is expressed in these words must refer to some object of which nothing is affirmed except that it has no corners." By contrast, consider the proposition 'What has no corners is round'. If this sentence is to be true, then "we must by no means think of any arbitrary something that has no corners, since a spirit would be such a thing, and since it is absurd to claim that a spirit is round. Hence, by the expression 'what has no corners', we here mean a *body* that has no corners."

4. These examples show that common linguistic expressions rarely show whether ideas are negative, and what kind of negative ideas they are. "We can see this only from an exact analysis and from their use."

NOTE 1: Some, among them Locke, have claimed that there are no negative ideas (*Essay*, III, 1, 4). Words such as 'nothing', 'infertility' etc. are said by them to designate the mere absence of ideas. "But if somebody uses the word 'infertility', then the idea of fertility is not only familiar to him, but presently in his mind. The word 'infertility' does not express the absence of that *idea*, but the absence of the *property* denoted by it. We mean something by every negative word, even by the word 'nothing'; otherwise we should have to say that these words are meaningless; the content of the thought that we have in connection with any of these words is the idea itself that is designated by them". There may be no object for an idea, but that does not mean that there is no idea.

NOTE 2: Some logicians consider negation to be a relation, such that an idea is affirmative or negative, depending on with what other idea it is compared. B. maintains that ideas are in themselves either affirmative or negative.

NOTE 3: It is not always obvious from the linguistic expression whether an idea is affirmative or negative. B. takes the idea of a right angle to be affirmative, that of an oblique angle to be negative, since the right angle but not the oblique one is equal to its supplement, and since the idea of equality is, in all likelihood, affirmative. Sometimes a concept is affirmative, even though it is expressed by a negative phrase, e.g. 'unlimited time', "since it is the time that includes *all* moments."

NOTE 4: Absolutely negative ideas are sometimes called 'infinite ideas' since an infinite number of them can be attributed to an object without the object being thereby determined. "Against this I wish to object that ideas that are absolutely negative actually have no object at all. Thus, what is said here holds at best of partially negative ideas (no. 3, b). . . . It is true that not even an infinite number of them, e.g. 'something that is not red', 'something that is not sweet', 'something that is not square', etc. completely determine the object to which they are attributed. However, we must not forget that there are also infinitely many affirmative ideas that also do not suffice for the determination of an object. For example, if the object to be determined is some existing finite individual, e.g. Socrates, then, as is well known, there are infinitely many affirmative predicates, e.g. 'Athenian', 'son of a midwife', 'sculptor', 'philosopher', 'husband of an argumentative wife', etc., after whose enumeration it is still undetermined which further attributes he had. Furthermore, if the object does not exist, but is some general concept like that of the magnitude $\sqrt{2}$, then it is easy to determine it through the indication of a pair of suitably chosen attributes, so that the answer to any further question about its attributes is already fixed. But it is equally simple to indicate infinitely many attributes that do not determine it. Thus, one could indicate the digits at all the even decimal places of $\sqrt{2}$, and after infinitely many indications of this type the magnitude of $\sqrt{2}$ is still so undetermined that we have to answer an infinity of other questions (namely for the digits of the odd decimal places) before the number can be completely determined in this way."

NOTE 5: Schulze (*Grundsätze der allgemeinen Logik*, 2nd ed., Helmstädt 1810, § 39) claimed that all negations can be united into one concept which is the concept of the complete nothing. B. points out that this concept would have to have the form 'something which is non-*A*, non-*B*, etc.'. "And since among its parts there must be contained all attributes which an object can have at all, for example the attribute that it is something, this concept is inconsistent, but the concept 'nothing' is not inconsistent." Hence the two concepts are not even equivalent, let alone the same.

NOTE 6: B. here criticizes Klein for opposing the distinction between affirmative and negative concepts and claiming that "affirmation and negation constitute the content of every concept". "Errors such as these take their origin from Spinoza's celebrated proposition *Omnis determinatio est negatio.* For if one thinks that the proposition 'this is a triangle' contains all the propositions 'this is not a quadrangle', 'this is not a pentagon', etc. then the difference between affirmative and negative propositions must soon disappear. But these are merely consequences of an erroneous opinion, which we have already encountered, and will meet again, namely that every attribute of an object must be represented in its idea (cf. § 64)."

NOTE 7: Krug is criticized for claiming that negative ideas are empty.

NOTE 8: Since every combination of ideas which is not a proposition is another idea, the expression 'no *A*', e.g. 'no man' also designates an idea. But while the idea 'no man', contains the ideas 'no' and 'man', in any propositional context the negation always combines with the following predicate, e.g. 'No man is *b*' is the same as 'every man is non-*b*'. Hence no special discussion of ideas of the form 'no *A*' is necessary.

§ 90. *Symbolic Ideas*

B. here introduces ideas whose main part (§ 58) is formed by the concept of an idea. Their general form is 'idea which has (the attribute) *b*'. He calls them ideas of ideas, or symbolic ideas. "If the attribute *b*, which is ascribed to an idea, is neither self-contradictory, nor an attribute that contradicts the nature of an idea, then the idea 'idea that has the attribute *b*' is real and has a referent (§ 66), i.e. there is an object (namely in the realm of ideas) which truly corresponds to this idea." Examples are ideas designated by such phrases as 'a simple concept', etc. Some symbolic ideas do not have objects.

CHAPTER 3

Distinctions between Ideas that Stem from their Relations to each other

§ 91. *No Two Ideas are Exactly Equal; Similar Ideas*

1. There can be many subjective ideas that have exactly the same content, but it would be absurd to speak of two exactly equal objective ideas, because "in this case we consider nothing but the idea in itself, and hence we cannot say that they are [exactly] equal except when all their recognizable properties (components, mode of composition, etc.) are identical. But if this is the case, then we cannot distinguish them, hence cannot claim that they are several in number."

2. There may be ideas that have so many common attributes that they are often confounded. These ideas shall be called 'similar ideas'.

NOTE 1: Logicians who have maintained that there are exactly equal ideas have generally confused one of the following: subjective with objective ideas, ideas with the signs for them, or equality of ideas with identity.

NOTE 2: "I should like to warn the novice from a confusion which often attaches to the words 'equal' and 'identical'. In my opinion, identity is the concept that arises in our minds when we observe one and the same object several times, provided we realize that it is the same object. Equality, on the other hand, is the concept that arises whenever we observe *various* objects and find that they are subsumable under the same objective ideas; this does, of course, not mean that *all* of the objective ideas under which they fall are the same, but only that some are. If every idea that we form of a given object also applies to another, then it is impossible, for this very reason, to *recognize* that there are two objects. And if it is altogether impossible to indicate an idea that applies only to one but not to the other of the two objects, then it is not even true that one is a different object from the other, for this proposition itself refers to one of the two objects by means of an idea which does not apply to the other. Thus, all equality is *partial*, and if we claim that two or more objects are equal, then, if we want to be precise, we must indicate in what respect they are equal, i.e. we must indicate the idea which applies to them all. The opposite of identity is plurality, the opposite of equality is inequality."

When identity or sameness is attributed to a real object that endures in time, different criteria are employed, depending on what kind of object it is. An object that has parts must retain most of them, or at least the most important ones. (This is the same watch although it has a new glass.) "Occasionally, as in the case of the ship of Theseus (which was constantly repaired by the Athenians), we mean by identity of an object A with another B (which consists of an entirely different substance) no more than that B was produced from A by means of series of imperceptible or unimportant alterations, none of which was big enough to call the object different from what it was before."

In cases where location is important, as with rivers, identity of location alone is taken into account. Organic beings are called identical if one originated from the other by series of mediating causes other than procreation.

NOTE 3: Some logicians have called ideas similar if they have any characteristics in common. This would have the undesirable consequence that all ideas are similar, because all of them share at least the characteristic that they are ideas. "But if, as I have suggested above, we call ideas similar only if they have so many common characteristics that we are in danger of confounding them, then it will admittedly often be unclear whether or not we should say that two ideas are similar. This vagueness, however, does not lie in the concept, but in the nature of the case, and the concept of similarity will continue to be useful since we can use it to call attention to the fact that here are two concepts that are likely to be confounded."

NOTE 4: B. here distinguishes his concept from the concept of similarity in mathematics. "In mathematics the term 'similar' [*ähnlich*] is used only when two objects have in common all those inner attributes which can be represented by pure concepts, in such a way that every pure concept that applies to one also applies to the other, provided only that the concepts are of inner attributes. For example, in geometry we call two circles similar because all inner attributes which can be represented by pure concepts which apply to one of them also apply to the other." If in one of them the ratio between diameter and circumference is as 1 : 3.14 . . ., then the same holds for the other. It is obvious that objects similar in this sense can be distinguished only because of their relations to other objects.

§ 92. *Relations between Ideas with Respect to their Content*

1. In § 56 the content of an idea was defined as the sum of its components, irrespective of order. Hence there may be different ideas with the same content. They are contrasted with ideas that share no part at all. Simple ideas that have the same content must be identical, because simple ideas and their content are one and the same thing.

2. Some ideas share some, but not all, components. B. calls them 'kindred ideas'. The larger the number of common parts, the "closer" the kinship. Kinship and similarity must be carefully distinguished. '*A*' and 'non-*A*' are akin, but not similar. On the other hand, "the idea of moral goodness is often confounded with, and hence similar to, the ideas of public utility, honesty, etc., although it is likely that they have not a single common component."

§ 93. *Relations between Ideas with Respect to their Scope*

The number of objects contained in the extension of two ideas can be equal or unequal. "In the case where both numbers are infinite, there may be as little justification in calling them equal as unequal. Hence in this third case the two extensions must be called 'not comparable'." An example of the third kind are 'sphere' and 'tetrahedron'. If the concept of an extension is extended in an obvious way so as to apply to whole classes of ideas, the concept of scope can be similarly extended. (Cf. also § 66.)

In a note B. points out that if the scope of one idea is finite, and that of another infinite, then their relation cannot be indicated by a pair of numbers, but the second is greater than the first. If both are infinite, then sometimes no comparison can be made (are there more triangles or more syllogisms?), but sometimes the relation can be exactly determined. For example, there are as many circumferences of circles as there are areas. "For, if the number of circumferences were finite, nobody would hesitate to say that there are as many circumferences as there are circular areas, since for each circumference there is a circular area. Hence nobody would hesitate to say that both concepts have the same scope. But this relation between their scopes does not change, no matter how much the number of referents of these concepts is increased. Thus, since there is no evidence to the contrary, we are justified in assuming that the scope of these two concepts can be represented by the same quantity, even though they are infinite." For similar reasons we should admit that the scopes of the concepts 'centre of an ellipse' and 'focal point of an ellipse' are as 1 : 2, while the corresponding relation between 'circumference of a circle' and 'diameter' is as 1 : ∞. These conclusions cannot be denied unless we are prepared to claim, e.g. that the concept 'triangle' does not have a greater scope than the concept 'right triangle'. However, "it is true that we cannot measure infinite quantities as such, and from this it appears that by 'scope' we here mean something other than the mere sum of the referents of an idea. I already said § 66, note (1) that a scope actually is any quantity assigned to the set of referents of an idea according to a rule, provided that it equals the sum of the quantities which the same rule assigns to the parts of that set. This definition justifies the above determinations sufficiently. The quantity of the set of referents of 'circumference of a circle' and 'area of a circle' cannot in itself be determined, but the relation between circumferences and areas of circles assures that the two quantities can be equated, since there is no reason to assume that they are unequal. Hence if we take the scope of one of these concepts as the unit, then we can assign 1 also to the other; and if we take the scope of the concept 'centre of an ellipse' as the unit, then the scope of the concept 'focal point of an ellipse' becomes 2."

If all this is true, then it disproves Kant's contention (*Logik*, § 13, A 152) that concepts can be compared with respect to scope only if one is subordinate to the other.

§ 94. *Relations between Ideas with Respect to their Objects*

1. Concerning the objects themselves which are represented by ideas, it appears that two ideas either have common objects or they do not. Both cases are important enough to deserve a special name. I therefore call ideas that have one or several objects in common *compatible*, and those that have not even a single common object I call *incompatible*. Thus the ideas 'red thing' and 'something with

a pleasant fragrance' are compatible, for both have certain objects in common, e.g. a rose. On the other hand, the ideas 'body' and 'surface' are incompatible, for no object of one of them is also an object of the other. Sometimes we find among a given set of ideas $A, B, C, D, \ldots$ the curious situation that only a certain number of them, e.g. any n, are compatible. Thus, the four ideas: 'roots of the equation $(x-a)(x-b)(x-c) = 0$', 'roots of the equation $(x-a)(x-b)(x-d) = 0$', 'roots of the equation $(x-a)(x-c)(x-d) = 0$', 'roots of the equation $(x-b)(x-c)(x-d) = 0$', are compatible in pairs.

2. If several ideas $A, B, C, D, \ldots$ are compatible with one another, then the *smaller* number of ideas $A, B, \ldots$ whose class forms only a part of the first class, must also stand in the relation of compatibility. If this is not the case, i.e. if there is no object that is jointly represented by $A, B, \ldots$ then it certainly also would be false that there is an object that is jointly represented by $A, B, C, D, \ldots$ Conversely, even if the smaller number of ideas $A, B, \ldots$ are compatible with one another, members of the larger class $A, B, C, D, \ldots$ to which $A, B, \ldots$ belong, may stand in the relation of incompatibility. For, even if the ideas $A, B, \ldots$ have a common object, this does not mean that it is also common to the remainder $C, D, \ldots$.

3. If a pair of attributive ideas a and b are compatible with one another, then their concreta A and B (§ 60) are also compatible. However, the converse does not hold. . . . For, if a and b are compatible, then there must be a common attribute x which is a as well as b; in this case, an object that has the attribute x is an A as well as a B, and A is compatible with B. But merely from the fact that A and B are compatible it does not follow that a and b are also compatible. It is indeed true that the object that is A as well as B must combine in it both attributes, a and b, but it does not follow that that attribute of it which belongs under a also belongs under b. Thus the abstracta 'prudence' and 'foresight' are compatible and consequently their concreta 'prudent' and 'foresightful' are compatible too. On the other hand, the concreta 'pious' and 'learned' are compatible though their abstracta, 'piety' and 'learnedness' are not, for no kind of piety can be called a kind of learnedness.

4. Just as an individual idea can stand in the relation of compatibility to another idea, a whole class of ideas $A, B, C, D, \ldots$ can stand in this relation or in the relation of incompatibility either to some other class $M, N, O, \ldots$ or to a single idea M. The first is the case if there is some object that falls under any one of the ideas $A, B, C, D, \ldots$ and at the same time under M or under one of the ideas $M, N, O, \ldots$, the second if this is not the case.

5. If a whole class of ideas *A*, *B*, *C*, *D*, . . . stands in the relation of *incompatibility* to a whole class of ideas *M*, *N*, *O*, . . . then any one of the ideas *A*, *B*, *C*, *D*, . . . must stand in this relation to any one of the ideas *M*, *N*, *O*, ... But when both classes stand in the relation of *compatibility*, then it is not necessary that each, but only that one of the ideas *A*, *B*, *C*, *D*, . . . stand in this relation to one of the *M*, *N*, *O*, ...

6. If a single idea *A*, or a whole class of several ideas *A*, *B*, *C*, *D*, . . . is compatible with *M*, *N*, *O*, . . ., and *M*, *N*, *O*, . . . is compatible with *R*, *S*, . . ., then it does not follow that *A* or *A*, *B*, *C*, *D*,... must be compatible with *R*, *S*, ... For the objects that *A*, *B*, *C*, . . . and *M*, *N*, *O*, . . . have in common may be different from the ones that *M*, *N*, *O*, . . . and *R*, *S*, . . . have in common. Similarly, if *A*, *B*, *C*, . . . is incompatible with *M*, *N*, *O*, . . ., and *M*, *N*, *O*, . . . is incompatible with *R*, *S*, . . . it also does not follow that *A*, *B*, *C*,... and *R*, *S*, . . . must be incompatible. . . .

§ 95. *Special Kinds of Compatibility:* a. *Inclusion*

1. If the concept of compatibility is taken in the sense of the preceding section, then there are several species of this relation that deserve special consideration. If a pair of ideas *A* and *B* stand in the relation of compatibility to one another, then it may be the case that not only one, but all of the objects under *A* also fall under *B*. If it is not assumed that this relation is mutual, i.e. if it is undetermined whether *B* has some objects that do not fall under *A*, then I wish to call this relation between *A* and *B* a relation of *inclusion*. I wish to say that the extension of the idea *B*, or briefly that *B*, includes *A*, and I call *B* the including and *A* the included idea. Thus I wish to say that the idea 'man' is included in the idea 'inhabitant of earth', since every referent of 'man' also falls under 'inhabitant of earth'.

2. It is clear how this relation can be extended so that we have a class of ideas on each side, or a class of ideas on one, and a single idea on the other. I shall say that *A*, *B*, *C*, *D*, . . . are included in the ideas *M*, *N*, *O*, . . . if every referent of one of the ideas *A*, *B*, *C*, *D*,... also falls under one of the ideas *M*, *N*, *O*, ...

3. If an idea *A* is included in another, *B*, then it cannot have a greater scope than *B*. For if *A* were to represent more objects than *B*, how could all *A* be represented by *B*? A similar law holds of classes of ideas.

4. If an idea *A* is included in *B*, and *B* is included in *C*, then *A* is also included in *C*. A similar law holds of classes of ideas.

NOTE: This relation of inclusion has already been discussed by some other logicians. Thus Maß says (*Grundriß der Logik*, Halle 1793, § 80) "A concept *a* includes another concept *b*, if all *a*'s are also *b*'s". This relation is of great importance since it is the relation in which the subject idea of any true proposition stands to the concretum that corresponds to the predicate idea. For, given that all propositions are of the form '*A* has *b*' or '*A* is *B*', it is required for their truth that the idea *B* include the idea *A*.

§ 96. b. *The Relation of Mutual Inclusion or Equivalence*

1. The definition of the concept of inclusion in the preceding section allows this relation to be mutual between two ideas *A* and *B*. *A* can be included in *B* and *B* in *A*. This is the case when all objects of *A* are objects of *B*, and all objects of *B* are also objects of *A*. Or, more briefly, if both ideas have precisely the same objects. This relation I call a *mutual* or *precise inclusion* or *equivalence*, and the ideas themselves I call *equivalent* ideas. The two ideas 'equilateral triangle' and 'equiangular triangle' are an example.

2. B. here considers whether there can be equivalent ideas that have the same content (without being identical). He answers in the affirmative, giving the examples 'a virtuous person who is prudent' and 'a prudent person who is virtuous'; and '2^4' and '4^2'.

3. There are no two simple ideas that are equivalent, for we should have no way of distinguishing them. On the other hand, it is not necessary that both ideas must be complex for a pair to be equivalent: one of them can be simple. For each simple idea there is an infinite number of equivalent complex redundant ideas, not only of the form 'Non-non-*a*' but also of the form '*a* which is *a*', etc.

4. In the above case, equivalent ideas were generated in such a way that one of them was redundant. Let us now consider some of the simplest cases of equivalent ideas that are not redundant. (a) First of all, it may be the case that all objects of a certain idea *A* also fall under the two ideas *B* and *C*, and that the latter have no other common object. In this case, '*B* which is *C*', and '*C* which is *B*', and 'something which is *B* as well as *C*', are equivalent to the idea *A* (where *A* may be simple). As an example, let *A* mean the morally good (or that which one ought to do), *B* that which is possible, and *C* that which advances common well-being. (b) Secondly, it can be the case that the idea *A* has certain objects in common with *B*, and the same objects also in common with *C*, where *B* and *C* are different ideas. In this case, the ideas '*A* which is *B*' and '*A* which is *C*' will be equivalent, as when *A* means a celestial body, *B* something that

is 1/50 of the size of our earth, and *C* something that illuminates the earth at night. The moon is the only object to which both these complex ideas refer. (c) Finally, it can be the case that *A* and *B* have certain objects in common which are also common to *C* and *D*, where *C* and *D* are otherwise quite different from *A* and *B*. In this case '*A* which is *B*' and '*C* which is *D*' are equivalent ideas. For example, let *A* be a way of life, *B* something that accords with the moral law, *C* a way to bliss, and *D* something that never fails, etc.

5. From the equivalent ideas *A* and *A′* we can generate a new pair of ideas by connecting both *A* and *A′* with the same other idea in the same way. We must not conclude, however, that the ideas thus generated will also be equivalent. Thus 2^4 and 4^2 are equivalent, and the two ideas 'root of the number 2^4' and 'root of the number 4^2' are generated from them in the same way, and yet they are in no way equivalent, since one has 2 as its object, and the other 4.

6. If *A* is equivalent to *B* and *B* is equivalent to *C*, then *A* and *C* are equivalent.

7. Let us denote the idea 'something that has (the attributes) *a*, *b*, *c*, *d*, . . .' by '[something] $(a+b+c+d+\ldots)$'. Let us furthermore denote (cf. § 60) the idea 'something that has (the attribute) *x*' by *X*. Then the following symbols represent equivalent ideas provided they have a referent: [something] $(a+b+c+\ldots)$, [*A*] $(b+c+\ldots)$, [*B*] $(a+c+\ldots)$ [*C*] $(a+b+\ldots)$.

8. If we want to extend the concept of equivalence to whole classes of ideas, then we shall have to say that the class of ideas *A*, *B*, *C*, *D*, . . . is equivalent to the class of ideas *M*, *N*, *O*, . . . if every object of one of the ideas *A*, *B*, *C*, *D*, . . . also falls under one of the *M*, *N*, *O*, . . . and conversely. Thus the two ideas 'a right triangle' and 'an oblique triangle' taken together are equivalent to the three ideas 'equilateral triangle', 'isosceles triangle' and 'scalene triangle', etc.

9. In order for the two classes *A*, *B*, *C*, *D*, . . . and *M*, *N*, *O*, . . . to be equivalent, it is not necessary that any one of the ideas *A*, *B*, *C*, *D*, . . . , taken by itself, should be equivalent to any one of the *M*, *N*, *O*, . . ., nor is it necessary that the sum of the scopes of the ideas *A*, *B*, *C*, *D*, . . . should equal that of the ideas *M*, *N*, *O*, . . . The first can be seen from the preceding example. The second is shown in the following example. The two ideas 'member of the series 1, 2, 3, . . . 10' and 'member of the series 2, 3, . . . 11' taken together are certainly equivalent to the following two 'member of the series 1, 2, 3, 4, 5' and 'member of the series 6, 7, . . . 11'. Yet the scope of the first idea is 10 and so is that of the second; thus the

sum of both scopes is 20. But the scope of the third idea is 5, and that of the last one 6, with the sum of them merely 11. . . .

§ 97. c. *The Relation of Subordination*

1. The relation of inclusion (§ 95) can be either mutual or not mutual. If an idea *A* is included in another, *B*, without *B* being included in *A*, then *B* must represent one or more objects in addition to those represented by *A*. This relation is called *subordination* and we say that *B* is *higher*, *A* is *lower*, that *A* is *subordinated* to *B* or *stands under B*. Thus the ideas 'man' and 'living being' stand in the relation of subordination; that is to say 'man' is the lower and 'living being' the higher idea, since every object that falls under the idea 'man' also falls under 'living being', but not conversely.

2. Several ideas *A*, *B*, *C*, *D*, . . . stand in the relation of *subordination* to one or several other ideas *M*, *N*, *O*, . . . if each of the objects represented by one of the former is also represented by one of the latter, but not conversely. The class *A*, *B*, *C*, *D*, . . . will be called the lower and *M*, *N*, *O*, . . . the higher one.

3. A higher idea must also have greater scope, but it is not the case that an idea with greater scope must also be higher. The same holds of whole classes.

4. If *A* is lower than *B*, *B* lower than *C*, then *A* is lower than *C*. The same holds for whole classes of ideas.

§ 98. d. *The Relation of Overlapping or Linking*

1. . . . If the objects of an idea *A* do not all fall under another idea *B* which is compatible with *A*, then a relation results which I call *overlapping* or *linking*. Thus I call the two ideas 'learned' and 'virtuous' overlapping since, aside from the objects which they share, each has some which do not fall under the other.

2. From the mere fact that of a pair of ideas *A* and *B* each stands in the relation of overlapping to a third, *M*, nothing can be concluded concerning the relation in which *A* and *B* stand to one another. They may exclude one another, they may themselves overlap, one may be subordinated to the other, or they may even be equivalent.

3. The first case, where certain ideas *A*, *B*, *C*, *D*, . . . overlap a certain other, *M*, but exclude one another is quite important. The reason is that through the mediation of *M* a certain *connection* between *A*, *B*, *C*, *D*, . . . is established, which would not otherwise be present, and knowledge of which may occasionally be of importance. Thus it may be important to know if two or more learned societies,

that do not have any members in common have each of them some members in some other such society. In such a case we call the ideas A, B, C, D, . . . *mediately linked* through M. If each of the ideas A, B, C, D, . . . is linked to the preceding one, but is incompatible with all earlier ones, then this series of ideas will be called a chain. If all remains as above, except that the last idea is again linked with the first, then the class of these ideas will be called a *closed chain*. The following seven ideas are an example: 'tones classed under c, d, e, f, g, a, b'; similarly the ideas 'red', 'orange', 'yellow', 'green', 'light blue', 'violet'.

4. An even more remarkable relation, which also occurs more often than the above, holds if the ideas A, B, C, D, . . . overlap in pairs in such a way that no two ideas share all objects, but any two have some common objects. In this case the idea of something that is A as well as B, C, etc. or [something]$(a+b+c+\ldots)$ is a referring idea in which none of the parts a, b, c, d, . . . is *redundant* (§ 69). This is the case because the ideas that result if we leave out any one of these parts e.g. [something]$(b+c\ldots)$, [something]$(a+c\ldots)$ are all wider than the idea [something]$(a+b+c\ldots)$. This relation could be called *collective overlapping*. An example are the ideas 'polygon', 'equilateral', 'equiangular', which we can combine to get the non-redundant concept of an equilateral, equiangular polygon.

5. It is obvious that the relation of overlapping is not confined to individual ideas but can be extended to classes of them. We shall say that the class of ideas A, B, C, . . . overlaps the class of ideas M, N, O, . . . if there are objects that fall under one of the ideas A, B, C, . . . as well as under one of the M, N, O, . . ., provided that there are others that are represented by each of these classes alone. . . .

§ 99. *Ideas of Absolutely Greatest and Smallest Scope, Highest and Lowest Ideas*

1. B. concludes that there cannot be an individual idea whose scope is greater than that of any other, because, according to § 96, we can find an equivalent idea for any given idea. Hence there is at most a *class* of ideas with greatest scope, which would also be the highest ones, since any higher idea would have a greater scope.

2. There are such ideas; examples are 'something', 'object in general'. "We say of an idea that it has an extension only if there are objects which it represents. Hence it is impossible that the extension of an idea should be any more extended than that which encompasses every object that there can be, and this extension belongs to the idea 'object' or 'something in general'." There are many equivalent ideas of greatest extent, e.g. 'not nothing', and all ideas of the form 'something that has the attribute *b*', where *b* is an

attribute that belongs to all objects without exception, such as self-identity, or the attribute of having only one of any pair of contradictory attributes. Hence there are infinitely many ideas of greatest scope, "but since they are all equivalent to one and the same idea, namely to the idea of something in general, they share one and the same extension. Hence there is only one extension that is of absolutely greatest scope, but there are many ideas that relate to it."

4. Are there also ideas with smallest scope? (They would also be the lowest.) B. maintains that singular ideas are of this sort (§§ 68 and 78).

5. While ideas of greatest scope all have the same extension, ideas of smallest scope do not.

6. Among general ideas, ideas of smallest scope would have to have exactly two objects, "since an idea whose scope is smaller than that, i.e. which has fewer than two objects, would no longer be a general idea." Ideas of the required sort can be formed by taking two singular ideas, *A* and *B*, and forming the idea 'one of the things *A* or *B*'. Other examples are 'sons of Isaak'.

NOTE 1: B. claims that his concept 'something' is different from Aristotle's substance, "since Aristotle merely wanted to classify *real* objects into species." More recent logicians claim that the highest concept is the conceivable; i.e. "everything that can be thought. But this restricts it to ideas and propositions. In my opinion, no other things belong to the conceivable [*zum Denkbaren*]". Also the concept of the 'possible', which is contained in the concept 'conceivable' (as is indicated by the suffix '-ble') is lower than 'something in general', because it does not represent ideas and propositions in themselves, "which not only have no existence, but cannot assume existence. Hence, strictly speaking, they do not belong to possible things. (Possible things are those that can become actual.) Of course, they also do not belong to the impossibilities, in the sense of the contradictory."

Some philosophers believe that there is a generic concept which is higher than, and includes both, something and nothing. B. rejects this on the grounds that nothing is not an object.

NOTE 2: There are only few logicians who admit that there are lowest ideas. B. considers arguments against this view: (a) The content of an idea can always be enlarged. Reply: true, but this does not always restrict the extension, so that an addition to the content of an idea may generate an equivalent idea. (b) Before a concept is restricted to the point that it fits only one or two objects, it must contain infinitely many parts. Reply: This is false (see §§ 64 and 68). (c) The law of continuity is violated if the addition of some part could restrict the extent of an idea from infinite to finite. Reply: Often the addition of some part cuts infinitely many objects away from the extension, and the subtraction of infinitely many objects from infinitely many objects often leaves a finite remainder. Example: the scope of the idea 'being' is infinite. If we add the part 'which does not have a cause of its existence', the scope is cut down to one object.

§ 100. *Ideas that are Adjacent in Scope or Height*

1. If an idea is narrower than *A* and wider than *B*, then it is said to have

a scope intermediate between *A* and *B*. If an idea is lower than *A* and higher than *B*, then it is said to be of intermediate height between *A* and *B*. If an idea *B* has smaller scope than an idea *A*, and there is no idea of intermediate scope between *A* and *B*, then *A* and *B* are of adjacent scope. A similar definition is given for '*A* and *B* are adjacent in height'.

2. If any idea *A* has a finite number of objects, *n*, and another idea *B*, has $n-1$ objects, then *A* and *B* are adjacent in scope. If it is the case that all *B* are *A*, then *A* and *B* are also adjacent in height.

3. There are also ideas *A* and *B* such that there are infinitely many ideas of intermediate scope between *A* and *B*.

Example: Since the fourth power of an integer is the square of some other integer, etc. it follows that the concept of an angle whose relation to its complement can be represented by the formula $1 : n^2$ is higher and of greater scope than the concept of an angle for which this relation can be represented by $1 : n^4$, which is higher and of greater scope than the concept of an angle for which this relation can be represented by $1 : n^8$, etc. Each of these concepts is higher and of greater scope than that of a right angle, for which $n = 1$, and of smaller scope and lower than that of an angle in general, because the latter includes angles which do not have the ratio of 1 to an integer between themselves and their complements.

§ 101. *Whether a General Concept can be Found for any Arbitrarily Selected Set of Objects*

A concept which represents at least objects α, β, γ, δ . . . is easily found. The concept 'something' will do the job in any case. It is more difficult to find a general concept that represents no more than these objects. An idea fulfilling this requirement is the distributive idea 'any part of the class *A*, *B*, *C*, *D*, . . ., where '*A*' represents α, 'B', β, etc. Such an idea always exists objectively, but may not be subjectively attainable, and it is not a pure concept. Whether a pure concept can be found in all cases is doubtful, but there is always the concept of such a concept, namely the symbolic idea (cf. § 90) 'a concept which represents all and only α, β, γ, δ. . . .' "This could be called the symbolic general concept of α, β, γ, δ . . ."

§ 102. *No Finite Set of Units Suffices to Measure the Scopes of all Ideas*

1. We have seen that there are Ideas whose extension includes only a finite number of objects, or only one. In order to measure the scope of these, we must choose as the unit of measurement an idea that likewise represents only a finite number of objects. It is most natural to choose for the unit the scope of ideas that have only one object. But we have already seen that there are ideas that represent an infinite number of objects. The scope of these can be measured only by taking as unit the scope of an idea that also has infinitely many objects. One might think that this would suffice to measure all

ideas of this kind. But this is not the case. Rather, it can be shown that there are infinitely many ideas which are of such a nature that one is infinitely surpassed in scope by the next. From this it follows that the unit that serves to measure one cannot be used to measure another, so that no finite set of units of measurement suffices to measure the scope of all ideas. It seems to me that the following example proves the truth of this assertion: Let us abbreviate the concept of any arbitrary integer by the letter n. Then the numbers $n, n^2, n^4, n^8, n^{16} \ldots$ express concepts each of which includes infinitely many objects (namely infinitely many numbers). Furthermore, it is clear that any object that falls under one of the concepts following n, e.g. n^{16}, also falls under the predecessor of that concept, n^8. It is also clear that very many objects that fall under the preceding (n^8) do not fall under the following (n^{16}). Thus of the concepts n, n^2, n^4, n^8, n^{16} each is subordinated to its predecessor. It is furthermore undeniable that the scope of any of these concepts is *infinitely larger* than the scope of the concept immediately following it. (And this holds even more for concepts that follow later in the series.) For, if we assume that the largest of all numbers to which we want to extend our computation is N, then the largest number that can be represented by the concept n^{16} is N. and thus the number of objects that it includes is equal or smaller than $N^{\frac{1}{16}}$ and likewise the number of objects that fall under the concept n^8 is equal or smaller than $N^{\frac{1}{8}}$. Hence the relation between the scope of the concept n^8 and that of the concept n^{16} is $N^{\frac{1}{8}} : N^{\frac{1}{16}} = N^{\frac{1}{16}} : 1$. Since $N^{\frac{1}{16}}$ can become larger than any given quantity, if N is large enough, and since we can take N as large as we want, and since we come closer and closer to the true relation between the scopes of the concepts n^8 and n^{16}, the larger we take N, it follows that the scope of the concept n^8 surpasses infinitely that of the concept n^{16}. Since the series $n, n^2, n^4, n^8, n^{16} \ldots$ can be continued indefinitely, this series itself gives us an example of an infinite series of concepts each of which is of infinitely greater scope than the following.

§ 103. *Special Kinds of Incompatibility among Ideas*

1. In § 94 the set of ideas $A, B, C, D, \ldots$ was said to be incompatible if there was no common object represented by each of them. Now the notion of *exclusion* is introduced. $A, B, C, D, \ldots$ exclude one another if they are incompatible in pairs. If a set consists of only two ideas A and B, then to say that A excludes B is the same as to say that A is incompatible with B.

2. The relation of exclusion can be extended to whole classes of ideas. The classes $A, B, C, D, \ldots; M, N, O, \ldots; R, S, T, \ldots;$ exclude each other if they are incompatible in pairs.

3. If an idea which is incompatible with A represents every object not represented by A, then it is called a contradictory of A, otherwise a contrary.

4. "It is evident that for every idea, as long as it has an extension, and as long as its extension is not of greatest extent, we can find another idea which is its contradictory." For every idea A, non-A is such a contradictory, provided that A has an extension (otherwise it could not be called an excluding idea) and provided it does not have the widest extension (otherwise its contradictory would have no extension). "But if an idea has one contradictory idea, then, according to § 96, it has infinitely many. . . . But all the ideas that contradict a given idea must be equivalent to each other. . . . This is not the case with contraries."

5. In an obvious way, the concepts of contradictoriness and contrariety can be extended to classes of ideas.

§ 104. *Coordinated Ideas*

1. If the ideas A, B, C, D, . . . are all subordinate to the idea X, and if they exclude each other, they are called ideas *coordinated under the idea X* or, in short, disjunct ideas. It follows from this and § 99 that all ideas that exclude each other can be envisaged as coordinated in a certain respect, because an idea to which they are all subordinate can always be found.

2. If the extension of a set of coordinated ideas A, B, C, D, . . . M is the same as that of the higher idea X, then their extensions are called the complementing parts of the extension X; we also say that they *exhaust* the latter. In the case where X is the idea of something in general, they are said to complement each other absolutely.

4. Not every class of ideas A, B, C, . . . M which is equivalent to some other idea X also exhausts the extension of the latter, since the ideas A, B, C, . . . M may not be exclusive.

7. If we can say that in some respect the difference between an object β and each of the two objects α and γ is smaller than the difference between α and γ, then we can call β an *intermediate* object between α and γ. A foot is intermediate between an inch and a yard. If the objects of a certain idea B are intermediate between the objects of A and C, then B can be called intermediate between A and C. If this is the case, A, B, and C must stand in the relation of exclusion. Furthermore, if β is envisaged as an intermediate object between α and γ, then we consider only those of its attributes which it shares with α and γ, or in which it differs from them less than from other objects. Hence an idea B is considered intermediate between two others A and C only to the extent in which all three are thought to be subordinate to the same higher idea, hence only in so far as A, B, and C are considered coordinated ideas. This shows sufficiently that intermediacy in the present sense must not be confounded with intermediacy in scope or height (§ 100).

8. Some objects are such that infinitely many objects lie between them; example: angles between 30° and 60°. "But if there are pairs of objects between which infinitely many intermediate objects can be inserted, then there must also be ideas between which there are infinitely many intermediate ideas."

9. "There are also pairs of objects, and hence pairs of ideas, between which there are no intermediate objects and intermediate ideas. Hence there

is probably nothing that could be considered an object intermediate between a rose and the truth that a square has only right angles, since the differences between these two objects do not allow of more or less. Similarly, there is no idea intermediate between the ideas 'right angle' and 'acute angle'; for even though we can find an intermediate angle between a right angle and any one acute angle, the inserted angle is not an object intermediate between the objects of 'right angle' and 'acute angle', but is itself acute."

In a note, B. attacks the so-called principle of logical continuity, due to Kant. He gives it in the formulation of Krug (*Logik*, Königsberg 1806, § 45, b): "between every higher and lower concept there must be a third which is the same with both and different from both, i.e. which is more closely related to each of them than they are among themselves." How could this be proved? "The concept 'real' is subordinate to 'possible', but I cannot think of a third concept that is intermediate between them."

§ 105. *Some Theorems*

This section contains theorems that belong to §§ 93–104. B. gives proofs which are here omitted.

1. If *X* is subordinate to both *A* and *B*, then *A* and *B* are compatible ideas.
2. If two ideas are incompatible, then there is no third idea which is subordinate to them both.
3. If two ideas are compatible, then all ideas higher than they are also compatible.
4. If two higher ideas are incompatible, then all ideas under them are also incompatible.
5. If two incompatible ideas *A* and *B* are compatible with *X*, then *X* is not subordinate to either of them.
6. If two ideas *A* and *B* overlap, then there is no idea *X* which is higher than *A* and lower than *B*: either *X* is higher than *B*, too, or overlaps *B*.
7. If *A* and *B* overlap, and *X* is lower than *A*, then *X* is either lower than *B*, or overlaps *B*, or is incompatible with *B*.
8. If *A* and *B* overlap, and *X* is incompatible with *A*, then *X* must be lower than, or incompatible with, *B*.
9. If *A* is higher than *B*, then non-*A* is lower than non-*B*.
10. No idea can be incompatible with both *A* and non-*A*: If it is incompatible with one, then it must be equivalent or subordinate to the other.
11. Two merely contrary ideas can be subordinate to the same higher idea.
12. Both of two contrary ideas can be incompatible with one and the same idea.
13. One and the same idea can be compatible with each of two contrary ideas, and also with each of two contradictory ideas.
14. If *B* is lower than *A*, then *A* and *B* are compatible.
15. If *A* is neither equivalent nor subordinate to non-*B*, then it must be compatible with *B*.
16. If *A* and *B* are compatible, then of the two ideas *A* and non-*B* and of the two ideas *B* and non-*A* neither includes the other.

17. Conversely, if inclusion holds between A and non-B, then A and B are incompatible.

18. If A and B both have referents, but neither includes the other, then A and non-B, B and non-A are compatible, and neither of non-A and non-B is subordinate to the other.

19. If A and non-B and B and non-A are compatible, then A and B are either incompatible, or they overlap.

20. If A and B are contraries, then A and non-B, and B and non-A are not equivalent; non-A and non-B are consistent in this case.

21. If B is lower than A, then non-B cannot be lower than A, except if A is an idea of greatest scope. If B is lower than A, then a contrary of B can be lower than A.

22. If A and B overlap, then it is possible that non-A is subordinate to B.

23. If the extension of two overlapping ideas A and B is smaller than the largest extension, then non-A and non-B also overlap.

§ 106. *Ideas of Kinds, Species, etc.*

1. If A is a general idea, then the idea of the form 'the class of all A' is called the idea of a kind, or species.

2. "Sometimes the idea A itself, i.e. any idea that has more than one object (any general idea) is called an idea of a kind or species, but this is inaccurate, since it is merely an idea out of which, through combination with certain other ideas, the idea of a species or kind can be formed. . . . The objects that fall under the idea A are not the same as the objects of the idea 'the class of all A'. For the latter has actually only one object, namely the whole that consists of the collection of objects that can be represented by A. Each of the objects that are represented by A is only a *part* of the whole which is the object of the class-idea of which A is a part. From this we see another difference between the idea A taken by itself and the idea 'the class of all A'. The former always refers to several objects, the latter only to one."

4. If A and B are equivalent, then 'the class of all A' is equivalent with 'the class of all B'.

5. If the idea A is higher than the idea B, then kind A is also called higher than kind B; kind B is then called a subordinate kind, or subspecies.

6. B. gives a long list of names that could be given to classes in a hierarchy.

7. The existence of a highest kind is guaranteed by the argument of § 99. There are also lowest kinds, compatible, exclusive, overlapping, contradictory, contrary, coordinate and intermediate kinds. They correspond to the respective ideas.

§ 107. *Opposing Ideas*

1. The relation of opposition between ideas is very important. I shall say that two ideas are opposed to one another, when actually only their objects are opposed. The opposition between objects, however, I conceive only in the strict mathematical sense, according to which before and after in time, above and below in space, credit

and debit, are opposed to each other. The definition of this concept, it seems to me, is the following: Let A be an idea which represents precisely α, and let B be an idea which represents precisely β, and let $p, q, r, \ldots$ be pure concepts. We shall call two objects α and β *opposed* to each other if there are some $p, q, r, \ldots$ such that with A they will form an idea $[A, p, q, r, \ldots]$ which represents precisely β and with B an idea $[B, p, q, r, \ldots]$ which represents precisely α.

An example will make this clearer. If two lines OR and OS proceed from the same point O in two different directions, and if they are so positioned that the distance between point M on OR and N on OS equals MN, namely the sum of its distances from the common origin O, i.e. if $MN = OM + ON$, then mathematicians call these lines opposed. Why? Because it is possible in this case to form an idea $[OR, p, q, r \ldots]$, which represents the direction OS, and which is such that an exchange of OS for OR yields an idea $[OS, p, q, r \ldots]$ which represents the direction OR. To form such an idea we only have to consider that for every given direction OR there is one and only one other directional OS, which proceeds from the same point O, such that the distance between the two points M and N equals the sum of their distances from O; for every other direction such as Os the distance $Mn < OM + On$. Hence the following idea will fulfil our requirements: 'a direction which proceeds from the same point as OR and which stands to OR in such a relation that the distance between a point on it and a point on OR equals the sum of the distances between these points and the point of origin of direction OR'. It is obvious that this idea has the form $[OR, p, q, r, \ldots]$, i.e. that its parts, with the exception of the idea OR (which may be mixed) are all pure concepts. Similarly, it is true that this idea does not apply to any other object except the direction OS. In the case of any other direction, e.g. Os, it holds that $Mn < OM + On$. Finally, it is obvious that this idea has the attribute that it can be changed from an idea of OS into an idea of OR as soon as we replace the part OR by OS. ... Hence it is obvious why the directions OR and OS must be called opposite under the given definition. It only remains to show that and why directions which form an angle, such as OR and Os, cannot be called opposite under this definition. In this case it is not possible to form from OR an idea $[OR, p, q, r \ldots]$ which applies only to Os, no matter how many pure concepts are used. For such an idea could be generated only through a determination of the relation between Os and OR. But it is obvious that the relation between Os and OR, if it is to be described only by pure concepts, also holds between an infinite number of other directions (namely all directions which are generated by turning *Os* around *OR* as an axis without changing the angle *ROs*).

2. Since I call a pair of ideas opposite only if they have a pair of opposite objects, it follows that each of them can only represent a single object; hence they are singular ideas; it also follows that they are incompatible with each other, more precisely that they are contrary ideas (§ 103). They obviously form only a special subclass of these. For to every given idea several contrary ideas can be found whose extensions differ from each other, while the several ideas which might be opposed to a given idea must all represent one and the same object.

NOTE 1: B. points out that most logicians have used the expression 'opposing ideas' for what he has called contrariety. Even if common usage is on their side, the concept introduced in this section is still of great importance, since it is used not only in mathematics, but in many other sciences. "If I am not mistaken, pleasure and pain, desire and aversion, benefit and harm, the morally good and the morally bad, merit and fault, praise and blame, command and prohibition, the true and the false, knowledge and error, wisdom and folly, the beautiful and the ugly, the sublime and the lowly, and many other such objects are opposed to each other in the truly mathematical sense if they are considered from a certain point of view. It is obvious, then, how there can be many different sciences in whose theorems the concept of opposition is either already present, or could be profitably used."

It is true that there are many objects which we consider opposites although they do not fulfil the above definition. But it turns out that there is always a point of view from which we can consider these objects, and under which they appear to be opposites. This also holds for contradictory ideas, if we consider the circumstance that either of them can be generated from the other in one and the same way. "In exactly the same way in which the idea non-*A* can be generated from *A*, the latter can be generated from the former (at any rate an idea that is equivalent to *A*). It may be for a similar reason that we call merely contrary ideas of the form '*A* which is *B*' and '*A* which is not *B*' opposites. We do this only in so far as we consider that each of them can be generated from the other in one and the same way, namely through insertion of a negation in the proposition which is attached to the main part (*A*). If we take large and small to be opposites, then we do this because we consider one of them larger and the other one smaller than a certain given measure. We can then think of one of them as generated through an addition, the other through a subtraction from this measure. But adding and subtracting are operations which can be called opposite.

It is most puzzling that we call such attributes as black and white, sweet and bitter opposites of each other. According to Aristotle, this is done because generally we call things opposites if they represent extreme differences among things of their kind. Hence, black and white are called opposites because among all colours they differ most from each other. The same holds for sweet and bitter among sensations of taste. I do not see why this reason, even if it holds, could generate an opposition between two things. I doubt whether white and black differ more than, for example, blue and yellow. Are there not cases where we do not quite know whether we should call something black or white while we never have any difficulty in deciding whether something is blue or yellow? I suspect that the indicated attributes are called opposite because we think that they are created by powers which cancel each other (hence which are opposed in the strict sense of the word). If we add white to black, the black is somewhat reduced, and conversely. On the other hand the mixture of yellow and blue produces a new colour."

NOTE 2: Kant, in his *Attempt to Introduce Negative Magnitudes into Philosophy*, has grasped the nature of mathematical opposition better than any other. However, the Kantian definition (of real repugnance) is not applicable to objects which do not have effects and consequences.

NOTE 3: "There is another relation between ideas that I shall mention only briefly, as it seems less important to me. The special kind of ideas which I have called ideas of relations in § 80 fall into pairs known as converse ideas. If an object *A* stands in a certain relation to another object *B*, a (usually different) relation can be indicated in which *B* stands to *A*. The ideas of these two relations are called converse, or *abstract converse* ideas. On the other hand, the objects *A* and *B* themselves, as a pair of objects of which the first stands to the second in one relation and the second to the first in the other, are called a pair of *concrete converses.* Any object *A* stands to its attribute *b* in the relation of 'having this attribute'. Conversely, the attribute *b* stands to the object *A* in the relation of 'being an attribute of it'. Thus the two ideas 'having as an attribute' and 'being an attribute (of something)' are a pair of converse ideas, indeed abstract converses. On the other hand, the idea 'something that has the attribute *b*' and 'the attribute *b* as something that belongs to an object' are a pair of concrete converses. Thus the concepts 'fatherhood' and 'sonhood' are abstract converses, while 'father' and 'son' are two concrete converses. Also the concepts 'cause' and 'effect', 'end' and 'means', 'part' and 'whole', 'substance' and 'attribute', 'higher idea' and 'lower idea', 'above' and 'below', 'before' and 'after' are all examples of concrete converses."

§ 108. *Ways in which the Relations Considered in §§ 93 ff. could be Extended to Ideas that do not have Referents*

According to their definitions, the relations between ideas that we have considered in § 93 ff. cover only ideas that have referents. It is nevertheless certain that we apply, even in common language, several of those relations to ideas which represent no object, or even to those which *cannot* represent any object, since they assign certain contrary attributes to it. Thus we do not hesitate to call the two ideas 'a mountain that is golden' and 'gold that forms a mountain' *equivalent*, even if we doubt that there is an object corresponding to these ideas. However, according to the definition of the relation of equivalence (§ 96) this can be said only if such a mountain exists. In the same way, mathematicians do not hesitate to call the concepts of a body that is bounded by five equal sides and of a body that is bounded by seven equal sides incompatible, although they know that there is neither a body of the first nor of the second kind. We also say that the concept of a *being* that does not have a single good attribute, and the concept of a *man* that does not have a single good attribute stand in the relation of *subordination*, and that the first concept is *of greater scope* than the second, so that if anybody could show though nobody can that there is no *being* of the first kind, he would at the same time have shown that there is no *man* of that kind. Here we attribute a relation of subordination in the very act of saying that these ideas do not have referents. This could not be the case unless these relations were taken in a certain *extended* sense. What is this extended sense? In § 69, we discussed the concept of redundancy. This concept was originally defined in such a way that it could be applied only to ideas with referents. But we discovered a means for a suitable expansion of it by considering certain constituents of the given idea i, j . . . to be *variable*. This method can be applied in the present case too. We can at once extend all of the definitions mentioned in §§ 93–107 to ideas without referents, if we are only permitted to envisage certain of their parts as variable. For then it would only be necessary to consider the infinitely many new ideas that are generated from the given ones if the variable parts i, j, . . . are replaced by any other ideas. The same relation that holds between these new ideas, *whenever they have a referent*, we also attribute to the given non-referring ideas. It is understood that this is the case only *relative* to the parts i, j, . . . which are taken to be variables. In particular, we shall say that two non-referring ideas A and B are *equivalent* with respect to the variable parts i, j, . . ., if the ideas that are generated when we put any other ideas in the place of i, j, . . . are *equivalent* in the narrower meaning of § 96 whenever they have referents. We shall say that A is *higher*, B *lower*, if the new ideas that are generated from A and B stand in the relation of subordination in the sense of § 97 whenever they have referents, etc. Accordingly, the two ideas 'mountain that is of gold' and 'gold that forms a mountain' shall be considered equivalent with respect to the variable parts 'mountain' and 'gold'. For all ideas that are generated if we replace these two parts by whatever other parts are equivalent in the sense of § 96, i.e. have the same objects, if they have any objects at all. In the same way we say that the idea 'A *being* that does not have a single good attribute' is higher than the idea 'a *man* that does not have a single good attribute', if we consider as variable the common constituent 'not having a single good attribute'.

This makes it evident that an idea that is generated from the first is always higher than one that is generated from the second in the sense defined in § 97 whenever we replace the variable part by something that gives us a referring idea; etc.

NOTE. The mathematician makes the most important use of this extension of these relations, especially in the theory of equations. According to their original concept, equations are nothing but assertions of the equivalence of two ideas. Thus to say that $4+5 = 11-2$ is to say that the idea '$4+5$' has the same objects as the idea '$11-2$'. But if we stay with this concept then we can never establish equations such as $2-2 = 0$, $\frac{1}{\sqrt{-1}} = -\sqrt{-1}$, etc. whose meaning can be easily explained according to the above.

CHAPTER 4

Distinctions among Ideas that Result from their Relations to other Objects

§ 109. *Correct and Incorrect Ideas of an Object*

"When we not merely believe that *A* is an idea of α, but when *A* is in fact an idea of α, we usually say, for the sake of emphasis, that *A* is a *correct* idea of α. Sometimes it is called a *real* or *true* idea of α. When we mistakenly assume that *A* is an idea of α, we occasionally say that *A* is an incorrect, false, or erroneous idea of α instead of saying that *A* is not an idea of α. Thus the idea 'something that has corners' is a correct idea of a cube, but an incorrect idea of a sphere. It is obvious that the correctness or incorrectnesss of ideas is not an internal, but only an external or relative attribute of them." We do not say that an idea is incorrect in itself, but only that it is incorrect *of an object*. Inconsistent ideas (§ 70) never have an object and hence are incorrect of every object.

In a note, B. points out that the words 'true' and 'false' should be reserved for propositions, and should not be applied to ideas.

§ 110. *Complete and Incomplete Ideas of an Object*

1. If *A* is a *correct* idea of object α, or, what comes to the same, if the proposition 'α is *A*' is true, then from this sentence either all, or only some of the attributes of α can be derived, where for this derivation nothing is used but truths of the form '*A* has the attribute *m*', '*A* has the attribute *n*', etc. If all attributes of α can thus be derived, then *A* is called a *complete* or *exhaustive* idea of its object; otherwise we say that A represents the object α only *incompletely*. Thus, I call the concept 'a being that is omnipotent' a complete concept of God, since this concept (although perhaps not for man, but in itself) suffices to derive from it all the attributes of God in the just indicated way, i.e. only by means of propositions of the form 'an almighty being must also be omniscient, independent, etc.'. For the same reason I call the idea 'father of Alexander the Great' a complete idea of King Philip of Macedon, while the concept 'King' and also the concept 'King of Macedon' are examples of incomplete ideas of that man. For it is impossible to deduce from them anything approaching all the attributes of the object that they are to represent.

2. We can see from these examples that it is not necessary that an

exhaustive idea of an object should contain the ideas of all the attributes of that object. Such an idea would be greatly redundant, and, since there is an infinity of attributes in any object, it would have to have an infinite number of constituents.

3. If an idea represents only one object, then it is always an exhaustive idea of that object (singular ideas, cf. § 68). For if there are some attributes of the object α that cannot be derived from the idea *A* in the indicated way, then a second object β can be thought which has all attributes that can be derived from *A* but lacks those that are present in α but cannot be derived from *A*. Thus the object β, too, would fall under the idea *A*, and *A* should not be called a singular idea.

4. Since all sensations are singular ideas, they are all exhaustive ideas of their objects.

5. No two real objects have precisely the same attributes. It also holds that no idea of a real object is exhaustive as long as it represents several objects, i.e. as long as it is not a singular idea of that object. For, in order to apply to several objects, it must leave undetermined those of their attributes in which they differ.

6. Whether an idea *A* is exhaustive cannot be determined by considering it alone, but only by comparing it with the object to which it is applied by means of the proposition 'this is *A*'. For, depending on the object to which it is applied, the same idea may or may not be exhaustive. Thus the idea 'the kind of figure with straight lines whose internal angles add up to two right angles' is an exhaustive idea of the kind *triangle in general*, but it is not an exhaustive idea of the kind *equilateral triangle*. For from this idea can be derived all attributes that belong to triangles in general, but not all of those that belong to equilateral triangles.

§ 111. *Essential and Inessential Attributive Ideas*

1. If the proposition '*A* has the attribute *b*' is true, then *b* is called a *correct* attributive idea of each object of *A*; otherwise it is called an incorrect one. . . .

2. If, in addition, the idea *A* is a pure concept, and if therefore the attribute which is expressed by *b* belongs to the objects of *A* by virtue of the pure concept under which we comprehend them, then I call *b* an *essential* attribute of them; otherwise an *inessential* attribute. Ideas of essential attributes are called essential attributive ideas; similarly for inessential attributes. The possession of certain limbs by an organic body is an essential attribute of it, since the proposition

'every organic being has limbs' is true, and since the idea 'every organic being' is a pure concept. Health and sickness on the other hand I call merely inessential attributes of an organic body since they do not follow from its concept, i.e. since neither the proposition 'every organic being is healthy' nor the proposition 'it is sick' is true.

3. According to the above definition, we cannot determine from a given attributive idea alone whether it is an essential or inessential attribute of some object. Rather, this depends in part upon the object whose attribute it is supposed to represent, and in part upon the concept that we form of this object. The same attribute will sometimes be called essential and sometimes inessential of the same object, depending upon the concept under which we subsume the object. Thus equilateralness is an inessential attribute of an object that we merely subsume under the concept 'quadrangle in general', but it is an essential attribute if we represent the same object as a square. In general, we can see that an object contains more esssential attributes the *smaller in scope* and *lower* the concept is under which we subsume it. For, if an idea is called smaller in scope and lower than another, then it must exclude some objects that the second includes. This can be the case only if it determines some attributes that are not determined by the other. Finally, if the concept under which we subsume an object is a singular one (§ 68) and thus an exhaustive concept of it (§ 110, no. 3), then by virtue of this concept all the attributes of this object are envisaged as essential to it. For from an exhaustive idea of an object all its attributes can be derived. On the other hand, as long as the concept under which we have subsumed an object is not exhaustive, there are attributes of this object which we cannot derive from our concept of it, and which therefore are envisaged as inessential. Thus if an attribute of an object can be represented by a pure concept, then there is a pure concept under which the object can be subsumed, such that the attribute becomes essential to the object. For, if 6 is the idea of that attribute, then at least the concept 'something that has the attribute *b*' fulfils the requirement: it represents the object, and the attribute *b* follows from it.

NOTE 1: B. indicates that he wishes to reserve the expressions 'necessary' and 'accidental' to objects that have real existence, while 'essential' and 'inessential' are to be applicable to such objects as propositions in themselves.

NOTE 2: In real objects, all necessary attributes are permanent, but the converse does not hold.

NOTE 3: "It may seem to some readers that the basis for my decision whether or not a certain attribute is essential of an object, namely the concept under which we subsume the object, is altogether too arbitrary and insignificant a circumstance. However, if we consider that no object is known to us save through the idea under which we have subsumed it, and that it cannot become an object for our enquiry except through this idea, then we must admit that for the purposes of logic it is neither an insignificant nor an arbitrary circumstance whether a given object is subsumed under one idea or another. As an object of our *knowledge* a thing is no more than what we represent in our minds, whenever we believe that we represent *it*. Thus, for the purposes of logic, its idea constitutes its *essence*."

§ 112. *Common and Peculiar Attributive Ideas; Indicators, Marks*

1. If a certain attribute of an object α does not belong to any other object, or at least not to another object of the same kind, then I call it, in the first case an *absolutely*, and in the second case a *relatively*, peculiar or exclusive attribute of α. Other attributes which do not belong merely to a certain object, but also to others, perhaps even of the same kind, I call common attributes. The concepts of peculiar and common attributive *ideas* are defined accordingly. Hence 'uncaused existence' is an attribute of God which does not belong to any other object, i.e. an absolutely peculiar attribute of God. Similarly, if it is true that among all creatures on earth man alone has earlobes, then their possession is a peculiar attribute of man relative to the creatures on earth. On the other hand possession of hands is an attribute of man which he has in common with several other creatures, for example apes. If *a* is the idea of a peculiar attribute of object α, then *A* (the concretum of *a*) is an exhaustive idea of α, since it represents nothing but α (§ 110).

2. Peculiar attributes of objects, if they are used to identify these objects, are called either absolute or relative indicators of them.

3. Obviously we cannot see from an attributive idea alone whether it is a common or a peculiar idea, or whether it is absolutely or only relatively peculiar. This depends on the object to which it is applied as well as on the kind to which the object is reckoned; finally, it also depends upon the attributes of all the other objects of this kind, or of objects in general.

4. Peculiar as well as common attributes can be either internal or external, either essential or inessential. Thus, omniscience is a peculiar and internal attribute of God, 'being accompanied by seven moons and a double ring' is a peculiar and external attribute of the planet Saturn. That all angles add up to four right angles is a peculiar

and essential attribute of a quadrangle; that everyone of its angles is a right angle is a peculiar but inessential attribute of it, etc.

5. Obviously, the indicator of an object, just like any other attribute, can be composed of several individual attributes. Thus the indicator that a certain object is a square consists of the sum of the following four individual attributes: the object must be a) a plane surface, b) it must be bounded by four straight lines all of which c) are of the same length and d) they must all meet with the same angle.

Individual attributes which can be used to identify an object either absolutely or within a certain kind I call marks; I call them sufficient marks if they suffice to identify it, and insufficient marks if they can do it only in connection with others. Hence the attributes that a certain object is a plane surface bounded by four straight lines etc. are marks, but insufficient marks, that this object is a square. If a mark is sufficient, then it is an indicator.

6. We can also (following several other logicians, for example Maaß and Schulze) divide marks into immediate and mediate ones. The latter become characteristics of an object only because they are marks *of* one of its characteristics. Thus, for example, the ability to speak is a mark of rationality, and this is a mark of man (at least among creatures on earth). Hence the ability to speak is also a mark of man, but a mediate one.

7. Finally, marks are generally divided into positive or affirmative, and negative. If this is understood to mean that affirmative marks should consist of affirmative concepts and the negative of negative concepts, then it will not be very useful. This distinction would be much more important if by positive marks or indicators of a certain kind of thing we understand attributes which belong exclusively to objects of this kind, and by negative marks attributes which belong to all objects of this kind but not exclusively to them. Given these definitions, the following inferences are permitted: where a positive mark is present, the thing is present; where a negative mark is absent, the thing is absent. But it will not be permissible to conclude from the absence of a positive mark the absence of the thing, and similarly from the presence of a negative mark the presence of the thing. In other words, the affirmative mark only justifies the affirmative assertion that the thing is there, while the negative mark only justifies the negative assertion that the thing is not there. . . . A negative mark is sometimes called a *condition*. Where the attribute a is a merely positive mark of things of kind X, the idea A must be a lower concept than X; and where the attribute b is merely a negative mark of X,

the idea B must be a higher concept than X. If y is an attribute which belongs to all and only X, then Y and X are equivalent concepts and the attribute y justifies an affirmative as well as a negative argument: from the presence of y we can conclude the presence of X and from the absence of y we can conclude the absence of X. It is obvious from this how such an exclusive indicator can be positive and negative at the same time.

§ 113. *Basic and Derived Attributive Ideas*

1. Attributes of an object whose ideas are constituents of the idea under which the object is subsumed are called basic attributes; all others are derived.

2. If the idea under which a certain object is subsumed is a pure concept, then all basic attributes of that object are essential.

3. Basic attributes do not have to be internal. For example, if a certain rose bush is subsumed under the idea 'centre rose bush at the centre window', then the attribute of being in the centre of the centre window is basic, but not internal.

4. Can a basic attribute be an indicator of its objects? Three cases must be distinguished: (a) If the idea in question is not a singular idea, but has several objects, then it is obvious that every part of that idea applies to several objects. Hence none of the basic attributes are peculiar to the object in question. (b) If the idea is a singular idea, but not redundant, then it also does not contain any proper part that does not apply to several objects. Hence, in this case too, none of the basic attributes are indicators. (c) A basic attribute can be an indicator only if its idea is part of a redundant singular idea.

5. Although the idea of any basic attribute of an object is part of the idea of that object, the converse does not hold. Not every part of the idea of an object is an idea of a basic attribute of that object. For example, the concept of an oblique angle has as parts 'supplement', 'not', 'equal', etc., none of which are basic attributes of oblique angles.

§ 114. *Ideas of Difference*

1. B. defines the difference between two objects α and β as "an attribute which one of them has but the other one lacks."

2. Whether a given idea m is the idea of a difference does obviously not depend on m alone, but on whether the propositions 'α has the attribute m' and 'β does not have the attribute m' are true or false.

3. As all differences are attributes, there will be internal and external differences (§ 80), essential and inessential differences (§ 111), general and singular differences (§ 112) and basic and derived differences (§ 113).

4. Is there a difference between any given two objects? B. points out that of two objects, α and β, α always has the attribute of being α and not β, while the converse holds for β. But this is only an external difference. Usually,

when this problem is posed, what is sought is an internal attribute, a property that one of them has but the other lacks. B. thinks that "such an inner difference must hold between any two objects that have real existence. In other words, I think that there are no two real objects that equal one another in all their internal attributes. (The reader will notice that I have referred to this theorem already several times under the name of the Leibnizian maxim *de identitate indiscernibilium.*) My reasons are as follows: Every finite substance is influenced by every other substance, no matter how far they are apart. Thus, even if we make the most extreme assumption, namely that the original attributes of two substances, i.e. the attributes that stem from the Creator Himself, are completely equal, an undisturbed equality of the two substances can continue only so long as both are surrounded by equal substances acting upon them from the same distances, or if there is an inequality in one of these parts, then this difference must be of such a sort that its effects are the same for both substances. But both of these assumptions are infinitely improbable, because of the infinitely many different cases that have the same probability. It is certain that at least in *most* cases there is not only one, but many, nay infinitely many, external and internal differences that can be truly asserted of two objects. The more closely we compare two people, or two trees, or any other two real objects, the more differences between them we discover."

5. No matter how large the number of differences between two objects α and β may be, it is often the case that there is one from which all the others can be derived. This is the case if the idea 'something that has the attribute *m*' is exhaustive of α, and 'something that does not have the attribute *m*' is exhaustive of β. In this case, all other differences must be derivable from these two ideas. Such a difference one could call an *exhaustive* or *determining difference.*

6. If the two objects α and β, which are determined by a certain difference, are individuals, we speak of an individual or numerical difference, if they are whole species or types, of a specific, generic, or class, difference. The difference between King Philip of Macedon and all other Philips is that he was the father of Alexander the Great. Hence this is an individual difference. Triangles differ from all other figures in that their internal angles sum to two right angles; this is a specific difference.

7. If the only difference between α and β is that α has *a*, and β has *b*, where both *a* and *b* are subordinate to *m*, we say that the difference between α and β is of kind *m* (especially if *m* is the next idea above *a* and *b*). Thus we say that red and blue larkspur differ only in colour.

8. Differences can be qualitative and quantitative.

In a note, B. maintains that many qualitative differences can be reduced to quantitative ones, e.g. 'visible to the naked eye' is a qualitative difference between macroscopic and microscopic objects, but it can be reduced to a difference in size. "I am inclined to think that the same can be done for many differences notably all differences between forces and perfections of merely finite substances. But I do not wish to extend this assertion to all objects, e.g. the infinite being or things that have no reality at all." B. does not think that the difference between a pound and a mile can be reduced to quantity.

APPENDIX

Earlier Treatments of the Subject Matter of this Part

§ 115. *Some General Remarks concerning the Difference between the Customary Presentation and My Own*

1. Many differences result from the fact that B. was not concerned with subjective ideas, but only with ideas in themselves. "Hence the division into clear and obscure, distinct and indistinct, possible, real, and necessary ideas, and other similar distinctions I could not discuss since they obviously apply only to subjective ideas."

2. The second distinction is that the discussion was extended beyond concepts to intuitions and mixed ideas.

§ 116. *Some Distinctions Made by Recent Logicians*

Most logicians begin the discussion of the subject of this part with the distinction between form and matter of concepts. Logic, they claim, is concerned only with the former. They thus seem to draw a sharp dividing line that was unknown to the ancients. "But in order to decide to what extent this boundary line is correct we must first know what precisely is meant by form and matter. But we do not get any satisfactory information. Kant, for example, merely says: "matter is the object, form the generality". Thus he merely replaces one word by another, and for a vague word he gives one that is even more vague." Other definitions are similarly unsatisfactory. B. claims that if logic were to refrain from considering the matter of concepts, i.e. of the parts of which concepts are composed, then it could not even distinguish affirmative and negative concepts; but this distinction is made in every handbook.

"What then is the truth that lies at the bottom of this whole claim? In my opinion it is this: that logic is not so much concerned with the definition of individual concepts (though it is concerned with the definition of some of them) as it is concerned with the determination of whole kinds of them, provided their peculiarities require a specific treatment in the sciences (§ 12)." But if this is all that is meant, then the distinction into form and matter does not delineate the contents of this part with the kind of precision that might be expected by somebody who is "blinded by the erudite twilight of the words 'form' and 'matter'. That logic does not have to describe every last idea but only certain kinds has been known for a long time."

2. "After our logicians have introduced the distinction of form and matter and have expelled the latter from the domain of logic, they usually assure us that there are neither more nor less than four viewpoints, from which all (formal) differences between ideas can be derived." These are quantity,

quality, relation, modality. (By modality is meant their relation to the thinking subject itself.) "Since the word quality simply means an inner attribute (property), and the word relation the same as an external attribute, the two headings *quality* and *relation* comprise, in my view, everything that can be said about an object. It is of course permissible to make a special heading for those attributes (whether internal or external) which have a magnitude, but it will only very rarely be appropriate. By introducing this distinction the most heterogeneous attributes will be placed together, while others, which are intimately related, will be separated. It would be very unreasonable, for example, if the geometer were to divide his theory of triangles into two parts, in one of which he speaks only of those attributes of a triangle which have a magnitude, e.g. of its area, the size of its angles, etc., while in the other division he would consider all the attributes of triangles which are not magnitudes, e.g. that the three altitudes meet in one point. . . . I cannot but confess that this manner of considering an object from these four points of view, which has become so popular in recent times, does not seem very logical to me; for the present investigation it seems completely out of place."

3. With respect to quantity most logicians find two kinds of quantity in every idea, namely an intensive quantity of content, and an extensive quantity of extension. The intensive quantity of content is simply the number of simple ideas which constitute the given idea. B. considers this concern unimportant; similarly, the magnitude or scope is only a relatively unimportant feature of extensions. Also, it is not clear why one of these magnitudes is to be called intensive, and the other one extensive. "In my opinion every magnitude of an object (e.g. of a line or surface) is to be called extensive if the object is compounded in such a way that the magnitude of its individual parts added together produces its own magnitude. This holds just as much for the magnitude of the content of an idea as for the magnitude of its extension."

4. Modern logicians are not in agreement concerning the viewpoint of quality. Here they distinguish clear and obscure, distinct and indistinct ideas. But these distinctions obviously apply only to subjective ideas and do not belong in the present discussion. Other logicians make a distinction with respect to quality into affirmative and negative ideas. This is unobjectionable; but the matter should not be presented "as if affirmation and negation were the only two differences between ideas from the viewpoint of quality".

5. The various distinctions which were advanced under the heading of *relation* are more defensible than the others. What can be said under the heading of *modality* does not belong into this part at all.

§ 117. *About the Five So-Called Predicables of the Ancients*

B. is of the opinion that the five predicables *genus*, *species*, *difference*, *property*, and *accident* should indeed be discussed in logic (as he has done in §§ 80, 108, 111, 114). But he does not think it necessary to put them at the very beginning, or even to discuss them all in one chapter. "They have been accorded this honour only because Aristotle uses them (as many others) without defining them first. His commentator found it necessary to draw them together in a special introduction as concepts with which we should be familiar in order to understand the *Organon*" (cf. Porphyry's *Introduction*, ch. 1).

1. B. notes that predicables are what he has called general concepts (§ 73). Hence the claim that there are five predicables is more correctly expressed by saying that there are five *kinds* of predicables.

2. According to the ancients, genus always has species under it and is based upon *essential* characteristics which objects belonging to several species have in common. But what is meant by 'essence' in this context? If the usual definition is followed, namely that the essence of an object is that which belongs to it necessarily, and which does not have its ground in something else, "then it is a very inconvenient restriction of the concept of a genus, to demand that it should contain only what is *essential* to several things. For then the concept of a merely accidental attribute, or of one that is grounded in something else, can never be made into a generic concept. Hence we can never say that morally good men and angels, etc. together form the genus of morally good beings; for moral goodness is not a necessary attribute of a being, much less one which does not presuppose any other." On the other hand if 'essential' here means the same as 'important' then this concept does not belong into the definition, since an unimportant genus is a genus just the same.

But it is likely that Aristotle meant by the essence of an object the same thing that was indicated in § 111, namely the class of all those attributes of an object which can be derived from the concept under which we represent it. "If we assume that he also decided to mean by genus only such general concepts as have lower kinds under them (namely their species), then it becomes obvious why he called a genus that which several essentially different things have in common: a general concept which has others under it represents only that which is common to the objects that stand under these subordinate concepts. And since we represent these objects through different concepts, they differ essentially from each other." However, this is unsatisfactory as a definition, since it cannot be reversed. "Not every concept which represents something which several different things have in common is a general concept that has others under it. For if the two concepts A and B are singular concepts and the concept M represented merely the two objects which fall under A and B respectively, then M is a concept which represents that which falls under two essentially different concepts; and yet it is not a genus in the classical sense, since A and B are not species. An example of this somewhat unusual case are the following concepts: 'Root of the equation $x-2=0$', 'Root of the equation $x-3=0$' and 'Root of the equation $x^2-5x+6=0$'. The first two are singular concepts and one of them has 2 as its object, the other 3; the third concept represents two objects, namely the numbers 2 and 3 at the same time; hence it contains something which 2 and 3, two essentially different objects, have in common; but that does not make it the concept of a genus, only the concept of an *infima species*."

3. It is correct to define the concept of a species as a general concept which is subordinate to that of a genus. But Porphyry and many scholastics make a mistake when they claim that things of the same species must differ from each other only in number.

4. It is a mistake to think that a difference is always a part of the concept of a species, such that this concept is compounded from this part and the

concept of the genus. This is often, but not always, the case. "We should in vain look for an addition through which the concept of the real is generated from that of the possible, although the two are related as species and genus. It was correctly maintained that a concept, *B*, which is subordinate to another, *A*, must differ from *A*. But it does not follow from this that *B* must consist of *A* and something else. For *B* can be different from *A* if it has no part at all in common with *A*, as is necessarily the case with simple ideas."

Aristotle, Porphyry and the scholastics make a sharp distinction between the concept of a genus and that of a difference. This is justified to the following extent: if the concept of a species is a composite of the concept of a genus and something else, then this "something else" is not a mere concept, but a complete proposition, and can therefore not be called a *generic concept*; hence if we define a spirit as a living being which is immortal, the generic concept is 'living being' and the "something else" is the proposition 'which is immortal'. But usually logicians mean by *difference* the predicate idea of this proposition, namely the concretum 'immortal'. And there is no reason why this concept cannot be envisaged as a generic concept.

5, 6, 7. In the sequel B. advances a number of criticisms of Porphyry's discussion of properties, and points out that the classical conception of accident is closely related to the one which he has advanced in § 111. He concludes the chapter by pointing out that all general concepts belong to one of the five indicated kinds of idea, but that the converse does not hold, namely that not every concept which falls under one of these five kinds is a general concept; this is based on his view that "not every idea of an attribute is a general concept. For if *b* is the idea of an attribute (e.g. omnipotent), which applies only to the one object which falls under the idea *B* (e.g. God), then *b* can obviously not be called a general concept. Similarly the idea of a difference does not always have to be a general concept."

§ 118. *Concerning the Categories and Post-Predicaments of the Ancients*

1. It seems that Aristotle meant by categories nothing but certain highest genera which stand under the absolutely highest genus of a something in general. "There is, however, something vague about this concept since it is not clear how far down from the highest concept of something in general we may go and still have a genus which is comprehensive enough to be called a *highest* genus." It is therefore not surprising if the various lists of categories do not agree. One could, for example, introduce only two categories, namely the possible and the impossible; or one could subdivide the possible into the real and the nonreal. The real could be divided into substance and adherence, or else into caused and uncaused. We could divide nonreal objects into propositions and ideas in themselves, etc. It is obvious that categories in this sense of the word are not exclusive ideas. One and the same object can fall under several categories, e.g. under the possible and the real, etc. Also, not all categories are simple concepts, nor are all simple concepts categories, e.g. 'not', 'has', etc. But since categories are generic concepts, they must all have an extension.

2. If we apply this to the categories of Aristotle, we must make the follow-

ing criticisms: (a) the categories do not appear in proper order, i.e. it is not noted how they are subordinated and coordinated. (b) Certain concepts of similar, even larger extensions, are not mentioned at all, although they are required to exhaust the domain of the concept 'something in general'. Thus the concepts of the possible and impossible, of the real and nonreal are lacking. (c) Aristotle seems to have thought only of the domain of existing things but not of the domain of the merely possible or even the not possible. (d) The differences between place and position, and quality and state do not seem important enough to make them into separate categories, etc.

3. In an appendix to the *Categories*, which does perhaps not even stem from Aristotle himself, certain concepts subordinate to the categories are listed, namely 'opposition', 'earlier', 'later' and 'simultaneous', and the various kinds of having. The scholastics called them post-predicaments and introduced a kind of opposition between them and the categories. This is unwarranted since there is no determinate boundary that limits the categories.

§ 119. *Concerning the Categories and Concepts of Reflection of Modern Philosophers*

B. criticizes the fact that Kant and his followers could not agree upon any definition of 'category', though all of them produced exactly the same table of categories. He objects to Kant's definition of categories as 'pure concepts of the understanding' or 'concepts which contain the subjective determination of all thought'. For, "if the categories are *concepts*, then they must have a certain inner distinction from others which does not depend upon our thinking. I should like to see their definition based upon this inner determination, rather than their relation to our understanding."

B. then gives several detailed reasons why the derivation of the table of categories from the table of judgments is unsatisfactory. He does not think, for example, that the concept of universality is contained in the so-called universal judgment, "for the expression 'all men are mortal' is equivalent to the expression 'man (as such) is mortal' (§ 57). It is obvious that such a judgment does not contain the concept of universality." Likewise, singular judgments do not contain the concept of unity (the idea 'Socrates' is quite different from the idea 'one'). Similarly forced is the derivation of the categories of relation from the corresponding forms of judgment. If, for example, the propositions 'distance has magnitude' or 'omnipotence is a property which only God has' are claimed to be categorical, it is difficult to see how the subject and predicate ideas of these judgments contain the concepts of substance and adherence or could even be subsumed under them.

The claim that in all four classes of categories the third can be derived from the first two holds, at best, for the category of limitation, since we could perhaps declare a limitation to be the affirmation of something and the negation of something else. In all other cases the derivations are quite opaque. The most likely one is the claim that the concept of necessity follows from that of possibility and reality, since the necessary is often said to be the real whose nonexistence is impossible. B. does not think that this is a valid derivation since the words 'necessary' and 'possible' are also applied to objects which do not have reality at all, e.g. to truths. "We shall see later that the correct definition of these concepts is the following: necessity is what follows from purely conceptual truths; possibility is that whose contradictory does not follow from a purely conceptual truth. Once this is seen, it will be admitted that the concept of possibility is not a constituent of the concept of necessity; rather it is even more complex than the latter."

B. then gives an extended criticism of the twice four concepts of reflection. They are supposed to come to the fore when the relation between subject and predicate in a proposition is viewed under the four viewpoints of quantity, quality, relation and modality. Given the opinion stated in § 116 no. 2, B. does not consider this much of a recommendation.

§ 120. *Concerning the Claim that Intension and Extension Stand in an Inverse Relation*

Ever since the *Logic of Port Royal,* this canon has been included in all handbooks of logic. Obviously, the expression 'inverse relation' is not to be taken in its strict mathematical sense that, for example, the extension would be halved if the number of constituents of a concept were doubled. But all novices tend to think that if an idea *A* has a greater content than *B*, such that *B* is a part of *A*, then *A* has a smaller extension than *B*, such that the extension of *A* is part of the extension of *B*. B. maintains that neither this proposition nor its converse is true. The proposition itself is false because the content of an idea can be increased without decreasing its extension. This can be done by making a concept redundant. For example the concept of a round sphere has a greater content than the concept of a sphere in general, although the extension of both concepts is exactly the same. It is also possible to add to the content of an idea in such a way as to *increase* its extension. Hence the concept of a man who knows all European languages has a smaller extension than the concept of a man who knows all living European languages. This example shows at the same time that the converse of the proposition is false.

Objects which fall under a certain idea *A* can have attributes whose ideas are not components of the idea *A*. Given that *b* is such an attribute which belongs to all objects that fall under *A*, then all objects which fall under *A* also fall under the idea *B* or 'something which has the attribute *b*'. Consequently *A* is either equivalent to *B* or subordinate to it; yet *A* does not have to be composed of *b* and some other component.

The doctrine was introduced by the author of the *Port Royal Logic* and is based on the assumption that every attribute of an object must be represented in its idea "but if all attributes which belong necessarily to an object (or at least their ideas) must be components of its idea, then the extension of an idea cannot be reduced except by enlarging its content. For, if the idea *B* is to be of smaller extent and lower than the idea *A* then necessarily the objects which are represented by *B* must have some special attributes in addition to those that are common to all the objects represented by *A*. . . . Hence one part of the canon is proved (namely the second proposition). But once it is assumed that the extension of an idea can be narrowed down only if its content is increased, there is an inclination to suppose that the former *always* becomes smaller when the latter is increased. And since this is indeed the case except for some minor exceptions of the kind listed above, it is understandable why these exceptions are either not taken into account at all or are not used as a counter argument."

B. admits that he owes his discovery of the above error to Kant's distinction between analytic and synthetic judgments "which could not have place if all attributes of an object had to be components of its idea."

PART II

Of Propositions in Themselves

§ 121. *Contents and Chapters of this Part*

Since the expression 'proposition in itself' has been sufficiently explained (§§ 19–23), it remains to discuss characteristics of propositions: first those common to all propositions, then those in which they differ; after that, relations among propositions and, finally, propositions that state relations between propositions.

CHAPTER 1

General Characteristics of Propositions

§ 122. *No Proposition in Itself is Real*

No proposition is real or exists; only mentally entertained propositions, or accepted propositions are real. "Propositions are merely the matter [*Stoff*] that is apprehended by the thinking individual in these thoughts and judgments."

§ 123. *Every Proposition Necessarily Contains several Ideas*

Every proposition is complex, ideas being its parts. Even apparently simple propositions such as 'come!' contain the concept of coming and of a certain obligation. Likewise the Latin '*sum*' contains the concept of existence and the idea 'I' as subject. Thus every proposition contains several ideas and may be called complex. "I shall call the sum of all its immediate and remote parts its *content*. Hence the proposition 'God has omniscience' is composed of the ideas 'God', 'has' and 'omniscience'. The sum of these three ideas I call the content of the proposition. The idea 'omniscience' is itself complex, being the idea of knowledge that extends over all truths. Thus it contains the ideas 'knowledge', 'truth', and others, all of which are considered parts of the content of the above proposition. Hence the content of a proposition stands to the proposition in the same relation in which the content of an idea stands to the idea (cf. § 56)."

In a note, B. points out that he takes the copula of a proposition to be part of its content, while most other logicians consider it part of the form.

§ 124. *Every Proposition can be Viewed as Part of another Proposition, even as Part of just an Idea*

Propositions can be indefinitely compounded, i.e. any of them can be constituents of other propositions and ideas. For example, we can form a class of propositions A and B; in thinking this class we have an *idea*, and if we assert something about this class, we generate a proposition of which A is a constituent.

§ 125. *Every Proposition is either True or False, and Remains that Way Always and Everywhere*

Apparent exceptions arise only when the linguistic expression of a proposition is ambiguous (cf. § 25).

§ 126. *Three Parts which are Obviously Contained in a Great Number of Propositions*

1. It must be admitted that many propositions, though perhaps not all of them, state something about certain objects. This holds in particular of true propositions. Thus, God is clearly the object of the proposition 'God is omnipotent', while the proposition 'all equilateral triangles are equiangular' is about infinitely many objects. But if a proposition is about certain objects, then it must contain an idea which represents or comprehends them. For it is only by virtue of such an idea that a proposition is about this object rather than another. We could not say that the proposition 'God is omnipotent' is about God, or has God as its exclusive object, if it did not contain an idea which has God as its only referent. The propositions 'all equilateral triangles are equiangular' is about all equilaterial triangles only by virtue of the fact that it contains an idea which represents all equilateral triangles. Now the idea in a proposition which represents the objects which the proposition is about I wish to call its *object idea*, *subject idea*, or *basis*.

2. It must also be allowed that at least some, if not all, propositions claim a certain attribute for the objects with which they deal. It is clear that these propositions must contain, in addition to their object ideas, an idea of the attribute which is claimed for the object. Finally, there must be a further idea through which these two ideas are connected, that is, an idea which indicates that the objects under consideration *have* that attribute. Thus in the idea 'God is omnipotent' or rather 'God has omnipotence' the concept 'God' occurs, God being the object of the proposition; secondly the concept of omnipotence is present as an attribute claimed for the object; and finally there is the concept of having. The idea of the attribute which is claimed for the subject I wish to call the *attribute* or *predicate idea*. The idea which connects the two, i.e. the concept of having, I am going to call the *connector* or *copula* of the proposition. Hence, subject, predicate, and copula are the three parts which are without doubt contained in a great many propositions.

§ 127. *Parts which the Author Takes all Propositions to have*

Closer consideration shows that *all* propositions have three parts, a subject idea, the concept of having, and a predicate idea, as indicated in the expression '*A* has *b*'. In the sequel, a number of propositions, or rather kinds of propositions, will be shown to have these parts, although this is often obscured by their linguistic form.

1. All grammarians admit that in every complete sentence, i.e. in every linguistic expression which stands for a proposition in itself, there must be a determinate verb; from this alone it follows that the concept of having occurs in every proposition. Every verb either contains the verb 'to have' as a part, or is identical with it. To prove this, I must make an assumption which will be readily granted. Every determinate verb which differs from the word 'is' can, without change in meaning, be replaced by its participle together with the word 'is'. Thus, '*A* does' means exactly the same as '*A* is doing'. But if the word 'is' occurs in a sentence, then the latter either has the form '*A* is' or '*A* is *B*', where *A* and *B* represent the other parts of the sentence. But in both cases it is easy to show that the proposition contains the concept of having. A proposition of the form '*A* is' (an existential proposition) as, for example, 'God is', clearly means the same thing as '*A* has existence'; hence the concept 'has' is contained in it; it likewise has an object, namely whatever is designated by the idea *A*, and with which the proposition deals; the attribute which is claimed for this object is existence. (A certain objection which might be advanced by specialists will be discussed in § 142.) It is even more obvious that propositions of the form '*A* is *B*' always have the sense which is indicated by the expression '*A* has *b*', where *b* is the *abstractum* which belongs to the *concretum* *B*. It will be admitted that the 'is' does not mean the same thing in these propositions as it does in existential propositions. That these propositions do not assert existence follows from the fact that such a proposition can be true even if the object *A* does not belong to things which have existence. Thus, for example, concepts in themselves do not have existence, but we do not hesitate to form the judgment 'the concept of a triangle is complex' since we do not intend to indicate by the 'is' that the concept of a triangle has existence. But it seems obvious to me that the indicated proposition has no other sense than 'the concept of a triangle has complexity'. The latter expression shows clearly that the copula in our proposition is none other than the concept indicated by the word 'has'. Propositions of the form '*A* is *B*' can always be more clearly and more correctly rendered as '*A* has (the attribute of being a *B*, or) *b*; hence all propositions have as their copula the concept of having.

But why is it that we use forms containing the verb 'is' ever so much more often than forms containing 'to have', and why do we have such difficulties if we want to use the latter? We can explain this from the fact that in the form '*A* has *b*' the predicate *b* is a mere *abstractum* (or attributive idea), while in the form '*A* is *B*' the place of *B* is taken by a *concretum*. But every language is far richer in signs for *concreta* than in designations for the corresponding *abstracta* which, where they exist, are often long and inconvenient words, or even combinations of several words. The reason why in all languages the *concreta* more often than the corresponding *abstracta* have their own, usually simpler, sign is this: the concrete ideas have for the most part objects which are evident to the senses and engage our attention more strongly than the *abstracta*; this has a double consequence: we have a greater urge to find designations for them and it is easier for us to reach agreement about the sense of these signs. Thus, for example, a body which attracts our attention by its shine, its beautiful yellow colour, its remarkable weight, its unusual malleability, etc. will soon be considered important enough to warrant a special designation, namely 'gold'. On the other hand, it seems rather superfluous to find a special designation for the concept of the totality of attributes by virtue of which this body is gold. Hence we have a sign for the *concretum* (gold), but for the *abstractum* (goldhood) we either have no special word at all or we designate this concept, if necessary, by a combination of several words, such as 'the constitution [*Beschaffenheit*] of gold'. It may well be the case that instead of the form '*A* has *b*' the form '*A* is a *B*' was invented because then the same sign which allows us to express the subject ideas in a proposition can also be used to designate the predicate.

2. However, there are many other reasons why we could doubt that the just-indicated form holds for all propositions. First of all, there are propositions which, aside from the subject idea, seem to contain only one further, apparently simple, idea, e.g. '*A* should', '*A* acts', '*A* wills', '*A* feels', etc. If the ideas of the words 'should', 'acts', 'wills', 'feels', etc. were indeed simple, then it would be established that not all propositions contain the concept of having together with two others. But I think that the indicated ideas are all complex, and that they contain the concept of having together with another concept, namely that of duty, efficaciousness, will, and sensation, respectively. I think that the above propositions, if they are to be expressed in such a way that their components become obvious, should be formulated in the following way: 'A has an obligation, has efficacy, will, a sensation, etc.'.

In general, I can accept only one of the following two opinions: either the concepts 'should', 'acts', etc. are, as I have just said, compounded from the concepts of duty, efficacy, will, sensation, etc. or, conversely, the latter are compounds of the former. But the first view is much more probable, since the derivation of the former concepts from the latter can be carried out in the indicated way by virtue of the concept of having, while the opposite case causes a lot of difficulties. One could think that the concept of an effect could be explained as the concept of something which is wrought; but it is obviously incorrect to construe the concept of will analogously as the concept of something which is willed, since what somebody wills is the object of his will, rather than the will itself. More than that, it seems to me that the attempted explication of an effect [*Wirkung*] does not give the abstract concept of an effect (efficacy), but only the concrete concept of that which is brought about.

The most plausible objection to my view is that the same relation which holds between the concepts of having and of an attribute also holds between the concepts of willing and will, and the concepts of acting and effect. Hence if the concepts of willing, acting, etc. are not simple, but are compounded from the concept of will, effect, etc. together with the concept of having (as I have just assumed), then it also seems to be the case that, analogously, the concept of having is not simple but compounded from the concept of an attribute and some other idea. For, if to will means the same as to have a will, if to act means the same as to have an effect, etc., then to have should mean the same as to have an attribute. But the latter is absurd, since the concept of having cannot possibly be the same as the concept of having an attribute; it can at most be equivalent to it; hence it would follow at most that the concepts 'wills' and 'has a will', 'acts' and 'has an effect', etc. though they are not one and the same, are at least equivalent.

I should like to point out, in reply, that the initial assumption of this objection, namely that the two concepts 'attribute' and 'to have' stand in the same relation to each other as the concepts 'will' and 'to will', 'effect' and 'to act', etc., is false. For though the concept of an attribute is not the same, it is at least equivalent to the concept of a something which is had, one cannot say about the concepts of a will, an effect, etc., that they are the same or even equivalent to the concepts of something that is willed, acted, etc.

3. But there are propositions of which it is even less plausible that they have the form '*A* has *b*'. Among them are the so-called hypothetical propositions of the form 'if *A* is the case, then *B* is the case'; similarly the disjunctive propositions of the form 'either *A* or *B* or *C*, etc.' I shall consider these forms of propositions in the sequel, and hope to persuade the reader that they do not form an exception to my rule.

4. A new objection could be occasioned by the fact that the verb

'to have', does not always have the same form, even in those sentences in which it occurs explicitly; rather, it varies depending on the person, number (and in some languages, gender) of the subject idea. From this one could draw the conclusion that it is not always the same concept which is designated by this verb in its various forms. But this objection vanishes once it is realized that these changes in form are due to the arbitrariness of language and are introduced merely for the sake of greater clarity, and perhaps variety, of expression; thus some languages do not have these inflections. Language inclines toward a certain kind of redundancy, which consists in the fact that oftentimes a certain concept is expressed repeatedly in the same sentence to make it more obvious. Such a redundancy occurs when, instead of 'I has *b*', we say 'I have *b*'; thus we use a certain alteration in the verb in order to indicate that the subject of the proposition is the speaker himself, which is actually already expressed in the symbol 'I'.

5. Through the forms of the verb 'to have', or the forms of any other verb which includes this concept, language allows us to express not only person and number of the subject, but also time determinations. Can we conclude from this that the true copula is formed not by the pure concept of having, but by a concept of having, combined with the determination of a time at which something is had? I reply that we express a time determination through the verb 'to have' even when we speak of objects which are not in time at all; for example we say 'every truth—has—an object with which it deals'. Hence we speak of it as if it took place at the present time. This shows that we must not conclude that since language connects the concept of having with time determinations, there is an essential connection between them. We have already remarked in §§ 45 and 79, and it will become more clear in the sequel, that these determinations belong essentially to the subject idea of a proposition. A proposition of the form 'the object *A*—has at time *t*—the attribute *b*', if its parts are to be clearly indicated, must be expressed in the following way: 'the object *A* at time *t*—has—(the attribute) *b*'. For it does not happen at time *t* that the attribute *b* is claimed for the object *A*; but the object *A*, inasmuch as it is thought to exist at time *t* (hence to have this determination) is claimed to have attribute *b*.

6. Once we are convinced of this, we will readily admit that certain other determinations which linguistic expression connects with the verb do not actually belong to the copula. I have in mind the determinations: 'often', 'rarely', 'always' and similar ones, likewise degrees of ***probability*** which we want to assign to the proposition. We say

'*A*—probably has—*b*'; '*A*—certainly has—*b*', etc. But it is obvious that these determinations are not concerned with the way in which the predicate *b* belongs to the subject *A*, but with a relation in which the entire proposition '*A* has *b*' stands to our cognitive faculty or to other propositions. Hence the proposition '*A* probably has *b*' means obviously the same as 'the proposition that *A* has *b*—has—probability'. Similar considerations hold for the determinations of *necessity* and *contingency*, which we also often connect with the copula of the proposition by saying '*A*—necessarily has—*b*', etc. The true sense of such propositions will be made clear later (§ 182), where I will show that they have the same copula as all other propositions. It will also be shown (§ 136) that even the concept of *negation* is not a component of the copula, but of the predicate idea, no matter how intimate its linguistic connection with the verb; the proposition '*A*—does not have—*b*' actually has the sense '*A*—has—lack of *b*'.

7. If every proposition contains as copula only the concept which is designated by the word 'has', then there is no doubt that there must be two further parts. For neither 'has', nor '*A* has', nor 'has *b*', taken by themselves, are expressions of a complete proposition. But from the definition given in § 48, it follows that parts *A* and *b* can only be either mere ideas, or whole propositions. Moreover, if the proposition in question is to be true, then it should be possible to prove that the two parts *A* and *b* must be ideas, indeed ideas with referents, and that *b*, in particular, must represent an attribute. But if it is not required that the proposition '*A* has *b*' is a truth (and at present we merely speak about attributes which hold for all propositions, even false ones), then I do not see why *A* and *b* must always be ideas, or even referring ideas, and why *b* must be an attributive idea. Can we not call every compound of the form '*A* has *b*' a proposition, no matter whether the signs '*A*' and '*b*' designate mere ideas, or ideas of a certain sort, or even entire propositions?

Since this is not a matter of great importance we may restrict the concept of a proposition to those cases where *A* and *b* are mere ideas, but we should in no case demand that these ideas should both have referents, and that the latter should be an attributive idea. Why should we not give the name of a proposition to the combination of ideas expressed by the following words 'a body which is bounded by five equal sides is not bounded by triangles'; if anybody actually connected these ideas, we should certainly claim that he has formed a judgment, even though there is no body that is bounded by five equal sides. According to § 66 it is often a matter of accident whether or not a given idea has an object; whether the idea 'golden tower' has a referent depends on whether it has pleased somebody to build a tower from gold. If it were a part of the nature of a proposition that its subject must be a referring idea, then the answer to the question whether a certain combination of ideas should be called a proposition would depend on the accidental circumstance whether the subject idea has a referent. The words 'a golden tower is valuable' would express a proposition only in case a golden tower had actually been built somewhere; if this had not been done, these words would not only fail to express a true proposition, they would express no proposition at all.

§ 128. *Attempted Definitions*

B. points out that if the concept of an idea is simpler than that of a proposition, then a proposition could be defined as the connection of any two ideas by means of the concept 'has'. "But from the circumstance that ideas are components of propositions, we cannot with certainty conclude that the *concept* of an idea must be simpler than that of a proposition." On the contrary, the orientation introduced in § 48 seems to contain the *definition* of the concept of an idea; an idea would then be "any component of a proposition which is not itself a proposition." Hence we cannot define propositions in terms of ideas.

§ 129. *Other Theories*

B. here examines and rejects sundry diverging theories concerning the composition of propositions: 1. That propositions have only two parts. 2. That at least existential propositions have only two parts (*A* is). 3. That hypothetical propositions, disjunctive propositions, and equations have a different composition from categorical propositions (he will examine this view later). 4. That the concept of having is not the proper copula. 5. That there are several kinds of copula. 6. That modalities are part of the copula. 7. That if a proposition deals with several objects, then it has several subject ideas. 8. That logical forms are merely subjective. "Only those who do not believe that in addition to things in themselves and our thinking there are also truths in themselves which we merely *grasp* in thought would be inclined to consider logical forms as something that merely adheres to our thought." 9. Various definitions of the parts of propositions, e.g. that the subject is the first part of a proposition, are examined and rejected.

§ 130. *The Extension of a Proposition is always the Same as the Extension of its Subject*

B. defines as the extension, the range, or the sphere of a proposition that characteristic by virtue of which it deals with these rather than some other objects. Hence, the extension of a proposition is the same as the extension of its subject. It must not be confounded with the scope which is merely the number of objects with which it deals.

§ 131. *Whether the Predicate Idea of a Proposition is Taken in its Full Extension*

B. answers in the negative, pointing out that when we say, e.g. 'Caius has intelligence', we do not wish to say that Caius has every kind of intelligence there is, for example, a well-developed, as well as a crude, a human, and an angelic, intelligence, etc. "The proposition '*A* has *b*' has no other sense than that every object under *A* has one of the attributes that fall under *b*; and if there are several of the latter, it remains undetermined, which of them belong to every *A*."

In a note, B. discusses the view, expressed by Aristotle [*De Interpr.*, 17ᵇ15 *Anal. Prior.*, 43ᵇ20], that no affirmative judgment is true, if its predicate is taken in general. For example, the proposition 'every man is every living being' is said to be indefensible. "I am somewhat astonished that many have thought to avoid this difficulty by remarking that the predicate in an affirmative proposition must be understood to be a particular. Hence, 'all men are living beings' is said to have the sense 'all men are some living beings'. Now I do not want to deny that the latter sentence is true under a certain interpretation, but I think that under this interpretation it is not the same proposition as the former. In the first sentence the expression 'all men' is understood to be distributive, i.e. it is equivalent with the expression 'every man' or 'man as such'. But if we say 'all men are some living beings' then, if what we say is to be true, the expression 'all men' must be taken as collective, i.e. as equivalent with the words 'the totality of men', or 'the class of all men'. Similarly, by the expression 'some living being' we must understand merely a class-concept, namely that of a set of living beings. But since this last idea, namely that of a class of living beings, has several objects, since, e.g. the totality of birds, or of fish, is a set of living things, the original question arises again: it is not clear what extension we ought to assign to the general idea which takes the place of the predicate. It follows that the change of the sentence has altogether failed to accomplish its purpose." He contends that the only proper analysis of the above proposition is 'every man—has—animation', where the predicate position is occupied by an attributive idea which is not taken in its entire extension.

CHAPTER 2

Differences between Propositions which Arise from their Internal Constitution

§ 132. *Simple and Complex Propositions*

B. maintains that the concept designated by the word 'to have' is simple but that the subject and predicate ideas of propositions are usually complex; but he holds that there are also propositions all of whose parts are simple. He will call these simple propositions, all others complex, and points out that there may be only very few of them, and that he is hard put to give even a single example: "this is due to the fact that it is very difficult to persuade oneself of the simplicity of an idea. But it doesn't follow from this that it is false that there are and must be such propositions, some of them true, others false".

In a note B. contends that the complexity of propositions is more than merely a matter of their linguistic expression.

§ 133. *Conceptual and Empirical Propositions*

No matter what anybody's theory concerning the parts of propositions may be, he can hardly deny that there are propositions, even true propositions, which consist entirely of pure concepts, without containing any intuitions [*Anschauung*]. The following propositions are obviously of this sort: 'God is omnipresent', 'gratitude is a duty', 'the square root of 2 is irrational', etc. We shall see in the sequel how propositions of this kind differ essentially from propositions which contain intuitions, especially if they are true. Hence I believe that a special designation for them is indispensable for the purposes of science, and I shall call them *conceptual propositions* or *conceptual judgments* and, if they are true, *conceptual truths*. All other propositions will contain one or several intuitions and may for this reason be called *intuitive propositions*, also *empirical* or *perceptual* propositions, etc. I shall therefore call propositions like 'this is a flower', 'Socrates was born an Athenian', empirical, since each of them contains at least one, perhaps several, intuitional ideas.

NOTE: The major reason why I consider the division into conceptual and empirical propositions so important is that truths which are put forth in a scientific treatise must be treated in entirely different ways, depending on whether they consist of pure concepts or contain intuitions, especially if it is a matter of their objective grounding, rather than mere persuasion. The ground of a purely conceptual truth can only lie in other conceptual truths, while the ground of an empirical truth can at least in part lie in objects to which its intuitions refer. But since the distinction between conceptual and empirical truths is so important, it would be very strange if it had escaped logicians altogether. This is indeed not the case; we might complain that this distinction is not established with sufficient clarity in the usual text books of logic, but we must admit that the earliest philosophers already knew it and discussed it frequently.

It is well known that Plato made a very important distinction between, on one hand, pure concepts (νοήσεις) and the Ideas which transcend all experience and, on the other hand, merely empirical ideas or intuitions (φαντασίαι). He demanded of a science (ἐπισήμη), especially of pure science (καθαρὰ), which is concerned with the unchangeable, that its doctrines should not be derived from experiences, but from pure concepts. He took pure thought to be concerned with pure concepts, namely with their analysis, connection, etc., without any regard to sensory perception (intuition). The only thing that is lacking is a special designation for the concept of propositions which are composed of pure concepts, and another for the concept of propositions which do not fulfil this condition. But we could not very well expect this from him, since he was not concerned with an exact determination of the parts of propositions, and since he was more concerned with concepts or ideas than with propositions. On the other hand, it is well known that Aristotle distinguished general propositions (πρότασις καθόλου; these are probably purely conceptual propositions) from others (as for example in *Anal. Post.* 75^b 21 ff.). He also insisted that one should not believe that such propositions could be derived from others which have as their subject merely changeable things (φθαρτὸν, empirical things, intuitions).

Locke (*Essay* IV,III,31; IV,IV,6; IV,IV,16, and elsewhere) not only distinguished conceptual and perceptual propositions with great clarity, he also differentiated between the sciences in which they are indigenous, and claimed that the former produce absolute certainty, the latter only probability. Although I cannot altogether agree with this, it seems to betray the correct view concerning the nature of the difference between conceptual and perceptual propositions; because all judgments of experience (they are the largest and most important part of the perceptual judgments) are only probability judgments, since all of them have a premise which is merely probable. On the other hand it would be a mere accident if a conceptual judgment were only probable; this case would arise only if we are not certain that we have not made an error in its derivation, or if we derive it from mere experience. . . .

In recent times the distinction between intuitions and concepts has been seen more clearly than ever before. One might have expected that this would have thrown more light upon the distinction between conceptual and empirical propositions; but I believe that the following circumstance detracted from a correct definition. The ancients already knew, but did not pay much attention to, a division of our cognitions into those whose correctness we ascertain merely through experience, and those which do not require any experience. Leibniz and Kant emphasized this distinction as extremely important; as it happens, it nearly coincides with the division of propositions into conceptual and empirical ones, since the truth of most conceptual propositions can be decided by pure thought, while propositions which contain an intuition can be judged only by experience. As a consequence, the essential difference between these propositions was not seen to lie in the character of their components, but in the method through which we ascertain their truth or falsity; hence the former were defined as propositions which can be known without any experience, while the latter require experience; for this reason they were called judgments *a priori* and *a posteriori* (cf. for example the Introduction to Kant's *Critique of Pure Reason*). I agree that the distinction which is here made is important enough to be retained; but I do not think that we should allow it to replace that other distinction which does not rest upon the relation between propositions and our cognitive faculty, but upon their inner characteristics, namely the distinction between those that are, and those that are not, composed of pure concepts. I even go as far as to claim that they had this distinction in mind without being clearly aware of it. For, if what they meant by judgments *a priori* were identified by the definition that they are cognitions which are independent of all experience, then it would not have been necessary to add to this definition a pair of other characteristics which allow us to recognize judgments *a priori*, namely *necessity* and *universality*. However, whether a proposition is strictly universal or not, and whether we can say that its predicate belongs to the subject by necessity, these are circumstances which depend upon the inner characteristics of the proposition itself and have nothing to do with its accidental relation to our cognitive faculty. Also, we can hardly doubt that Kant, since he said explicitly that all mathematical propositions are judgments *a priori*, counted among them propositions which we do not presently know, e.g. a formula which allows us to drive all prime numbers; on the other hand, the answer to the question what the inhabitants of Uranus are doing just now he would have reckoned among the empirical propositions, although there are no experiences which lead to a decision of this question. Beck (*Logik*, § 67) states explicitly that a judgment can be objectively *a priori* although it is subjectively only *a posteriori*, and I do not think that much can be said against this. But it follows from it that the circumstance whether or not a judgment is *a priori* is considered an objective attribute of the judgment, which belongs to the judgment itself, and hence should be defined in an objective way, and not in a way which depends upon its relation to our cognitive faculty. However, the two characteristics of *universality* and *necessity* are unsuitable for this purpose; this follows from the fact that they can be applied only to true propositions. Moreover, most logicians consider the proposition 'some numbers are prime numbers' as particular and the proposition 'every finite being is fallible' as problematic; but both are purely *a priori* propositions. I hope to show in § 182 that, if we want to define the concept of *necessity*, the distinction between *a priori* truths and others must be presupposed.

§ 134. *Abstract and Concrete Propositions*

According to § 127, the predicates of all propositions are abstract, but their subjects may be either abstract or concrete (cf. § 60). If the subject of a proposition is concrete, then the proposition itself will also be called concrete; similarly for abstract subjects. If the subject idea of a proposition is an intuition, then the proposition is neither concrete nor abstract (cf. § 60).

§ 135. *Propositions Containing Ideas of Classes*

1–5. If the subject ideas of propositions are ideas of classes of the kind described in §§ 82–88, the propositions in which they occur are named accordingly.

6–7. Propositions of the form 'Caius has wisdom and integrity' are called propositions with a collective predicate, if the predicate is taken in its collective sense, otherwise they are called propositions with distributive predicates.

8–10. Depending upon the nature of the predicate, propositions are called assertions of a multiplicity (e.g. 'everybody has several faults'), of universality (as in 'Caius has all the virtues of a good father') or assertions of exceptions (as in 'Caius has all virtues of a good father except for thrift').

11. B. proposes to consider propositions in which ideas of classes form merely part of either the subject or predicate.

12. He first discusses assertions of equality, i.e. propositions like 'attribute *m* is common to the objects *A*, *B*, *C*, *D*, . . .'. He proposes the following as (preliminary) normal form for all such assertions: 'the relation of attribute *m* to the objects *A*, *B*, *C*, *D*, . . . is the relation of a common attribute to its objects'.

13. He explicates *propositions of difference* analogously, i.e. 'the relation of *A*, *B*, *C*, *D*, . . . to *F*, *G*, *H*, . . . is the relation between things which have *m* (*A*, *B*, *C*, *D*, . . .) to others, which lack it (*F*, *G*, *H*, . . .)'.

14. If a certain attribute *m* belongs exclusively to several objects *A*, *B*, *C*, *D*, . . . then a proposition which asserts this is called a *determination* or *determinative* proposition. It has the general form 'the relation between the attribute *m* and the objects *A*, *B*, *C*, *D*, . . . is the relation between an exclusive attribute and its objects'.*

15. Propositions asserting mathematical similarities in the sense of § 91, note 4, are called assertions of similarity.

In a note, B. points out that in propositions of the form '*A* is equal to *B*' and '*A* is similar to *B*' the *A* does not form the subject and the *B* the predicate. This follows from the fact that equality and similarity can be asserted to hold among more than two objects.

§ 136. *Propositions with Negative Ideas*

1. B. discusses propositions with negative subject ideas (cf. § 89), e.g. 'what is not corporeal does not have colour'. The subjects of these propositions always have referents, so long as the negated idea is not the idea of something in general.

2. He next considers the distinction between affirmative and negative propositions and wonders whether the negation belongs to the copula of the latter or to the predicate. He points out that in all languages (so far as he knows) the negation always occurs in close proximity of the verb, which would seem to indicate that it belongs to the copula. But this argument loses its plausibility once we realize that the signs for probability, necessity, and contingency occur in similar positions "yet it is certain that they do not state a relation between the subject and the predicate of a proposition, but an attribute which belongs to the proposition itself. Hence we can see that from this circumstance alone we can conclude nothing." On the other hand, the following would seem to indicate that the negation does not belong to the copula: from the proposition 'every *A* does not have *b*' (*Omne A non habet b*) we should conclude only that not every *A* is a *B*, but not that not a single *A* is a *B*. However, if the 'not' belonged to the copula and determined the relation between the subject of the proposition ('every *A*') and the predicate *b*, then the second interpretation would be the only correct one: *Omne A—non habet—b* would mean that the attribute *b* must be denied of every *A*. Generally, if we want to say *b* does not belong to any *A*, we say 'no

* In § 500, vol. IV, B. defines a determination [*Bestimmung*] as a proposition which asserts that a certain attribute *b* applies exclusively to objects falling under *A*, so that *A* and *B* are equivalent ideas. ED.

A has *b*'. Even if the subject idea has only a single object we often take the negation out of the copula and move it into the predicate, e.g. 'Caius—has—no wit'. This example shows that in all those cases where the negation was thought to belong to the copula it is actually part of the predicate. "All propositions which we express as '*A* does not have *b*', or which we could express in this form, simply assert the *lack* of attribute *b*. To show its constituents more clearly we can express it as '*A*—has—lack of b'. But if, as I suppose, the concept of negation never belongs to the copula of the proposition, and if in all propositions whatsoever this copula is the pure concept of having, how is it possible that there are propositions which *ascribe* an attribute to an object and others which *withhold* such an attribute? Is it not that this ascribing and withholding indicates a difference in the copula? If it is furthermore true that the concept of negation, in all propositions which are called negative, belongs essentially into the predicate, why is it that we have such difficulties when we want to incorporate the signs for the negation into the linguistic expression of the predicate and why can we not find words to express our thoughts correctly? My reply is that withholding differs from ascribing only in the fact that it is an ascription of an attribute which is thought to be the lack of some other attribute. Clearly, the lack of an attribute is itself an attribute, and the lack of the lack of an attribute is that attribute itself. But we rarely have words for an ordinary attribute (i.e. for the *abstractum*) hence we should not be astonished that a special word for the mere lack of an attribute is extremely rare. This explains the difficulty of putting into words a proposition which asserts a purely negative attribute. Language invents words only for ideas which occur frequently and for which we need a designation. But we are rarely forced to think an object merely as something which lacks a certain attribute; while we think it as an object which lacks a certain attribute we remember others which it does have, and in most cases it is much more useful that we should think both, namely the presence of attributes *b*, *c*, together with the lack of *a*, than to think merely the lack of *a*. Hence we invent a word for the former kind of idea but not for the latter. For example, we find people who cannot see, and this lack appears important enough to be given a special name, namely blindness. But it is by no means the mere lack (the mere absence) of the ability to see which we designate by this name (for we do not call a stone blind even though it has no ability to see), but this word indicates the absence of vision together with the presence of (inadequate) visual organs."

According to § 89, a complex idea is non-referring, if it consists of nothing but the concept of negation together with some other idea. For example, the connection of the idea 'something' with negation results in the idea 'nothing', which is empty. On the other hand, the idea 'a lack of *b*' represents an attribute, and hence has a referent. It follows that the idea 'lack of *b*' contains some other constituent idea besides *b* and the concept of negation. B. surmises that this third idea is the idea of an attribute, and that the best way of expressing the attribute 'lack of *b*' is the expression 'attribute non-*b*'. Hence the normal form of a negative proposition is said to be '*A*—has—the attribute non-*b*'.

3. In affirmative propositions the distinction between propositions that have a collective predicate and those that have a distributive predicate is not very important (cf. § 135, no. 6 and 7). In negative propositions, this distinc-

tion is crucial: '*A* has—not the class of attributes *b*, *b'*, *b''* . . .' means that *A* does not have *each* of the attributes *b*, *b'*, *b''* . . ., but it is not denied that it may have *some* of them. On the other hand, '*A* does not have a single one of the attributes *b*, *b'*, *b''* . . .' is the correct rendition of a negative sentence with distributive predicate and differs sharply from the first case.

§ 137. *Various Propositions about Ideas:* a. *Assertions that an Idea has Reference*

If a proposition deals with an idea, then its subject idea must be the idea of an idea; hence a *symbolic* idea (§ 90). An important kind of proposition about ideas are propositions which assert that a certain idea has a referent: 'idea *A*—has—reference'. Sometimes an assertion of reference occurs in other linguistic forms, "for example 'there is an *A*', 'there is a God', 'there is a supreme moral law', 'there are bodies which are bounded by four equal sides', etc. In sentences of this sort the words 'there is' are not always meant to indicate the real existence of the subject. We can see this from the fact that we use these words even with objects which do not have real existence, for example in the case of the supreme moral law, which is a mere truth in itself and hence nothing that exists." Propositions of the form 'a certain *A* has *b*' or 'certain *A* are *B*' or 'some (several) *A* are *B*' are fundamentally nothing but assertions of reference. "The only way I know to explicate these expressions is this: 'the idea of an *A* which is also *B* (or which has the attribute *b*) is a referring idea'. Hence the following expression indicates the logical parts of these propositions with sufficient clarity: 'the idea of an *A* which has the attribute *b*—has—reference'. It might be objected that it is unlikely that my explication and the linguistic expression which it explicates both indicate the same thought, since they are composed of entirely different words. I do not wish to quarrel about this matter; we only need to admit that the sense connected with the original formulation is equivalent to the sense of my explication, i.e. that if one of them is true then the other is also. This can hardly be denied (cf. § 173)."

§ 138. b. *Denials that an Idea has Reference*

That an idea has reference can be denied not only in sentences of the form 'the idea *A*—has—no reference', but also in propositions of the form 'there are no *A*'. Sentences of the form 'no *A* is *B*' can be interpreted in two ways, namely, 'the idea of an *A* which is also a *B* (or which has the attribute *b*) does not have reference' or 'every *A* is a non-*B*'. "If we, as is often the case, take the expression 'no *A* is a *B*' in the sense that we do not presuppose that there is an *A*, then it is not even equivalent to the expression 'every *A* is a non-*B*'. For, if the idea *A* does not have an object, then the complex of ideas indicated by the words 'every *A* is a non-*B*', does perhaps stand for a proposition, but never for a true one, since there is no object with which it deals. In such a case the proper sense expressed by the words 'no *A* is a *B*' is the first mentioned, namely 'the idea of an *A* which is also a *B* does not have reference'. For example, a geometer, wanting to show that there is no solid bounded by five equal sides, might start from the propositions 'no solid bounded by five equal sides is bounded by triangles, by quadrangles, etc.'; but he does not want these forms of words to indicate that there *is* a solid which is bounded by five equal sides, since he is about to prove that such a solid is impossible. He merely wants to say that the idea of a solid which is bounded by five equal sides does not have a referent when we think of these sides as triangles; similarly, if we think them as quadrangles, etc. From this he finally concludes that the idea of such a solid has no referent at all."

§ 139. c. *Propositions which Determine the Extension of an Idea*

B. explicates sentences of the form 'there are several *A*', etc. as 'the idea of a multiplicity of *A*'s (or the idea of a class each of whose members is *A*) has reference'. Propositions of this sort he calls 'claims that an idea is general'.

2. Propositions in which we claim that an idea *A* is not general are of the form 'the idea of a multiplicity of *A*'s does not have reference'.

3. Such a denial that an idea is general does not imply that it has at least some referents, though some of them, i.e. singular ideas, do. Examples of propositions claiming that an idea is singular are 'there is only one God', 'there is only one supreme moral law', etc. Propositions of this form are explicated in this way: 'the idea *A* is a singular idea', i.e. 'the idea *A* has reference and has the attribute that the idea of a multiplicity of *A*'s does not have reference'.

§ 140. d. *Propositions concerning the Relations between several Ideas*

B. here discusses propositions which make claims concerning *compatibility* (cf. § 94), *inclusion* (cf. § 95), *equivalence* (cf. § 96), *subordination* (cf. § 97), *overlapping* (cf. § 98), *exclusion* (cf. § 103), *contradictoriness* (cf. § 103, no. 3), and *coordination* (cf. § 104) of ideas and introduces an appropriate terminology.

§ 141. *Propositions which Deal with other Propositions*

B. considers propositions which have other individual propositions as their object. The most important of these are *affirmations* and *denials*; their normal forms are, respectively 'the proposition *A*—has—truth' and 'the proposition *A*—has—no truth'. "Denials are frequently expressed by simply adding the word 'not' to the copula or the verb. . . . Where *A* designates a proposition I shall frequently designate its denial, or the proposition that *A* is false, by '*Neg.A*'."

§ 142. *Propositions Asserting Actuality*

The propositions so far discussed may occur in almost any science. There are others which are indigenous to particular sciences, but are used in so many different ways that it is reasonable to mention them in a treatise on logic.

Into this class belong, in particular, all propositions which assert or deny existence, being, or actuality, and which are commonly called existential propositions, e.g. 'I am', 'God is', 'Truths in themselves do not have existence', etc. We know from § 127 what I take the form of these propositions to be. I take the general form of the affirmative ones to be '*A*—has—existence', and of the negative ones '*A*—has—no existence' (or '*A* has the attribute of non-being'). With affirmative existential propositions, I take the concept of existence to form the entire predicate idea, with negative ones it forms at least part thereof.

NOTE 1: The greatest possible confusion reigns in the assessment of these kinds of proposition. Kant in his *The Only Possible Basis for a Demonstration of the Existence of God* (*Der einzig mögliche Beweisgrund zu einer Demonstration des Daseins Gottes*, Königsberg 1763, A 7, *Akademieausgabe*, vol. II, p. 73) asserts "that the concept of actuality [*Dasein*] can never be predicate but only subject in a judgment, that we speak improperly when we say that a thing has actuality [*Dasein*] instead of saying: There is an existing thing (an actuality) which has such and such attributes". According to Herbart (*Lehrbuch zur Einleitung in die Philosophie*, Königsberg 1813, p. 56 ff., and *Allgemeine Metaphysik*, vol. 2, Konigsberg 1829, § 204) existential propositions have no subject at all, while in his *Hauptpunkte der Metaphysik* (Göttingen 1807, § 1) he says that the proposition 'being is' offends against itself since the 'is' contains being as a predicate. Hence he goes on to assert the proposition 'being is *not*'. . . . I take *being* [*Sein*] or actuality [*Wirklichkeit*] precisely as what language makes it out to be, namely an attribute; whoever denies this confuses (I believe) actuality with substance. By substance I mean an actuality which is not an attribute of another actuality; hence I admit that we cannot truly predicate the putative *abstractum* of substance (substantiality) of any object. For it is part of the concept of substance that there is no property of this kind. But it is not the same with actuality, which I consider to be a mere attribute, not only of a substance itself but also of each of its attributes, since every attribute of an actual thing is itself actual. And since every attribute of an object can be ascribed to it in a judgment of the form '*A* has *b*', why not the attribute of actuality?

This much is certain: in propositions in which we deny, rather than assert, the actuality of an object, the concept of actuality appears in the predicate. For hardly anybody will claim that in the proposition 'truths in themselves do not have actuality' the concept of actuality appears in some part other than the predicate. But if this concept occurs in the predicate in the case of negative existential propositions, must we not admit that affirmative propositions also contain it in their predicate? This is particularly obvious in the case of existential propositions in which the subject whose actuality is asserted is represented by a simple *intuition*. Nobody will imagine that in the proposition 'this is' or 'this has existence' some idea other than the idea 'this' forms the subject. But if it is the subject of the proposition, then the concept of actuality can appear nowhere but in the predicate.

I do not want to deny that there are some propositions which look like existential propositions, but which are more correctly envisaged as having some other concept than the concept of actuality as their predicate. Some of the propositions which are commonly expressed by '*A* is' or 'there is an *A*', where the idea *A* contains the concept of actuality as its main part, are of this sort. For, if by *A* we merely mean a certain actuality which has the attributes a', a'', . . . then the proposition '*A* is', if it is interpreted in the same way in which I interpret existential propositions proper, would have the following sense: 'the actual which has attributes a', a'' . . . has actuality'. While this is not false, it is hardly what we mean when we say the words '*A* is'. Since we use these words synonymously with 'there is an *A*', it follows from § 136 that this proposition is an assertion of reference, which has the sense 'the idea of an actuality, which has attributes a', a'' . . . has a referent'.

NOTE 2: B. here surmises that the concept expressed by the words 'existence' [*Dasein*], 'being' [*Sein*], and 'actuality' [*Wirklichkeit*] is a simple concept.

§ 143. *Propositions which Deal with Psychological Phenomena*

1. There is nothing with which we are more closely acquainted and concerned than with the events in our inner self, the so-called psychological phenomena through which we become aware of everything else. No wonder that there are several sciences which are concerned with these phenomena either in general, or with certain classes of them. Hence propositions which describe them are of great importance. By a psychological phenomenon I mean any effect which is brought about by a mind [*Seele*] (any simple being as such); I distinguish six kinds of them; five of them occur in the substance which brings them about, while the last consists in the effect of the mind upon surrounding substances. The first kind are what I have called *subjective ideas* (§ 48). I consider *judging* another such phenomenon (cf. § 34). All, or at least most, ideas are accompanied by a

certain pleasure or uneasiness; these I call a *sensation* which they cause within us, and I think that all sensations arise from ideas (though perhaps not always ideas of which we are clearly conscious). Whether the concept of a sensation is simple, or what parts it has, I do not dare decide. If we judge that a certain object would cause a pleasant sensation within us, then the idea of this object brings about a certain effect in our minds, which we call *wishing*, *desiring*, *wanting*. I believe that we must make a clear distinction between wishing and *willing*, for it is not true that we always wish what we will, but that we often will what we think we should, although it is very unpleasant. Though willing differs from wishing in this way, it also differs from *acting* or *doing*. By acting I mean any change which is brought about through our will either in our own mind or in certain substances which are different from us, such as the organs of our body, and through them in other surrounding objects. It is not always the case that the result that we want (or, as we say, intend) is brought about by our will. When we fail in this fashion, we say that we wished for the result but did not attain it. In this case we do not take the word 'to wish' in the same sense as above, but synonymously with 'to will'.

2. When we merely state that such a phenomenon, e.g. an idea, or a judgment, exists, without determining the being in which it appears or which brought it about, then the proposition in question is simply the statement of the reference of a certain idea (§ 137). The following proposition is of this kind: 'It is desired that a machine be built which allows man to fly through the air.' This means the same as 'the idea of a desire that somebody invent a machine . . . has reference'. Propositions of this kind, which state the presence of a desire or wish that somebody should do something, are generally called *problems* [*Aufgaben*]. Actually only the desire itself should be called a problem but the proposition which expresses this desire should be called the problem proposition. . . .

3. If we do not merely speak of the existence of a certain psychological phenomenon, but of somebody (a certain mental being) in whom this phenomenon occurs, or by whom it is brought about, then propositions of the following forms occur: '*A*—has—the idea *B*'; '*A*—has (or forms)—the judgment *C*'; '*A*—has—the sensation *D*'; '*A*—has—the wish (or the desire) *E*'; '*A*—has—the will *F*'; '*A*—has (or brings about)—the action *G*'. These expressions show what I take to be the logic of these propositions.

§ 144. *Ethical and Related Propositions*

B. surmises that the concept indicated by the word 'ought' or 'obligation' is simple when it occurs in such propositions as 'you ought not to lie', 'you ought to be charitable'. "The concept of obligation in my sense applies only to actions, or rather to mere *resolutions* [*Willensentschließungen*] of rational beings, but to them without exception, in such a way that for every resolution a certain attribute can be indicated which it *ought* to have. I further wish to remark that I take the concept of obligation in such a wide sense that it holds of every resolution which can be termed morally good, whether it is a definite duty or merely meritorious, so that we can say of both kinds that they ought to be performed. Thus I say, for example, that one ought not to lie, which is a duty; and I also say that we ought to be charitable, which is not a duty, but merely meritorious . . . When there is no obligation which interdicts a certain action, then we say that it is *permissible*. Hence the permissibility of an action is the non-obligatoriness of its omission." The general form of moral propositions is '*A*—has—an obligation to do *B*'. For such a proposition to be true *A* must be a rational being, and *B* must be the idea of a certain action, or rather of a certain resolution. The general form of propositions of permission are '*A*—has—no obligation to do non-*B*'. The general form of propositions which simply state that an action is obligatory or permissible, without specifying an agent, is 'the idea of action *X* as an action which is obligatory (or permissible) has a referent'.

In a note, B. points out that the definition of obligation as moral necessity does not show that the concept of obligation is complex, since the word 'necessity' is here not taken in its proper sense, and the concept of obligation is still buried in the word 'moral'.

§ 145. *Interrogative Propositions*

1. By an interrogative proposition or question, B. means any assertion of a desire that a certain truth, some of whose attributes are identified, should be produced. Hence 'what attributes does God have?' is a question, similarly 'I desire to learn certain truths which contain attributes of God', even though the latter does not have the grammatical form of a question.

2. Hence questions are a kind of problem (§ 143) "they differ from other problems in that they demand that a proposition, indeed a true proposition, and one which has the attribute specified in the question, be produced. But this producing can be nothing but the stimulation of this proposition in the mind of a thinking being, either by linguistic expression or some other method. For we can desire only that something should come about, or not come about, hence we can desire only something that is actual or can take on actuality. But propositions in themselves do not have existence and cannot acquire existence; hence only their appearance in the mind or their expression in language can become actual at a given time. Hence one of these must be meant when somebody states his desire to have a certain truth indicated to him."

3. B. distinguishes several types of questions, namely (1) Questions which ask for the truth or falsity of a stated proposition, e.g. 'is the human soul immortal?'; (2) Questions which indicate a subject idea and leave it to us to find predicates such that true propositions result; e.g. 'what are the attributes of right triangles?'; (3) Questions that indicate a predicate idea where we have to find subjects such that true propositions result; e.g. 'who was the

first person to climb the Mont Blanc?'; (4) Questions which ask for a method to achieve a given purpose; e.g. 'how does one draw an equilateral triangle?'

4. Some questions can be answered by any of several different propositions; for others there is only one proposition that can serve as an answer (cf. examples 3.1 and 3.3 above). B. calls the latter *determined* and the former *undetermined*.

5. "Obviously, there can also be questions which require attributes that cannot be found in any truth. The following question is of this kind: 'who caused God?', if these words are to indicate a desire to find a truth of the form '*a* is the cause of God'. Such questions have been called *impossible*, *imaginary* and *absurd*."

NOTE 1: B. here points out that questions cannot be defined in terms of ignorance. "We ask a question if we want to express our desire that somebody should indicate a certain truth; we can do this even if we know very well the truth after which we inquire, for example when we ask in the Socratic manner."

It might be objected that a question cannot be defined in terms of the desire to learn the truth, since we sometimes answer questions merely by giving names; e.g. 'what is the name of this flower?'. "We can easily reply to this. In many cases the linguistic expression which satisfies the questioner does not contain a complete proposition but only a single idea; in other cases it contains a complete proposition which, however, is not true. Nonetheless what the questioner finds out from these expressions, or is supposed to find out from these expressions, is always a complete and true proposition. If we answer the above question with the single word 'rose', then, by virtue of the context, we express an entire proposition, namely 'this flower is called rose'."

NOTE 2: B. claims that not all problems [*Aufgaben*] are also questions, e.g. the problem of forgiving one's enemies; however, it is true that with many problems we naturally think of certain questions, namely *how* (by what means) the problem can be solved. "It must be realized, however, that such means cannot always be indicated, since there must be certain effects which can be achieved without means. If the effect which is demanded by a problem is of this kind, then the question by what means the problem can be solved is absurd, but it does not follow that the problem itself is absurd; rather, it may be very important, even unavoidable and its execution may be quite simple."

§ 146. *Referring and Non-Referring Propositions, Singular and General Propositions*

B. gives these names to propositions if their *subjects* have the indicated attributes.

§ 147. *The Concept of the Satisfiability of a Proposition*

It is a familiar fact that propositions can be divided into *true* and not true; the latter are also called *false*. This fact is so well known and has so often been assumed in this book that it will suffice if I merely touch upon it here.

It cannot be denied that every given proposition is either true or false and never changes: either it is true forever, or false forever, except if we change some *part* of it, and hence consider no longer the same but some other proposition (cf. § 125). We do this frequently without being clearly aware of it; this is one of the reasons why it seems as if the same proposition could sometimes be true and sometimes false, depending on the different times, places or objects to which we relate it. Thus we say that the proposition 'The keg of wine costs 10 Thaler' is true at this place and time, but false at another place or time. Likewise the proposition 'This flower has a pleasant fragrance' is said to be sometimes true and sometimes false, depending on whether the 'this' is made to refer to a rose or a stapelia, etc. However, it is obvious that it is not really one and the same proposition which exhibits different truth-values, but that we are concerned with several propositions which have in common only that they can be generated from one and the same proposition, provided we envisage certain of its parts as variable, and replace them with different ideas. (In the manner employed in §§ 69 and 108 with respect to ideas.) In the first example it is the tacitly assumed condition of time and place which produces a true proposition if it is of one kind, and a false one if it is of another. In the second example we vary the idea denoted by the word 'this', if we relate the proposition first to a rose, and then to a stapelia; hence we have not one, but two, essentially different propositions before us.

These examples show that we often take certain ideas in a given proposition to be variable and, without being clearly aware of it, replace these variable parts by certain other ideas and observe the truth values which these propositions take on. Since we do this anyhow, it is worth the effort to undertake this procedure with full consciousness and with the intention of gaining more precise knowledge about the nature of such propositions by observing their behaviour with respect to truth. Given a proposition, we could merely inquire whether it is true of false. But some very remarkable properties of propositions can be discovered if, in addition, we consider the truth values of all those propositions which can be generated from it, if we take some of its constituent ideas as variable and replace them by any other ideas whatever. Let us, for example, consider the proposition 'The man Caius is mortal', and let us envisage the idea 'Caius' as arbitrarily variable; hence we put in its place any other idea, e.g. 'Sempronius', 'Titus', 'rose', 'triangle', etc.

If we do this, it becomes obvious that the new propositions which are thus generated are all true, without exception, as long as they have reference, i.e. as long as their subject-idea has a referent. For, if we replace 'Caius' by ideas which designate real men, e.g. 'Sempronius', 'Titus', we always get true propositions. On the other hand, if we take some other idea, e.g. 'rose', 'triangle', then we obtain a proposition which not only lacks truth, but which does not even have reference. It can easily be seen that this would not have been the result if the initially given proposition had been different. For instance, if the proposition had been 'the man Caius is omniscient', the exact opposite would have happened. Each of the propositions generated through an exchange of the idea 'Caius' by any other idea whatever would have lacked truth. However, had the proposition originally read 'The being Caius is mortal', then some of the propositions generated in this manner would have been true and others false, some of the latter nevertheless having reference. In this case, the outcome depends upon whether we replace 'Caius' by the idea of a being which is mortal or by some other idea. All this should make it clear that the truth-values of all the propositions which can be generated from a given proposition through the assumption of one or more arbitrarily variable parts may be envisaged as a quality which makes better known to us the nature of the original proposition itself.

Most propositions by far are such that the propositions which evolve from them are neither all true nor all false, but some of them are true and others are false. Here the question arises *how many* are true and how many false, or what the relation is between one set and the other (or the total set). If we allow a replacement of the ideas which we consider variable by any other idea whatever, then the totality of the true as well as false propositions which can be generated from a given proposition will be infinitely large. Let i' be an idea which has been introduced into the proposition A instead of the variable idea i, and which makes A true (or false); any idea equivalent to i' will in most cases, though not in all, do the same. But, according to § 96, there are infinitely many of such equivalent ideas. Thus if we do not somehow limit the kinds of ideas which may be put into the place of a variable one, we will rarely be in a position to determine the relation between the number of true or false propositions and the number of all propositions which can be generated in this way. On the other hand, if we restrict the choice, and stipulate that no two equivalent ideas may replace the ideas $i, j, \ldots$, which are treated as variables (equivalent in the sense of § 96 or in the wider sense of § 102), if we furthermore require that only those ideas be chosen which generate referring propositions, then the number of ideas which are still permitted, hence also the number of propositions which can thus be generated, will be significantly reduced. Thus, under these restrictions, we will more often succeed in assigning numerical values to the relation between true or false propositions on the one hand, and the sum of all propositions which can be generated on the other. For instance, consider the proposition 'The ball marked no. 8 will be among those drawn at the next lottery'. Let us stipulate that only '8' be considered a variable idea, and that it be replaced only by ideas which are not equivalent to previously chosen ones, and which form a referring proposition. The total number of propositions which can thus be generated will be 90, if the lottery contains 90 numbers, because we can replace the '8' in the given proposition only by any one of the numbers 1 to 90. Every other number or idea put in this place will turn the subject of the proposition, i.e. the idea 'The ball marked no. X', into a non-referring idea.

This example also shows that the stipulation through which we decrease the number of propositions that can be generated from a given proposition can often be drawn up so that it singles out only propositions which interest us in some way and which we want to collect and count. Thus we are not at all interested in all the different propositions which can be generated from the above by replacing the idea '8' by sundry equivalent ones. However, it is useful to know how many different propositions appear if one proceeds under the given limitation, and it is especially useful to know the relation between the true propositions which appear and the sum total of all of them. This relation determines the degree of probability which the proposition takes on under certain circumstances. I therefore want to give a special name to the concept of the relation of all true propositions to the total of all propositions which can be generated by treating certain ideas in a proposition as variables and replacing them with others according to a certain rule. I wish to call it the *satisfiability* [*Gültigkeit*] of the proposition. The degree of satisfiability of a proposition is expressed by the relation between the number of the true propositions and the total number of propositions which are generated when certain ideas contained in the original proposition are considered variable and exchanged for others according to a certain rule. The degree of this satisfiability can then be represented as a fraction, where the numerator is to the denominator as the number of true propositions to the total number of propositions. Thus, for example, the degree of satisfiability of the above proposition is 5/90, i.e. 1/18, if five balls are drawn at the same time; because then there are precisely 5 true propositions out of a total of 90.

It is evident that the satisfiability of a proposition will vary, depending on which and how many of its ideas are considered variable. Thus, for instance, the proposition 'This triangle has three sides' remains true so long as only the idea 'this' is considered variable, provided that all ideas put in its place will generate referring propositions. However, if in addition to 'this', we also consider the idea 'triangle' or, instead of both, the idea 'side' as a variable, then the degree of satisfiability of the proposition will turn out to be entirely different. Thus, in order to give a proper estimate of the degree of satisfiability of a proposition, there must always be an indication which of its ideas are to be considered variable.

Let the proposition A be such that all the propositions which can be generated from it are true, if the ideas $i, j, \ldots$ alone are considered variable, and if only referring propositions may be formed. Then the degree of satisfiability of A with respect to $i, j, \ldots$ is the largest possible, i.e. it equals 1, and we can call the proposition *universally* or *fully* satisfiable. If, on the other hand, all propositions developed from A are false, then the degree of satisfiability of A is the smallest possible, i.e. it equals 0. We can therefore say that it is a *universally* or *absolutely* non-satisfiable proposition. Universally satisfiable propositions could also be said to be true by virtue of their kind or form, universally non-satisfiable propositions false by virtue of their kind or form, where by *kind* is meant the sum of all propositions, which differ from A only in the ideas $i, j. \ldots$

If the proposition A is universally satisfiable or non-satisfiable with respect to the ideas $i, j, \ldots$, then the denial of A, or the proposition 'A is false', is universally non-satisfiable or satisfiable respectively.

In § 66 we demonstrated that the extension of an idea must be considered one of its inner attributes. It can be similarly demonstrated that satisfiability is an inner attribute of a proposition.

§ 148. *Analytic and Synthetic Propositions*

1. I showed in the preceding section that there are universally satisfiable as well as non-satisfiable propositions, given that certain of their parts are considered variable. It was also shown that propositions which have either of these properties on the assumption that

i, *j*, . . . are variable, do not retain this status if different or additional ideas are taken as variable. It is particularly easy to see that no proposition could be formed so as to retain such a property if *all* its ideas were considered variable. For if we could arbitrarily vary all constituent ideas of a proposition, we could transform it into any other proposition whatever, and thus could turn it into a true, as well as a false, proposition. But it would be important enough to deserve notice if a proposition contained even a single idea which could be arbitrarily changed without altering the truth or falsity of the proposition; i.e., the propositions which could be obtained from it through the arbitrary alteration of this one idea would either all be true or all false, provided only they have reference. Borrowing this expression from Kant, I allow myself to call propositions of this kind *analytic*. All other propositions, i.e. all those which do not contain any ideas which can be changed without altering their truth or falsity, I shall call *synthetic*. For example, I call the following propositions analytic: 'A depraved man does not deserve respect' and 'A man may be depraved and still enjoy continued happiness'. The reason for this is that both contain a certain idea, namely 'man', which can be exchanged for any idea whatever, for instance 'angel', 'being', etc., yet the former remains always true, the latter always false, provided only that they continue to have reference. On the other hand, in the propositions 'God is omniscient', and 'Any triangle has two right angles', I could not point out a single idea which could be arbitrarily changed with the result that the first always remains true, the second false. Hence the two last propositions are examples of synthetic propositions.

2. The following are some very general examples of analytic propositions which are also true: '*A* is *A*', 'An *A* which is a *B* is an *A*', 'An *A* which is a *B* is a *B*', 'Every object is either *B* or non-*B*', etc. Propositions of the first kind, i.e. propositions cast in the form '*A* is *A*' or '*A* has (the attribute) *a*' are commonly called *identical* or *tautological* propositions.

3. The difference between the last mentioned analytic propositions and those under no. 1 lies in the following: In order to appraise the analytic nature of the propositions under 2, no other than logical knowledge is necessary, since the concepts which form the invariable part of these propositions all belong to logic. On the other hand, for the appraisal of the truth and falsity of propositions like those given in no. 1 a wholly different kind of knowledge is required, since concepts alien to logic intrude. This distinction, I admit, is rather unstable, as the whole domain of concepts belonging to logic is not

circumscribed to the extent that controversies could not arise at times. Nevertheless, it might be profitable to keep this distinction in mind. Hence propositions like those in no. 2 may be called *logically* analytic, or analytic in the *narrower* sense; those of no. 1, analytic in the *broader* sense.

NOTE 1: In order to determine whether a proposition which is given as a certain linguistic expression is analytic or synthetic, more is required than a cursory inspection of its words. A proposition may be analytic, perhaps logically analytic, or even identical, though its literal phrasing does not make this immediately apparent. Conversely, a proposition may sound exactly like an analytic or even an identical one, and may still be synthetic in its meaning. Thus it may not be immediately obvious that the proposition 'Every effect has a cause' is in fact identical, or at any rate analytic; for by 'effect' we always mean something which is brought about by something else, and the phrase 'to have a cause' means as much as 'to be brought about by something else'; thus the above proposition merely means 'Whatever is brought about by something else is brought about by something else.' The same holds of the proposition 'If A is larger than B, then B is smaller than A', 'If $P = Mm$, then $M = \frac{P}{m}$' and many others. On the other hand, many propositions which are conversationally used as proverbs sound altogether analytic or even tautological, without actually being so. For instance, we say frequently, 'What is bad is bad'; taken literally, this is indeed an empty tautology. Yet, what we actually think in speaking these words might be something quite different, and might also be different under different conditions. For example, we may want to indicate by these words that we cannot but call bad anything that appears bad to us. Somebody else may intend to call to our attention that it is a vain effort to extenuate the bad since it is found out sooner or later, etc. Likewise, the proposition that Leibniz quotes (*Nouv. Ess.* IV, VIII, 4) as an example for identical (or rather analytical) propositions "that are not without their use" namely 'Even a learned man is a man' is not analytical if it is interpreted in a sense that makes it useful. For then it is interpreted as 'Even a learned man is fallible'. A similar point holds of Leibniz' second example: "What may happen to one, may happen to anyone." Similarly, Pilate's saying "What I have written, I have written." is not a tautology, but had the meaning 'What I have written, I will not change'. The observation that apparently identical propositions are not always in fact identical has already been made by others, especially *Reusch* (*Systema Logicum*, Jena 1760, § 435).

2. B. criticizes several views concerning identical judgments. The theory closest to his own is that of Wolff and Maaß who define an identical proposition as one where the same idea forms the subject as well as the predicate. "There is only one reason why I cannot accept this definition unaltered: in my opinion the proposition '*A* is *A*' does not have the idea *A* for its predicate, but the corresponding *abstractum*." The views that propositions with equivalent subject and predicate ideas are identical he regards as "completely erroneous"; the same holds for the assumption that all equations are identical propositions.

3. Some philosophers wanted to pass all our thinking off as nothing but a class of identical (or, at least, analytic) judgments; others went to the opposite extreme and did not even want to grant the name of *propositions* to identical propositions. Maimon (*Versuch einer neuen Logik*, Berlin 1794, p. 55) did the latter for the reason that such propositions cannot be the ground of others, which may be quite correct; but why deny them the name of propositions for this reason? Similarly, Keckermann's *Systema Logicae*, Hanover 1600, *I*, 2, *sect.* 1, says "Identicals are not per se propositions", etc.

4. The ancient logicians already knew a distinction, more or less like mine, between analytic and synthetic propositions. Thus, Aristotle (*Soph.*, 165^b16 f.) indicated repeating oneself as an error, into which the art of the Sophists is meant to lead us. Locke (*Ess.* IV, VIII, 1) introduces the concept of trifling propositions; he defines them as propositions that do not inform us. Among them he counts (a) all identical ones, (b) all those in which part of a complex idea is asserted of an object of such an idea. He obviously had analytic judgments in mind, and almost gives a better explanation of them than Kant, as we shall see later. But Locke made a far-reaching error when he added that all propositions where the species is the subject and the genus the predicate fall into this category. Not every concept of species is compounded from the concept of the genus.

Schmidt has reminded us that Crusius (*Weg zur Gewißheit*, Leipzig 1747, § 260), already saw the difference between analytic and synthetic judgments in the same way as Kant. Though it is true that this distinction has been touched upon before, Kant must be credited with first having firmly grasped and suitably used it. But it seems to me that the explanations of this distinction, whether in Kant's own, or in the writings of others, do not fully meet the requirements of logical strictness. For example, Kant says in his *Logik* (§ 36, A 173) "A proposition is called analytic, if its certainty depends upon the identity of the concept of the predicate with the notion of the subject." Yet this applies at most to identical propositions. In the *Critique of Pure Reason* (Intro. § 4, etc., B 11 ff.), Kant asserts that in analytic judgments the predicate is contained (in a hidden manner) in the subject, or that the predicate does not lie outside the subject, or that it is a constituent of the subject.... These are, in part, only figurative modes of speech which do not analyse the concept to be explicated, in part expressions which permit of too many interpretations. Everything stated here, namely that the predicate idea is nothing but a repetition of a constituent of the subject idea, that the predicate idea is contained in the subject, that it lies, or is thought, in the subject, etc. can be said of propositions which nobody will want to call analytic, for instance 'The father of Alexander, King of Macedon, was King of Macedon' or 'A triangle similar to an isosceles one is isosceles', etc. This shortcoming could be avoided if we were to say ... that in an analytic judgment the predicate forms an essential part of the subject or (what comes to the same) that it forms one of its essential characteristics; the latter are constitutive parts of the subject, i.e. parts which occur in the concept of the subject. However, this explanation is fit only for one kind of analytical propositions, namely those of the form '*A* which is *B* is *B*'. But should there really be no others? Should we not count among analytic judgments '*A* which is *B* is *A*' and 'Everything is either *B* or not *B*'? Generally, it seems to me that none of these explications sufficiently emphasizes what makes these propositions important. I believe that this importance lies in the fact that their truth or falsity does not depend upon the concepts of which they are composed, but that it remains the same irrespective of the changes to which some of their concepts are subjected, provided only that reference of the proposition is not destroyed. This is the reason why I gave the above definition, although I know that it makes the concept of these propositions somewhat wider than it is commonly thought to be; propositions like those under no. 1 are not commonly considered analytic. I furthermore thought it expedient to give a wide enough range to the concepts of analytic and synthetic propositions so as to include not only true, but also false, propositions.

Whatever definition one wants to subscribe to, I don't think it will be necessary to admit that the difference between analytic and synthetic judgments is merely *subjective*, and that the same judgment is sometimes analytic and sometimes synthetic, depending on what concept we have of the object to which the subject idea refers. Still, this has been maintained by many logicians.... Thus Maaß says (in

Eberhard's *Phil. Mag.*, vol. II, 1790) "We can define a triangle as a figure in which the sum of all angles equals two right angles. Then the proposition 'In any triangle the sum of all angles equals two right angles' becomes analytic, although it must be considered synthetic under the common definition." I have a different opinion about this matter. By a proposition I do not mean a mere connection of *words* which assert something, but the *meaning* itself of what is asserted. Therefore, I do not admit that the proposition 'In a triangle, etc.' remains the same if different concepts are connected with the word 'triangle'. I am just as far from admitting this as I am from saying that 'Euclid was a famed mathematician' is one and the same judgment whether by 'Euclid' we mean the man who taught mathematics in Alexandria under Ptolemy Soter, or Euclid of Megara, disciple of Socrates. It is true that in Maaß' example both ideas refer to the same object, in my example to two different objects. Still, if propositions consist of different ideas they must be considered different, even if they are about the same object.

§ 149. *Propositions with Transponible Parts*

1. If a proposition contains parts which can be transposed or interchanged without changing the truth value, the proposition will be said to have transponible parts. "Given that the two persons, Caius and Titius love each other, it follows that the proposition 'Caius loves Titius' is not only true but remains true if we interchange the two ideas C. and T. and form the proposition 'Titius loves Caius'."

2. If the parts of a proposition are complex, more than one pair of parts may be interchangeable. "Hence if we claim that a proposition has transponible parts, we shall have to indicate the parts in question. Given for example that the son of Caius loves the daughter of Titius, and the son of Titius loves the daughter of Caius, the proposition 'the son of Caius loves the daughter of Titius' remains true if we interchange the ideas C. and T. but not if we transpose the ideas 'son' and 'daughter'."

CHAPTER 3

Distinctions among Propositions which are Based upon their Relations to each other

§ 150. *No Two Propositions are Exactly Equal; Similar Propositions*

The considerations given here are analogous to those of § 91 concerning ideas in themselves.

§ 151. *Relations among Propositions with Respect to their Content*

B. here defines identity of content of propositions, and kinship among propositions. The definitions are analogous to those given in § 92 for ideas.

§ 152. *Relations among Propositions with Respect to their Extension*

According to § 130, the extension of a proposition is identical with the extension of its subject. Hence the extension-dependent relations among propositions can be reduced to analogous relations among ideas.

§ 153. *Relations among Propositions with Respect to the Extension of their Predicate Ideas*

B. points out that important relations among propositions are occasionally founded upon the extensions of their predicates, but he does not give any details.

§ 154. *Compatible and Incompatible Propositions*

1. The most important relations among propositions come to light when we follow the method introduced in § 147, and envisage certain ideas contained in them as variable; we then consider the new propositions which are generated, if these ideas are replaced by any other ideas whatever, and observe what truth values they take on.

2. We know already that almost any proposition in which we replace certain ideas by different, arbitrarily chosen, ideas will sometimes turn into a true, and sometimes into a false, proposition. Let us

compare several propositions *A*, *B*, *C*, *D*, . . . and consider as variable certain ideas *i*, *j*, . . ., which they have in common. The question arises whether there are any ideas which can be put into the place of *i*, *j*, . . . and which are of such a nature that they make *all* of the above propositions true *at the same time*. If this question must be answered in the affirmative, then I wish to call the relation between propositions *A*, *B*, *C*, *D*, . . . a relation of *compatibility*, and the propositions *A*, *B*, *C*, *D*, . . . will be called *compatible* propositions. If the question must be answered in the negative, i.e. if there are no ideas whose substitution for *i*, *j*, . . . will make the propositions *A*, *B*, *C*, *D*, . . . all true, I say that they stand in a relation of *incompatibility*, and the propositions themselves I call *incompatible*. Thus I call the following three propositions compatible with each other: 'this flower is red', 'this flower has a pleasant fragrance', 'this flower belongs into the twelfth class of the system of Linné', provided the idea 'this flower' is viewed as arbitrarily variable. For, if I replace it by the idea 'rose' then all three propositions become true. On the other hand, the following three propositions I call incompatible if only the ideas 'finite being', 'man', and 'omniscience' are viewed as variable: 'no finite being is omniscient', 'man is a finite being', and 'some man is omniscient'. For, no matter what ideas are used for replacement, we do not succeed in simultaneously turning all three propositions into truths; as soon as two of them are true, the third becomes false.

3. It follows from the above definition that the relation of compatibility and incompatibility holds mutually (between propositions).

4. It is also obvious that there is great similarity between this relation among *propositions* and the relation between *ideas* which I have given the same name in § 94, especially if the amplification in § 108 is taken into account. With ideas, the crucial question was whether or not a certain object is indeed represented by them; the corresponding question for propositions is whether or not they are true. Just as I have called ideas compatible or incompatible with each other, depending on whether or not they have certain referents in common, so I call propositions compatible or incompatible, depending on whether or not there are certain ideas which makes all of them true.

5. A class of propositions *A*, *B*, *C*, *D*, . . . can appear as either compatible or incompatible, depending upon which ideas are considered variable. Hence the two propositions 'a lion has two breasts' and 'a lion has two wings' are compatible if we envisage the idea 'lion' as variable: if we replace it by the idea 'bat', both propositions become true. But if the idea 'two' is the only one that may be altered,

then the two propositions present themselves as incompatible, since there is no idea that may be put in its place which makes both propositions true. It is quite obvious that, if we are permitted to increase indefinitely the number of variable ideas in a given class of propositions, then these propositions will always appear compatible. For, if we can change arbitrarily many, perhaps all, of the ideas that occur in a given proposition, then we can change every proposition into any other one, hence doubtlessly also into a true proposition. Thus, if we say of a given class of propositions *A*, *B*, *C*, *D*, . . . that they are compatible or incompatible, we must indicate, for the sake of precision, *in what respect*, i.e. in relation to what variable ideas *i*, *j*, . . . this is meant.

6. All truths are compatible with each other no matter which of their ideas are considered variable: they are already true by virtue of the ideas of which they originally consist.

7. Hence of any given set of incompatible propositions at least one must be false, but several, even all of them, may be false.

8. In a set of compatible propositions, there may also be false ones, even all of them may be false. The fact that certain propositions are false, given the ideas of which they originally consist, does not prevent them from becoming true if certain other ideas are introduced. It is obviously necessary, however, that each of these false propositions should contain at least one of the ideas which are considered variable, since otherwise it would not change and hence not turn into a truth.

9. Even if the *n* propositions *A*, *B*, *C*, *D*, . . . are incompatible with respect to ideas *i*, *j*, . . ., there may still be a relation of compatibility with respect to the same ideas *i*, *j*, . . . among any subset of these propositions e.g. between any $(n-1)$, $(n-2)$. . . . For, even if no idea can be found whose substitution for *i*, *j*, . . . will make all *n* propositions *A*, *B*, *C*, *D*, . . . true, there may still be a part of these propositions, for example $(n-1)$ or $(n-2)$ of them, which can be made true at the same time. Thus the three propositions 'all *A* are *B*', 'all *B* are *C*', 'no *A* is *C*' are not compatible with respect to the three ideas *A*, *B*, *C*, though any two of them are compatible with respect to these ideas.

10. Conversely, if a part of the propositions *A*, *B*, *C*, *D*, . . ., for example *A*, *B*, . . ., are not compatible with each other with respect to certain ideas *i*, *j*, . . ., then the total class is also not compatible with respect to the same ideas. For if there were ideas whose substitution for *i*, *j*, . . . made *A*, *B*, *C*, *D*, . . . true at the same time, then *A*, *B*, . . . would be compatible.

11. If certain propositions are compatible with each other with respect to the ideas *i*, *j*, . . ., then they are also compatible with respect to the more inclusive set *i*, *j*, *k*, *l*, . . . which contains the former. If certain propositions are incompatible with respect to the larger set of ideas *i*, *j*, *k*, *l*, . . ., then they are also incompatible with respect to the smaller set *i*, *j*. . . . Conversely, it does not follow that, if a number of propositions are incompatible with respect to ideas *i*, *j*, . . ., they are also incompatible with respect to *i*, *j*, *k*, *l*, . . .; and if they are compatible with respect to ideas *i*, *j*, *k*, *l*, . . ., it does not follow that they are also compatible with respect to *i*, *j*. . . .

12. From the fact that propositions *A*, *B*, *C*, *D*, . . . as well as propositions *G*, *H*, *I*, *K* . . . are compatible with propositions *M*, *N*, *O*, . . . with respect to ideas *i*, *j*, . . . it does not follow that propositions *A*, *B*, *C*, *D*, . . . and *G*, *H*, *I*, *K*, . . . are compatible with each other with respect to these same ideas. It could very well be the case that there are certain ideas whose substitution for *i*, *j*, . . . makes propositions *A*, *B*, *C*, *D*, . . . and *M*, *N*, *O*, . . . true at the same time, while there are certain other ideas which make *G*, *H*, *I*, *K*, . . . and *M*, *N*, *O*, . . . true. But there may be no ideas at all which make *A*, *B*, *C*, *D*, . . . and *G*, *H*, *I*, *K*, . . . true at the same time. For example, both propositions 'all *A* are *B*' and 'no *A* is *B*' are compatible with the proposition 'all *A* are *C*' with respect to the three ideas *A*, *B*, *C*; nonetheless, the first two propositions are not compatible with each other with respect to the same ideas.

13. Similarly, it does not follow that if propositions *A*, *B*, *C*, *D*, . . . as well as *G*, *H*, *I*, *K*, . . . are incompatible with propositions *M*, *N*, *O*, . . . with respect to certain ideas *i*, *j*, . . ., then *A*, *B*, *C*, *D*, . . . and *G*, *H*, *I*, *K*, . . . are incompatible with each other. . . .

14. From the fact that certain propositions *A*, *B*, *C*, *D*, . . . are compatible with respect to *i*, *j*, . . . it does not follow that their negations, i.e. the propositions *Neg.A*, *Neg.B*, *Neg.C*, *Neg.D*, . . . (§ 141) are compatible with respect to the same ideas. For it could be the case that one of the propositions *A*, *B*, *C*, *D*, . . ., e.g. *A*, is made true not only by those ideas which make all the others, *B*, *C*, *D*, . . . true, but, in addition, also by all ideas through which one of them, e.g. *B*, is made false, and hence the proposition *Neg.B* is made true. In this case there is no idea which would make propositions *Neg.A* and *Neg.B* true at the same time. Thus the two propositions 'some *A* are *B*', and 'it is false that all *A* are *B*' are compatible with respect to ideas *A* and *B*, but the two propositions which result from their denial, namely 'it is false that some *A* are *B*' and 'all *A* are *B*' are obviously incompatible with respect to those ideas.

15. From the fact that the negation of any one of the propositions *A*, *B*, *C*, *D*, . . . is compatible with the remaining propositions with respect to ideas *i*, *j*, . . . it does not follow that the negations of any *two* or more of these propositions are compatible with the remainder with respect to the same ideas. . . .

16. All propositions whose predicates are viewed as variable are compatible with each other no matter what their subjects are, so long as the latter have reference. For, if each of these propositions has a distinct predicate idea, which differs from that of the others, but is viewed as variable, then it will be easy to give each of these propositions a predicate which makes it true. All we have to do is find for each of them severally the idea of an attribute which all objects represented by its subject have in common. But if some or all of them have one and the same predicate idea, then we have to find an idea of an attribute which the totality of objects represented by their various subjects have in common. Such attributes can always be found, since even the most divergent objects have certain attributes in common.

17. All propositions whose *subjects* are different from each other, and are viewed as variable, are compatible with each other, no matter what their predicates are, so long as these predicates are proper attributive ideas. For, if this condition is fulfilled, we can always find ideas with which to replace the subjects of these propositions to make them true. We only need to find for each proposition the idea of one of those objects which have the attribute indicated by the predicate.

18. Propositions which have the same subject, which is also envisaged as the variable idea in them, are compatible with each other if the *concreta* corresponding to their predicates are compatible, and are incompatible if these *concreta* are incompatible. For, if the *concreta* of their predicate ideas are compatible, then there are certain objects which have united in them all the attributes which are asserted in these propositions. Hence, we can make all of these propositions true if we replace their subject idea by an idea that represents exclusively some such object. Hence the propositions are compatible. On the other hand, if those *concreta* are not compatible with each other, then there is no object which has all of the attributes asserted in the given propositions. Hence there is no idea of an object with which we could replace the common subject idea so as to make all propositions true.

19. For any given proposition, once the ideas are designated which are to be considered variable, we can indicate infinitely many other propositions that are incompatible with it and, unless it is a formally

false proposition, also infinitely many propositions that are compatible with it. For, all propositions which are formed according to the following rule will be incompatible with proposition *A*: 'proposition *A* is false', 'that proposition *A* is false, is true', etc. In this way infinitely many propositions can be formed which are incompatible with *A*, for there obviously can be no idea which will make proposition *A* and at the same time any of these other propositions true. Furthermore, if *A* is not formally false, then there is an infinitely large number of propositions that are compatible with it, for each truth which does not contain the variable parts *i*, *j*, . . . will be compatible with *A*, since no change of *i*, *j*, . . . will change it.

20. If a proposition is false and does not contain the ideas *i*, *j*, . . ., which we consider variable in a certain class of propositions, it is incompatible with these propositions, no matter what their nature. Since it does not contain ideas *i*, *j*, . . . it remains unchanged, no matter what ideas are put in their place. Hence it will never become true and is therefore incompatible with those other propositions.

21. Hence, if a certain proposition *F* is compatible with certain others *A*, *B*, *C*, *D*, . . . although it does not contain a single one of the variable ideas *i*, *j*, . . . then it must be a true proposition.

22. Whatever the relation between propositions *A*, *B*, *C*, *D*, . . . with respect to ideas *i*, *j*, . . ., that same relation must obtain between propositions *A'*, *B'*, *C'*, *D'*, . . . with respect to ideas *i'*, *j'*, . . ., where the propositions *A'*, *B'*, *C'*, *D'*, . . . have resulted from propositions *A*, *B*, *C*, *D*, . . . merely by replacing ideas *i*, *j*, . . . with *i'*, *j'*, . . ., and since ideas *i'*, *j'*, . . . are to be envisaged as variable, another replacement of the ideas *i'*, *j'*, . . . by the ideas *i*, *j*, . . . will recover the propositions *A*, *B*, *C*, *D*, . . . from *A'*, *B'*, *C'*, *D'*, . . . and through the continued exchange of ideas *i*, *j*, . . . with whatever other ideas, we can generate any proposition which we could have gotten by starting from *A*, *B*, *C*, *D*. . . . Hence the same relations obtain in both cases.

In a note, after discussing some other matters, B. criticizes some Kantians who had claimed that propositions are compatible if they can be thought together in the same consciousness. He points out that it must be possible to judge the compatibility of two propositions from their inner attributes, without considering their relation to our understanding.

§ 155. *Special Kinds of Compatibility:* a. *The Relation of Deducibility*

1. The indicated similarity between propositions and ideas with

respect to the relations of compatibility and incompatibility is close enough so that the same subsections can be made in the discussion of propositions as in the discussion of ideas. Let us, to begin with, consider the relation of *compatibility*.

2. If we assert that certain propositions *A*, *B*, *C*, *D*, . . . *M*, *N*, *O*, . . . stand in the relation of compatibility with respect to ideas *i*, *j*, . . ., then we assert, according to the above definition, no more than that there are certain ideas whose substitution for *i*, *j*, . . . turns all of those propositions into true ones. So far, we have not concerned ourselves with the question whether aside from these ideas, which turn propositions *A*, *B*, *C*, *D*, . . . *M*, *N*, *O*, . . . all into true propositions, there may be others which make some, but not all, of them true. Obviously, this problem is of some importance. Let us consider, first of all, the case that among the compatible propositions *A*, *B*, *C*, *D*, . . . *M*, *N*, *O*, . . . the following relation obtains: all ideas whose substitution for the variable ideas *i*, *j*, . . . turns a certain part of these propositions, namely *A*, *B*, *C*, *D*, . . . into truths, also have the characteristic of making a certain other part of these propositions, namely *M*. *N*, *O*, . . . true. This special relation which we think between propositions *A*, *B*, *C*, *D*, . . . on the one hand and *M*, *N*, *O*, . . . on the other is of special importance, since it puts us in a position to infer the truth of *M*, *N*, *O*, . . ., once we have recognized the truth of *A*, *B*, *C*, *D*. . . . I wish to give the name of *deducibility* [*Ableitbarkeit*] to this relation between propositions *A*, *B*, *C*, *D*, . . . on one hand and *M*, *N*, *O*, . . . on the other. Hence I say that propositions *M*, *N*, *O*, . . . are *deducible* from propositions *A*, *B*, *C*, *D*, . . . with respect to variable parts *i*, *j*, . . ., if every class of ideas whose substitution for *i*, *j*, . . . makes all of *A*, *B*, *C*, *D*, . . . true, also makes all of *M*, *N*, *O*, . . . true. Occasionally, since it is customary, I shall say that propositions *M*, *N*, *O*, . . . *follow*, or can be *inferred* or *derived*, from *A*, *B*, *C*, *D*. . . . Propositions *A*, *B*, *C*, *D*, . . . I shall call the *premises*, *M*, *N*, *O*, . . . the *conclusions*. Finally, since this relation between propositions *A*, *B*, *C*, *D*, . . . and *M*, *N*, *O*, . . . has great similarity to the relation between including and included ideas, I shall allow myself to call propositions *A*, *B*, *C*, *D*, . . . *included*, and *M*, *N*, *O*, . . . *including* propositions.

3. From the assumption that all ideas whose substitution for *i*, *j*, . . ., makes propositions *A*, *B*, *C*, *D*, . . . true also makes *M*, *N*, *O*, . . . true, it does not follow that the converse must hold, i.e. that all ideas which make *M*, *N*, *O*, . . . true also make *A*, *B*, *C*, *D*, . . . true. Therefore, the relation of deducibility is not necessarily mutual. For example, every pair of ideas whose substitution for *A* and *B*

makes the proposition 'all *A* are *B*' true also makes the proposition 'some *A* are *B*' true; hence the latter is deducible from the former. But conversely, not every pair of ideas whose substitution for *A* and *B* makes the proposition 'some *A* are *B*' true, also makes the proposition 'all *A* are *B*' true. Hence this proposition is not deducible from the other.

4. Assume that propositions *M*, *N*, *O*, . . . are deducible with respect to *i*, *j*, . . . from propositions *A*, *B*, *C*, *D*, . . ., and that proposition *A* does not contain any of these ideas. In this case *A* may be deleted and we may assert that *M*, *N*, *O*, . . . are deducible from *B*, *C*, *D*, . . . alone with respect to ideas *i*, *j*. . . . Under these circumstances the proposition A must be true; it must remain true, no matter what ideas are put in the place of *i*, *j*, . . .; thus as soon as propositions *B*, *C*, *D*, . . . are all true propositions *A*, *B*, *C*, *D*, . . ., and hence also *M*, *N*, *O*, . . . are also true.

5. If certain propositions *M*, *N*, *O*, . . . are deducible from certain others *A*, *B*, *C*, *D*, . . . and one of the former is false, then there must also be a false one among the latter. For if *A*, *B*, *C*, *D*, . . . were all true, *M*, *N*, *O*, . . . would also have to be true, since otherwise it would not be the case (namely for ideas *i*, *j*, . . . themselves) that each class of ideas whose substitution for *i*, *j*, . . . makes *A*, *B*, *C*, *D*, . . . true must also make *M*, *N*, *O*, . . . true.

6. If all propositions which are deducible from *A*, *B*, *C*, *D*, . . . with respect to certain ideas *i*, *j*, . . . are true, then *A*, *B*, *C*, *D*, . . . must themselves be true. For among the various propositions which are deducible from *A*, *B*, *C*, *D*, . . . regardless of the character of *i*, *j*, . . ., we find the proposition '*A* is true', '*B* is true', '*C* is true', etc. Hence if all propositions which are deducible from *A*, *B*, *C*, *D*, . . . are true, then these must also be true. But if they are true, then propositions *A*, *B*, *C*, *D*, . . . must themselves be true.

7. The negation of a given proposition *A*, namely *Neg.A*, i.e. the proposition '*A* is false' is never deducible from *A*, no matter what ideas *i*, *j*, . . ., which occur only in *A*, are envisaged as variable. For no class of ideas which make proposition *A* true can also make the proposition '*A* is false' true.

8. All conclusions *M*, *N*, *O*, . . . which follow from certain propositions *A*, *B*, *C*, *D*, . . . with respect to ideas *i*, *j*, . . . are compatible with respect to *i*, *j*, . . . with all propositions, with which the premises *A*, *B*, *C*, *D*, . . . are compatible with respect to *i*, *j*. . . . For, if propositions *A*, *B*, *C*, *D*, . . . are compatible with propositions *A′*, *B′*, *C′*, *D′*, . . . with respect to ideas *i*, *j*, . . . and if *M*, *N*, *O*, . . . are deducible from *A*, *B*, *C*, *D*, . . . with respect to *i*, *j*, . . . then there

are certain ideas whose substitution for *i*, *j*, . . . will make propositions *A*, *B*, *C*, *D*, . . . as well as *A'*, *B'*, *C'*, *D'*, . . . true. However, as soon as *A*, *B*, *C*, *D*, . . . are true, *M*, *N*, *O*, . . . also become true. Hence there are certain ideas whose substitution for *i*, *j*, . . . makes propositions *A'*, *B'*, *C'*, *D'*, . . . as well as *M*, *N*, *O*, . . . true at the same time.

9. Propositions which are not compatible are never conclusions from propositions which are compatible with respect to the same variable ideas. For if they were such conclusions, they would have to be compatible according to no. 8.

10. If certain conclusions are incompatible, then their premises must also be incompatible with respect to the same variable ideas. For, if the premises were compatible, then, according to no. 9, the conclusions would also have to be compatible.

11. But it is quite possible that conclusions are compatible when their respective premises are incompatible with respect to the same variable parts. Assume that propositions *M*, *N*, *O*, . . . are conclusions from propositions *A*, *B*, *C*, . . . with respect to ideas *i*, *j*, . . . and that *M'*, *N'*, *O'*, . . . are conclusions from propositions *A'*, *B'*, *C'*, . . . with respect to the same ideas; this merely requires that each class of ideas whose substitution for *i*, *j*, . . ., makes propositions *A*, *B*, *C*, . . . true, should also make propositions *M*, *N*, *O*, . . . true; and that each class of ideas that make propositions *A'*, *B'*, *C'*, . . . true should also make propositions *M'*, *N'*, *O'*, . . . true. But it is not required that, conversely, propositions *A*, *B*, *C*, . . . and *A'*, *B'*, *C'*, . . . should become true whenever *M*, *N*, *O*, . . . and *M'*, *N'*, *O'*, . . ., respectively, are true. Now if propositions *M*, *N*, *O*, . . . are true more often than *A*, *B*, *C*, . . . and *M'*, *N'*, *O'*, . . . more often than *A'*, *B'*, *C'*, . . ., then it is possible that there are certain ideas whose substitution for *i*, *j*, . . . makes propositions *M*, *N*, *O*, . . . and *M'*, *N'*, *O'*, . . . true at the same time, while there are no such ideas for *A*, *B*, *C*, . . . and *A'*, *B'*, *C'*. . . . Thus the two propositions 'Caius is miserly' and 'Caius is profligate' are not compatible with each other if only the idea 'Caius' is envisaged as variable. But from the first of them we can deduce, with respect to the same idea, the conclusion 'Caius is not generous', and from the second we can deduce the conclusion 'sooner or later Caius will no longer be able to be generous'; and these two propositions are quite compatible.

12. No proposition which is not universally satisfiable with respect to ideas *i*, *j*, . . . (§ 147) is deducible from both, the individual proposition *A* as well as its negation *Neg.A*. For, if the proposition *M* is not universally satisfiable, then there are certain ideas whose

substitution for *i*, *j*, . . . will make it false. But if it is to be deducible from *A*, then each idea which makes it false must also make *A* false (cf. no. 5), and if it is to be deducible from *Neg.A* as well, it must also make *Neg.A* false. Hence one and the same idea would have to make both *A* and *Neg.A* false at the same time, which is absurd.

13. But if there are several premises from which a certain proposition *M* is to be deducible, e.g. *A*, *B*, *C*, *D*, . . ., then it is quite possible that *M* is also deducible from the negation of some or even all of them. For, in this case, we cannot conclude from the falsity of *M*, i.e. from *Neg.M*, the negation of each of the individual propositions *A*, *B*, *C*, *D*, . . ., but, according to no. 5, we may conclude only that they are not all true. The following is an example of a pair of propositions where the same proposition can be deduced both from them and from their negations. From the two propositions 'every *A* is a *B*' and 'it is false that every *A* is a *C*', where only ideas *A*, *B*, *C*, . . . are variable, we can draw the conclusion that the ideas *B* and *C* are not equivalent [*Wechselvorstellungen*]. The negations of these two propositions are 'it is false that every *A* is a *B*' and 'every *A* is a *C*'. These two propositions allow us to draw the same conclusion.

14. If *M* is compatible with propositions *A*, *B*, *C*, *D*, . . . with respect to ideas *i*, *j*, . . ., then its negation *Neg.M* is not deducible from these propositions with respect to the same ideas. For if *Neg.M* were deducible from *A*, *B*, *C*, *D*, . . . with respect to *i*, *j*, . . . then every class of ideas whose substitution for *i*, *j*, . . . makes all of *A*, *B*, *C*, *D*, . . . true would also make *Neg.M* true, hence *M* false. Consequently *M* cannot be compatible with *A*, *B*, *C*, *D*, . . . with respect to the same ideas.

15. If propositions *A*, *B*, *C*, *D*, . . . are compatible with each other with respect to ideas *i*, *j*, . . . but incompatible with *M*, then the proposition *Neg.M* is deducible from them with respect to the same ideas. For if propositions *A*, *B*, *C*, *D*, . . . are compatible with each other with respect to ideas *i*, *j*, . . . then there must be ideas whose substitution or *i*, *j*, . . . makes all of these propositions true. But since *M* is supposed to be incompatible with these propositions, the very same ideas which make all of *A*, *B*, *C*, *D*, . . . true, must make *M* false, and hence must make proposition *Neg.M* true. Therefore *Neg.M* is deducible from *A*, *B*, *C*, *D*. . . .

16. If we delete any one of the premises *A*, *B*, *C*, *D*, . . . of a conclusion *M*, for example *A*, and replace it by the negation of *M*, *Neg.M*, then we can deduce the negation of the missing proposition, i.e. *Neg.A*, from the class of propositions *B*, *C*, *D*, . . . and *Neg.M*, provided only that *Neg.M* is compatible with the remaining premises.

For, if *Neg.A* were not deducible from the indicated propositions, it would not become true whenever they are true, i.e. there must be cases in which propositions *B*, *C*, *D*, . . . *Neg.M* and proposition *A* are true together, which is absurd. For, as soon as *B*, *C*, *D*, . . . and *A* are true together, *M* must also be true; hence, in this case, *Neg.M* cannot be true.

17. If propositions *M*, *N*, *O*, . . . are deducible with respect to the same ideas *i*, *j*, . . . from the class of propositions *A*, *B*, *C*, *D*, . . . and X, as well as from the class of propositions *A*, *B*, *C*, *D*, . . . and *Neg.X*, then they are also deducible from propositions *A*, *B*, *C*, *D*, . . . alone with respect to the same ideas. For each class of ideas whose substitution for *i*, *j*, . . . makes propositions *A*, *B*, *C*, *D*, . . . true will also make propositions *M*, *N*, *O*, . . . true, no matter whether it makes *X* true or false.

18. The matter would be quite different if one of the above classes contained the negation of more than one proposition from the other class. For, if *M* is deducible from the class of propositions *A*, *B*, *C*, *D*, . . . *X*, *Y*, as well as from the class of propositions *A*, *B*, *C*, *D*, . . . *Neg.X*, *Neg.Y*, it does not follow that *M* is deducible from *A*, *B*, *C*, *D*, . . . alone. For there could be certain ideas which make propositions *A*, *B*, *C*, *D*, . . . and only one of the two, *X* or *Y*, true. These ideas could make *M* false, and it could still be the case that it is deducible from both, namely the class of propositions *A*, *B*, *C*, *D*, . . . *X*, *Y*, as well as the class of propositions *A*, *B*, *C*, *D*, . . . *Neg.X*, *Neg.Y*.

19. If propositions *M*, *N*, *O*, . . . are deducible from *A*, *B*, *C*, *D*, . . . with respect to ideas *i*, *j*, *k*, . . . then they are also deducible from the same propositions with respect to the fewer ideas *j*, *k*, . . . (which are part of the former), provided that propositions *A*, *B*, *C*, *D*, . . . are compatible with each other with respect to the fewer ideas *j*, *k*. . . . For, if this is the case, there are certain ideas whose substitution for *j*, *k*, . . . makes propositions *A*, *B*, *C*, *D*, . . . all true, and as soon as they become true, propositions *M*, *N*, *O*, . . . also become true. Therefore *M*, *N*, *O*, . . . are deducible from *A*, *B*, *C*, *D*, . . . with respect to the fewer ideas *j*, *k*. . . .

20. Conversely, if propositions *M*, *N*, *O*, . . . are deducible from propositions *A*, *B*, *C*, *D*, . . . with respect to the fewer ideas *i*, *j*, . . ., then they are not necessarily deducible with respect to the larger class of ideas *i*, *j*, *k*, . . . (which contains the former), although propositions *A*, *B*, *C*, *D*, . . . would certainly remain compatible on this assumption (§ 154, no. 11). For if, in addition to variable ideas *i*, *j*, . . ., we take certain others *k*, . . . to be variable, then the number

of true propositions which can be formed from the given proposition *A, B, C, D, . . .* can increase considerably, and it is therefore possible that the *M, N, O, . . .* do not always become true whenever *A, B, C, D, . . .* are true. Thus the propositions 'Caius is mortal' is deducible from the proposition 'Caius is a man' provided that 'Caius' is the only proposition which is envisaged as variable in both of them. But if the idea 'man' is also envisaged as variable, then the relation of deducibility no longer holds, which is obvious when we replace 'man' by 'simple being'.

21. It is not the case that any proposition *M*, let alone any arbitrary class of propositions *M, N, O, . . .* can be placed into a relation of deducibility with another individual proposition *A* or a class of propositions *A, B, C, D, . . .* simply by stipulating that certain of their ideas *i, j, . . .* should be considered variable. Given, for example, that the two propositions '*A* has *b*' and '*C* has *d*' have no common constituent aside from the idea 'has', then it is obvious that, no matter which ideas we envisage as variable, these two propositions will never stand in a relation of deducibility, since the ideas which are placed in one of them are altogether independent of the ideas which will appear in the other.

22. If propositions *M, N, O, . . .* are deducible from propositions *A, B, C, D, . . .* with respect to ideas *i, j, . . .*, and propositions *P, Q, R, . . .* deducible from propositions *F, G, H, . . .* with respect to the same ideas; and if *A, B, C, D, . . .* are compatible with *F, G, H, . . .* also with respect to the same ideas, then the class of propositions *M, N, O, . . . P, Q, R, . . .* is deducible from the class of propositions *A, B, C, D, . . . F, G, H. . . .* For, if propositions *A, B, C, D, . . .* and *F, G, H, . . .* are compatible with respect to ideas *i, j, . . .* then there are ideas whose substitution for *i, j, . . .* will make propositions *A, B, C, D, . . .* and *F, G, H, . . .* true together; but these very same ideas will also make propositions *M, N, O, . . .* and *P, Q, R, . . .* true. Hence the latter are deducible from the former.

23. Even in the case where propositions *M, N, O, . . .* are deducible from propositions *A, B, C, D, . . .* with respect to certain ideas *i, j, . . .* and propositions *P, Q, R, . . .* are deducible from propositions *F, G, H, . . .*, with respect to certain other ideas *k, l, . . .* which are either altogether, or at least in part, different from *i, j, . . .*, then the class of propositions *M, N, O, . . . P, Q, R, . . .* is deducible from the class of propositions *A, B, C, D, . . . F, G, H, . . .* with respect to the class of ideas *i, j, k, l, . . .*, provided that none of the ideas of *k, l, . . .*, which are not also contained in *i, j, . . .* appear in propositions *A, B, C, D, . . .* and none of the ideas of *i, j, . . .* which are not also

contained in *k*, *l*, . . . occur in propositions *F*, *G*, *H*, . . ., and provided also that propositions *A*, *B*, *C*, *D*, . . . *F*, *G*, *H*, . . . are compatible with respect to ideas *i*, *j*, *k*, *l*. . . . For if this is the case, then there are certain ideas whose substitution for *i*, *j*, *k*, *l*, . . . makes all of *A*, *B*, *C*, *D*, . . . *F*, *G*, *H*, . . . true. And since none of the ideas of *k*, *l*, . . . that differ from the *i*, *j*, . . . occur in propositions *A*, *B*, *C*, *D*, . . . and none of the ideas of *i*, *j*, . . . that differ from *k*, *l*, . . . occur in propositions *F*, *G*, *H*, . . ., the stipulation that all of *i*, *j*, *k*, *l*, . . . should be variable will not result in any true propositions among *A*, *B*, *C*, *D*, . . . in addition to those generated if only *i*, *j*, . . . are variable; but in all these cases, *M*, *N*, *O*, . . . became true as well. Similarly, no true propositions in addition to those that appeared when only *k*, *l*, . . . were considered variable will be generated from *F*, *G*, *H*, . . .; but in all these cases *P*, *Q*, *R*, . . . become true as well. It follows that, as soon as all of *A*, *B*, *C*, *D*, . . . *F*, *G*, *H*, . . . become true, all of *M*, *N*, *O*, . . . *P*, *Q*, *R*, . . . will also become true. Hence the latter are deducible from the former with respect to *i*, *j*, *k*, *l*. . . .

24. If, with respect to certain ideas *i*, *j*, . . ., propositions *M*, *N*, *O*, . . . are deducible from propositions *A*, *B*, *C*, *D*, . . . and propositions *X*, *Y*, *Z*, . . . are deducible from propositions *M*, *N*, *O*, . . . and *R*, *S*, *T*, . . . with respect to the same ideas, then propositions *X*, *Y*, *Z*, . . . are also deducible from propositions *A*, *B*, *C*, *D*, . . . *R*, *S*, *T*, . . . with respect to the same ideas. For, if propositions *M*, *N*, *O*, . . . are deducible from *A*, *B*, *C*, *D*, . . . with respect to ideas *i*, *j*, . . ., then every class of ideas whose substitution for *i*, *j*, . . . makes all of *A*, *B*, *C*, *D*, . . . true, also makes *M*, *N*, *O*, . . . true. Hence every class which makes all of *A*, *B*, *C*, *D*, . . . *R*, *S*, *T*, . . . true, also makes all of *M*, *N*, *O*, . . . *R*, *S*, *T*, . . . and hence (because of the deducibility of *X*, *Y*, *Z*, . . . from *M*, *N*, *O*, . . . *R*, *S*, *T*, . . .) also *X*, *Y*, *Z*, . . . true.

25. Even if propositions *M*, *N*, *O*, . . . are deducible from *A*, *B*, *C*, *D*, . . . with respect to ideas *i*, *j*, . . . while propositions *X*, *Y*, *Z*, . . . are deducible from *M*, *N*, *O*, . . . *R*, *S*, *T*, . . . with respect to ideas *k*, *l*, . . ., which are either altogether or in part different from *i*, *j*, . . ., propositions *X*, *Y*, *Z*, . . . are still deducible from propositions *A*, *B*, *C*, *D*, . . . *R*, *S*, *T*, . . . with respect to the class of ideas *i*, *j*, . . . *k*, *l*, . . ., provided only that none of the ideas of *k*, *l*, . . . that differ from *i*, *j*, . . . occur in propositions *A*, *B*, *C*, *D*, . . . and none of the *i*, *j*, . . . that differ from *k*, *l*, . . . occur in *M*, *N*, *O*, . . . *R*, *S*, *T*, . . . For, if none of the ideas *k*, *l*, . . . that differ from *i*, *j*, . . . occur in *A*, *B*, *C*, *D*, . . ., then the stipulation that all of *i*, *j*, . . . *k*, *l*, . . . should be variable will not make true any of the propositions *M*, *N*, *O*, . . . *R*, *S*, *T*, . . . that are not already made true on the stipulation

that only *k*, *l*, . . . should be variable. Hence under this stipulation, if all of propositions *M*, *N*, *O*, . . . *R*, *S*, *T*, . . . are true, propositions *X*, *Y*, *Z*, . . . are also made true. Therefore *X*, *Y*, *Z*, . . . are deducible from *A*, *B*, *C*, *D*, . . . *R*, *S*, *T*, . . . with respect to *i*, *j*, . . . *k*, *l*. . . .

26. Let proposition *M* be deducible from premises *A*, *B*, *C*, *D*, . . . with respect to ideas *i*, *j*. . . . If *A*, *B*, *C*, *D*, . . . are such that none of them, nor even any of their parts, may be omitted, with *M* still deducible from the remainder with respect to the same ideas *i*, *j*, . . ., I call the relation of deducibility of proposition *M* from *A*, *B*, *C*, *D*, . . . ***irredundant*** [***genau bemessen***], ***precise***, or ***adequate***. A relation of deducibility that is not irredundant is said to be ***redundant***. Thus, the relation of deducibility between the two premises 'all α are β' and 'all β are γ', and the conclusion 'all α are γ', where ideas α, β, γ are to be considered variable, is irredundant: we cannot leave out a single constituent of the first two propositions, let alone an entire proposition, if the proposition 'all *α* are *γ*' is still to be deducible from the remainder, always given that ideas α, β, γ are the variables. On the other hand, the relation of deducibility between the same two premises and the conclusion 'some β are α' I call redundant, since it continues to hold if we only use the first of the above premises. If we deduce the conclusion 'all α are δ' from premises 'all α are β' and 'all β and γ are δ', the deduction is likewise redundant, for this conclusion can also be drawn if the premises 'all β and γ are δ' is replaced by the simpler proposition 'all β are δ'.

27. If a relation of irredundant deducibility holds between propositions, then neither the conclusion nor any of the premises may be a formally true proposition. The conclusion cannot be formally true, for such a proposition does not have the truth of its premises as a condition for its own truth. None of the premises can be formally true, for if a formally true proposition is deleted from the premises, the conclusion is still deducible from the remainder (cf. no. 4).

28. If the relation of deducibility with respect to ideas *i*, *j*, . . . between premises *A*, *B*, *C*, *D*, . . . and the conclusion *M* is irredundant, then the negation of this conclusion, *Neg.M*, must be compatible with any [proper] part of the premises with respect to the same ideas *i*, *j*. . . . For if *Neg.M* were incompatible with any part of the propositions *A*, *B*, *C*, *D*, . . ., e.g. with *B*, *C*, . . ., then, according to no. 15, the proposition *Neg.Neg.M*, hence *M* itself, would be deducible from propositions *B*, *C*, . . . alone. Therefore the relation of deducibility between *A*, *B*, *C*, *D*, . . . and *M* would be redundant (cf. no. 26).

29. If a relation of deducibility is irredundant with respect to

ideas *i*, *j*, . . ., then no premise must be deducible from the remaining premises with respect to the same ideas. For, if premise *A* were deducible from the remaining premises *B*, *C*, *D*, . . . then the whole class of propositions *A*, *B*, *C*, *D*, . . . would be deducible from propositions *B*, *C*, *D*, . . . with respect to the same ideas; consequently proposition *M*, which is deducible from *A*, *B*, *C*, *D*, . . . would also be deducible from *B*, *C*, *D*, . . . (cf. no. 23). Hence the relation of deducibility between *A*, *B*, *C*, *D*, . . . and *M* would not be irredundant.

30. If a relation of deducibility is irredundant, then the negation of any single premise must be compatible not only with all other premises, but also with the added negation of the conclusion *M*, given that the same ideas are variable throughout. For, if the proposition *Neg.A* were not compatible with propositions *B*, *C*, *D*, . . ., *A* would be deducible from them; hence, according to no. 29, *A*, *B*, *C*, *D*, . . . would not be suitable premises in a relation of irredundant deducibility. Furthermore if the propositions *Neg.A*, *B*, *C*, *D*, . . . were not also compatible with the negation of the conclusion, i.e. *Neg.M*, then, according to no. 15, *Neg.Neg.M*, i.e. *M* itself, would be deducible from *Neg.A*, *B*, *C*, *D*. . . . But since *M* is also to be deducible from *A*, *B*, *C*, *D*, . . ., the truth or falsity of *A* must be altogether indifferent for *M*; hence *M* must be deducible from *B*, *C*, *D*, . . . alone; therefore the relation of deducibility between *A*, *B*, *C*, *D*, . . . and *M* would not be irredundant.

31. It also holds for irredundant deducibility that the negation of two or more of the premises can be incompatible with the remainder, given the same variable ideas throughout. Let us abbreviate the three propositions 'all α are β', 'all β are γ' and 'all α are γ' by *A*, *B*, and *C* respectively, then the relation of deducibility between the three premises *Neg.A*, *Neg.B*, and *Neg.C* and the conclusion 'the class of the three propositions *Neg.A*, *Neg.B*, *Neg.C*, consists of nothing but true propositions' is irredundant, if α, β and γ are envisaged as variable; for we cannot delete any of the three premises, nor any of their parts, and retain the conclusion. Nonetheless, the negation of the first two premises, i.e. the affirmation of propositions *A* and *B*, is incompatible with the third, i.e. *Neg.C*.

32. If the relation of deducibility between premises *A*, *B*, *C*, *D*, . . . and conclusion *M*, also the relation between premises *M*, *R*, *S*, *T*, . . . and the conclusion *X* is irredundant with respect to the same ideas *i*, *j*, . . ., it does not follow that the relation of deducibility which holds between premises *A*, *B*, *C*, *D*, . . . *R*, *S*, *T*, . . . and the conclusion *X* (cf. no. 24) must also be irredundant. For example,

the relation of deducibility between the premises 'all α are β', 'all β are γ' and the conclusion 'all α are γ', also the relation of deducibility between premises 'all α are γ', 'all γ are β' and the conclusion 'all α are β' are without doubt irredundant. But the relation of deducibility between the same premises 'all α are β', 'all β are γ' and 'all γ are β' and the conclusion 'all α are β' is not irredundant.

33. But there can be no doubt that *sometimes* when the conclusion *M* is irredundantly deducible from premises *A*, *B*, *C*, *D*, . . . and the conclusion *X* from premises *M*, *R*, *S*, *T*, . . ., the relation of deducibility between conclusion *X* and premises *A*, *B*, *C*, *D*, . . . *R*, *S*, *T*, . . . is also irredundant. If, in the preceding example, we replace the last premise 'all γ are β' by the premise 'all γ are δ', and replace the conclusion 'all α are β' by the conclusion 'all α are δ', then all three of the indicated deductions will be irredundant. Whenever it is the case that the relations of deducibility between *A*, *B*, *C*, *D*, . . . and *M*, *R*, *S*, *T*, . . . and *X* and the resulting relation between *A*, *B*, *C*, *D*, . . . *R*, *S*, *T*, . . . and *X* are all irredundant, I wish to call the latter relation *compounded* from the other two. A relation of deducibility which is not compounded in this way I wish to call *simple*.

34. There must be simple relations of deducibility. It is very likely that, for example, the relation of the two premises, 'all α are β', 'all β are γ' to the conclusion 'all α are γ' is simple. It is unlikely that anybody can indicate a proposition which is irredundantly deducible from any one of these premises or from both of them together, and which is such that the above conclusion can be irredundantly deduced from it alone, or together with a second one.

35. If two propositions '*A* has *x*' and '*B* has *x*' have the same predicate, which is also the only variable idea in them, then the second is deducible from the first if the subject idea of the first, *A*, stands in the relation of inclusion to the subject idea of the second, *B* (§ 95). If this is not the case, then they don't stand in this relation. For, if idea *A* includes idea *B*, hence if every *B* is an *A*, then every idea whose substitution for *x* makes proposition '*A* has *x*' true, must also make proposition '*B* has *x*' true. Conversely, if *A* does not include *B*, hence if there is some *B* which is not *A*, there must also be an attribute which applies exclusively to this *B*. Let us call this attribute *b'*; then the attribute non-*b'* is an attribute which holds of all *A*, but not of all *B*. Hence, if we replace *x* by 'attribute non-*b''*, proposition '*A* has *x*' will be true, but proposition '*B* has *x*' will be false.

36. If two propositions '*X* has *a*', and '*X* has *b*' have the same subject, which is also the only variable idea in them, then the second proposition is deducible from the first if the idea *B* (the *concretum*

belonging to b) includes idea A; failing this, the relation of deducibility does not hold. For, if idea B includes idea A, hence if every A is also a B, then every idea whose substitution for X makes the proposition 'X has a' true, also makes the proposition 'X has b' true. On the other hand, if B does not include A, hence, if there is some A which is not also a B, then there will be some idea which applies exclusively to this A. Let this idea be A'; then the idea A', when put in the place of X, will make the proposition 'X has a', but not the proposition 'X has b', true.

NOTE 1: The relation of deducibility between propositions which I have here described is too obvious and important for the discovery of new truths to have escaped logicians altogether. Rather, the development of this concept (in the usual chapter on arguments) forms the most important part of the elements of logic. On the other hand, it seems to me that the nature of this relation has not always been properly grasped. Where it was comprehended, it was not discussed with sufficient generality, or no precise definition was given. It seems to me that deducibility of propositions from each other is a relation which holds *objectively*, i.e. regardless of our faculties of representation and understanding, and it should be discussed accordingly. This was not generally done; deducibility was described as a relation between *judgments* (i.e. thought and accepted propositions), and it was said that this relation consists in the fact that the acceptance of one proposition brought about the acceptance of another. It also seems to me that the relation of deducibility should not be confounded with what I shall later call the ground-consequent relation [*Abfolge*]; this relation, I believe, does not hold between propositions in general, but only between *truths*. This distinction has so far not been made since it was not considered necessary to distinguish propositions and judgments in themselves from their appearances in the mind (from judgments and cognitions). Also, it was generally assumed that an immediate relation of deducibility could hold only between two, or at most three, propositions, namely two premises and one conclusion. In my opinion there is an indefinitely large number of both, premises and conclusions, even where the relation of deducibility is *irredundant*. Finally, among the *definitions* of this concept found in other books one of the best is that of Aristotle (*Anal. Priora*, 24^b18, and *Topica* 100^a25, etc.): "a syllogism is a discourse (λόγος) in which, certain things being stated, something other than what is stated follows of necessity from their being so". Since there can be no doubt that Aristotle assumed that the relation of deducibility can also hold between false propositions, the 'follows of necessity' can hardly be interpreted in any other way than this: that the conclusion becomes true *whenever* the premises are true. Now it is obvious that we cannot say of one and the same class of propositions that one of them becomes true *whenever* the others are true, unless we envisage some of their parts as variable. For propositions none of whose parts change are not sometimes true and sometimes false; they are always one or the other. Hence when it was said of certain propositions that one of them becomes true as soon as the others do, the actual reference was not to these propositions themselves, but to a relation which holds between the infinitely many propositions which can be generated from them, if certain of their ideas are replaced by arbitrarily chosen other ideas. The desired formulation was this: as soon as the exchange of certain ideas makes the premises true, the conclusion must also become true. . . .

NOTE 2: In this note, B. attempts to justify the terms 'deducibility' [*Ableitbarkeit*] and 'inclusion' [*Umfassen*].

§ 156. b. *Relations of Equivalence*

If the relation of deducibility holds reciprocally between propositions *A*, *B*, *C*, *D*, . . . and *M*, *N*, *O*, . . . with respect to the same ideas *i*, *j*, . . . then I say that these propositions stand in the relation of equivalence. In other words, if every class of ideas whose substitution for *i*, *j*, . . . makes all of *A*, *B*, *C*, *D*, . . . true also makes all of *M*, *N*, *O*, . . . true, and, conversely, when every class of ideas whose substitution for *i*, *j*, . . . makes all of *M*, *N*, *O*, . . . true also makes all of *A*, *B*, *C*, *D*, . . . true, then I say that propositions *A*, *B*, *C*, *D*, . . . and *M*, *N*, *O*, . . . stand in the relation of equivalence with respect to the ideas *i*, *j*. . . . Thus I say that the proposition 'every *A* has *b*' is equivalent to the two propositions 'the idea *A* has a referent' and 'the idea of an *A* which does not have *b* does not have a referent', provided that ideas *A* and *b* are considered variable; for, on this assumption, the last two propositions are deducible from the first, and the first is deducible from them.

2. If propositions *A*, *B*, *C*, *D*, . . . are *together* equivalent to propositions *M*, *N*, *O*, . . ., also taken together, then according to this definition it is not required that for each individual proposition in *A*, *B*, *C*, *D*, . . . there should be an equivalent proposition in *M*, *N*, *O*. . . . It is not even necessary that the number of propositions *A*, *B*, *C*, *D*, . . . should be the same as the number of propositions *M*, *N*, *O*, . . ., as can be seen from the above example.

3. If propositions A, B, C, D, . . . are all true, then the equivalent propositions M, N, O, . . . must also be all true. Conversely, if the former are not all true, then the latter are also not all true.

4. If propositions A, B, C, . . . taken together, are equivalent to propositions A', B', C', . . ., also taken together, with respect to ideas i, j, . . ., and if propositions A, B, C, . . . stand in any one of the relations so far considered to certain other propositions M, N, O, . . . then the equivalent propositions A', B', C', . . . stand in the same relation to M, N, O, . . . with respect to the same ideas. This holds with respect to the relations of compatibility, incompatibility, equivalence, and deducibility, no matter whether M, N, O, . . . is deducible from A, B, C, . . . or A, B, C, from M, N, O. This can be shown in the following way: it is assumed that propositions A, B, C, . . . are equivalent to propositions A', B', C', . . . with respect to ideas i, j, . . ., and it is also assumed that i, j, . . . are the variable ideas in the relation between A, B, C, . . . and M, N, O. . . . Hence every class of ideas whose substitution for i, j, . . . in A, B, C, . . . and M, N, O, . . . makes all of A, B, C, . . . true also makes all of A', B', C', . . . true, and any class of ideas which fails to do so with A, B, C, . . . can also not accomplish it with A', B', C'. . . . And since the kind of relation between A, B, C, . . . on the one hand and M, N, O, . . . on the other depends entirely upon this circumstance, it is obvious that the same relation holds between A'. B', C', . . . and M, N, O. . . .

5. If propositions A, B, C, . . . are equivalent to A', B', C', . . . and propositions D, E, F, . . . equivalent to D', E', F', . . . with respect to the same ideas i, j, . . ., and if propositions A, B, C, . . . are compatible with propositions D, E, F, . . . with respect to the same ideas, then propositions A, B, C, . . . D, E, F, . . . taken together, are equivalent to propositions A', B', C', . . . D', E', F', . . ., also taken together, and again with respect to the same ideas. For, if propositions A, B, C, . . . are compatible with propositions D, E, F, . . . with respect to ideas i, j, . . . , then there are ideas whose substitution for i, j, . . . makes all of A, B, C, . . . D, E, F, . . . true. But these very same ideas will also make all of A', B', C', . . . D', E', F', . . . true, since all ideas which make A, B, C, . . . true also make A', B', C', . . . true, and all ideas which make D, E, F, . . . true also make D', E', F', . . . true. In a similar way we can prove that all ideas whose substitution for i, j, . . . makes all of A', B', C', . . . D', E', F', . . . true, also makes all of A, B, C, . . . D, E, F, . . . true. It follows that the two classes stand in the relation of equivalence to each other.

6. If propositions *A*, *B*, *C*, . . . are equivalent to propositions *A′*, *B′*, *C′*, . . . with respect to ideas *i*, *j*, . . . and if propositions *M*, *N*, *O*, . . . are deducible from propositions *A*, *B*, *C*, . . . with respect to the same ideas, then *A*, *B*, *C*, . . . *M*, *N*, *O*, . . . taken together stand in the relation of equivalence to propositions *A′*, *B′*, *C′*, . . . with respect to the same ideas. For, the totality of propositions *A*, *B*, *C*, . . . *M*, *N*, *O*, . . . are made true by all and only those ideas which can make propositions *A′*, *B′*, *C′*, . . . true.

7. Merely from the fact that propositions *A*, *B*, *C*, . . . are equivalent to propositions *A′*, *B′*, *C′*, . . . with respect to ideas *i*, *j*, . . . it does not follow that propositions *M*, *N*, *O*, . . . and *M′*, *N′*, *O′*, . . ., which are deducible one set from the first, the other from the second of these two classes, also with respect to *i*, *j*, . . , are also equivalent with respect to the same ideas. For, if *M*, *N*, *O*, . . . is merely deducible from *A*, *B*, *C*, . . . but not equivalent to it, then *M*, *N*, *O*, . . . can become true more often than *A*, *B*, *C*. . . . Now if *M′*, *N′*, *O′*, . . . is equivalent to *A′*, *B′*, *C′*, . . . or becomes true more often, but with different ideas than *M*, *N*, *O*, . . ., then *M*, *N*, *O*, . . . and *M′*, *N′*, *O′*, . . ., are in no way equivalent. Thus the two propositions 'this figure is an equilateral triangle' and 'this figure is an equiangular triangle' are equivalent with respect to the variable idea 'this'. But two propositions which are deducible from them namely 'this figure is equilateral' and 'this figure is equiangular' are not at all equivalent with respect to the idea 'this'.

8. Assume that the sum of propositions *A*, *B*, *C*, *D*, . . . is equivalent to the sum of propositions *A′*, *B′*, *C′*, *D′*, . . . with respect to ideas *i*, *j*, . . . and that a part of the first sum, e.g. *A*, *B*, . . ., is itself equivalent to a part of the second sum, e.g. *A′*, *B′*, . . . with respect to the same ideas. It does then not necessarily follow that the remaining part *C*, *D*, . . . and *C′*, *D′*, . . . are equivalent with respect to the same ideas. For, propositions *C*, *D*, . . . may be merely deducible from *A*, *B*, . . ., and propositions *C′*, *D′*, . . . from *A′*, *B′*. . . .

9. If two propositions *A* and *A′* are equivalent with respect to ideas *i*, *j*, . . . and if neither of them is formally true, then propositions *Neg.A* and *Neg.A′* are also equivalent with respect to the same ideas. For, if one of the two propositions, e.g. *A*, is not formally true, then there are certain ideas whose substitution for *i*, *j*, . . . make it false, hence make the proposition *Neg.A* true. But these very same ideas must also make proposition *A′* false and hence *Neg.A* true, since otherwise *A* would not be deducible from *A′*. It follows in the same way that every class of ideas which makes *Neg.A′* true also makes proposition *Neg.A* true.

10. The same does not hold for whole classes of propositions. If the class of propositions *A*, *B*, *C*, . . . is equivalent to the class of propositions *A′*, *B′*, *C′*, . . ., then it does not follow that the class of propositions *Neg.A*, *Neg.B*, *Neg.C*, . . . is equivalent to the class *Neg.A′*, *Neg.B′*, *Neg.C′*, . . . with respect to the same ideas. Thus the two propositions 'all *A* are *B*' and 'all *B* are *A*', taken together, are obviously equivalent to the following proposition 'every object of one of the two ideas *A*, *B* is also an object of both of them', where ideas *A*, *B* are considered variable. On the other hand, the two propositions which result from the denial of the first two, namely 'it is false that all *A* are *B*' and 'it is false that all *B* are *A*' are not equivalent to the proposition 'it is false that every object of one of the two ideas *A*, *B* is also an object of both of them'. For the latter proposition becomes true without the first two becoming true together, e.g. if idea *A* is higher than *B*, which makes the first proposition true, but the second false.

11. If two propositions '*A* has *x*', and '*B* has *x*' have the same predicate, which is to be considered the only variable idea in them, then they are equivalent if, and only if, their subject ideas *A* and *B* are equivalent (cf. § 96). For, if the ideas *A* and *B* are equivalent, then every object which falls under one of them also falls under both; hence any attribute whose substitution for *x* makes one proposition true, will also make the other true. But if *A* and *B* are not equivalent, then one of them, e.g. *A*, must represent an object which is not represented by *B*. Hence there must be some attribute which is peculiar to this object. Let this attribute be *a′*, then the attribute non-*a′* will apply to all *B*. Hence if we substitute the idea 'attribute non-*a′*' for *x*, then the proposition '*B* has *x*', but not the proposition '*A* has *x*', becomes true.

12. If two propositions '*X* has *a*' and '*X* has *b*' have the same subject, which is to be the only variable idea in them, then they are equivalent if the ideas *A* and *B* are equivalent, otherwise not. For, if ideas *A* and *B* are equivalent, hence if every object of one of them also falls under the other, then every idea whose substitution for *X* makes one proposition true, must also make the other true. But if *A* and *B* are not equivalent, then there is some object which falls under one of them, e.g. *A*, which does not fall under the other, *B*. There will then also be an idea which applies exclusively to this object. Let this idea be *A′*. If we replace *X* by the idea *A′* then the proposition '*X* has *a*', but not the proposition '*X* has *b*' becomes true.

13. From the mere fact that two propositions *A* and *A′* are composed in an analogous way out of parts (propositions or individual

ideas) which are equivalent to each other, it does not follow that they themselves are equivalent. Thus the two propositions 'an equilateral triangle is also equiangular' and 'an equiangular triangle is also equilateral' are in no way equivalent.

In a note, B. discusses the definitions of equivalence offered by a number of his contemporaries and predecessors.

§ 157. c. *The Relation of Subordination*

1. Let us assume that a relation of deducibility between propositions *A*, *B*, *C*, *D*, . . . and *M*, *N*, *O*, . . . is not reciprocal, as in the preceding section, but holds only in one direction, e.g. such that propositions *M*, *N*, *O*, . . . are deducible from *A*, *B*, *C*, *D*, . . . but not conversely, with respect to certain variable ideas *i*, *j*, . . . or (what comes to the same), if every class of ideas whose substitution for *i*, *j*, . . . makes propositions *A*, *B*, *C*, *D* . . . true, also makes propositions *M*, *N*, *O*, . . . true, but not conversely. In such a case we call the relation between propositions *A*, *B*, *C*, *D*, . . . on one hand and *M*, *N*, *O*, . . . on the other a relation of *subordination.* Because of the similarity of this relation with the relation between ideas discussed in § 97, I wish to call propositions *A*, *B*, *C*, *D*, . . . the *subordinated* or *lower*, or, if this is objectionable, propositions of *smaller degree of satisfiability* or *more limited* propositions, or propositions which *claim more.* On the other hand, I call propositions *M*, *N*, *O*, . . . *superordinated*, or *higher* or *propositions* with a *greater degree of satisfiability*, or propositions which *claim less.* It might be even less objectionable to say that propositions *M*, *N*, *O*, . . . are unilaterally deducible from propositions *A*, *B*, *C*, *D*, . . .; the latter are called the unilateral premises, the former the unilateral conclusions. Thus the proposition '*A* is *C*' is deducible from the two propositions '*A* is *B*', '*B* is *C*', provided ideas *A*, *B*, *C*, . . . are envisaged as variable. But conversely, the first two propositions '*A* is *B*' and '*B* is *C*' are not deducible under the same conditions from '*A* is *C*'. Thus a relation of subordination holds between the propositions '*A* is *B*' and '*B* is *C*' on one hand and '*A* is *C*' on the other. I call the first two propositions subordinated, and the last proposition superordinated with respect to them; I attribute a lower degree of satisfiability to the former than to the latter.

2. If propositions *M*, *N*, *O*, . . . are unilaterally deducible from propositions *A*, *B*, *C*, . . . with respect to ideas *i*, *j*, . . . then there are always certain propositions which are compatible with *M*, *N*, *O*, . . . without being compatible with *A*, *B*, *C*, . . . with respect to

the same ideas *i*, *j*. . . . Since there are ideas whose substitution for *i*, *j*, . . . makes all of *M*, *N*, *O*, . . . true without making all of *A*, *B*, *C*, . . . true, there will be ideas which will make true all of *M*, *N*, *O*, . . . together with one of the propositions *Neg.A*, *Neg.B*, *Neg.C*. . . . Let *Neg.A* be one such proposition; then *M*, *N*, *O*, . . . and *Neg.A* are compatible. But it is obvious that *Neg.A* and *A*, *B*, *C*, . . . are not compatible.

3. If propositions *M*, *N*, *O*, . . . are unilaterally deducible from propositions *A*, *B*, *C*, *D*, . . ., and propositions *R*, *S*, *T*, . . . are also unilaterally deducible from propositions *M*, *N*, *O*, . . . with respect to the same ideas *i*, *j*, . . ., then propositions *R*, *S*, *T*, . . . are unilaterally deducible from propositions *A*, *B*, *C*, *D*, . . . with respect to the same ideas. That *R*, *S*, *T*, . . . is deducible from *A*, *B*, *C*, *D*, . . . follows from § 155, no. 23, and that this deducibility is merely unilateral follows from the fact that there are certain ideas which make all of *R*, *S*, *T*, . . . true without making all of *M*, *N*, *O*, . . . true, and these cannot make all of *A*, *B*, *C*, *D*, . . . true, for if they did, *M*, *N*, *O*, . . . would also have to become true.

4. If two propositions '*A* has *x*' and '*B* has *x*' have one and the same predicate, which is to be considered variable, then the second is unilaterally deducible from the first if, and only if, its subject idea, *B*, is subordinate to the subject idea of the first, *A*. For, if idea *B* is subordinate to *A* then, according to § 155, no. 35, the proposition '*B* has *x*' is deducible from the proposition '*A* has *x*', but not conversely. Hence, if the proposition '*B* has *x*' is to be deducible from the proposition '*A* has *x*', but not vice versa, then, according to § 155, no. 35, the idea *A* must include *B*, but *B* must not include *A*; hence *B* must be subordinate to *A*.

5. If two propositions '*X* has *a*' and '*X* has *b*' have the same subject, which is to be the only variable part of them, then the second is unilaterally deducible from the first if, and only if, the idea *A* is subordinate to the idea *B*. This follows from § 155, no. 36.

§ 158. d. *The Relation of Overlapping*

1. It remains to consider the case where there is a relation of compatibility between propositions *A*, *B*, *C*, . . . and *M*, *N*, *O*, . . ., but in such a way that neither propositions *M*, *N*, *O*, . . . are deducible from *A*, *B*, *C*, . . . nor, conversely, *M*, *N*, *O*, . . . from *A*, *B*, *C*, . . . with respect to the same ideas *i*, *j*, . . .; in other words, where there are ideas whose substitution for *i*, *j*, . . . makes all of *A*, *B*, *C*, . . . and *M*, *N*, *O*, . . . true, but others which make only *A*, *B*, *C*, . . .

but not all of *M*, *N*, *O*, . . . true, and still others which make only *M*, *N*, *O*, . . . but not all of *A*, *B*, *C*, . . . true. This relation has great similarity with the relation which I have called, in the case of ideas, *overlapping* or *linking* (§ 98). Hence it is reasonable to give it the same name in the case of propositions, unless the designation *independence* is preferred.

2. Every pair of propositions of the form 'every *X* has *y*', and 'every *Y* has *x*', where the only variables are the ideas *x* and *y*, form a pair of independent propositions: there are certainly ideas whose substitution for *x* and *y* will make these two propositions true together; this holds, for example, whenever *X* and *Y* are equivalent ideas. But there are also other ideas whose substitution for *x* and *y* will make only one proposition true. For, if we choose for *x* and *y* a pair of attributes, such that *Y* is higher than *X*, then the proposition 'every *X* is *Y*' is true and the proposition 'every *Y* is *X*' is false. But if we choose to replace *x* and *y* in such a way that *X* is higher than *Y*, then the proposition 'every *Y* is *X*' is true and the proposition 'every *X* is *Y*' is false.

3. If a pair of propositions '*A* has *x*' and '*B* has *x*' have one and the same predicate, which is to be their only variable, then they stand in the relation of independence if and only if the ideas *A* and *B* are either linked or incompatible. For, according to § 154, no. 16, both these propositions can become true if *A* and *B* have reference. In order that any one of them can become true by itself it is necessary that the ideas *A* and *B* do not stand in the relation of subordination, since otherwise one proposition would be deducible from the other (cf. § 155, no. 35). Hence ideas *A* and *B* must be either linked or incompatible.

4. If two propositions '*X* has *a*' and '*X* has *b*' have one and the same subject, which is to be their only variable part, then they stand in the relation of independence if, and only if, the ideas *A* and *B* are linked. For, if ideas *A* and *B* are linked, then there are objects which fall under both of them. Hence if we replace *X* by an idea which applies exclusively to such objects, both propositions will become true. But there are also objects which fall under one, but not under the other, of these ideas, and an idea which applies exclusively to such objects will make one proposition true but not the other. But if the ideas *A* and *B* are not linked, then it is either the case that one or both of them are non-referring, and then one or both of these propositions are formally false; or else the ideas *A* and *B* are incompatible with each other and then, according to § 155, no. 18, so are the propositions; or one of these ideas is subordinate to the other

and then, according to § 155, no. 36, one of the two propositions is deducible from the other.

5. The members of some, but not all, pairs of independent propositions can become simultaneously false with respect to certain variable ideas. An example of a pair of independent propositions which can never be both false are the following 'every *X* has *a*' and 'it is false that every *X* has the attributes $a+b$', where the idea *X* is the only variable, and where *a* and *b* designate a pair of attributes which are sometimes, but not always, found together. Both of these propositions become true if we replace *X* by an idea which refers exclusively to things which have *a* but not *b*. The first proposition alone will become true if we replace *X* by an idea which applies only to objects which have attribute *b* in addition *a*. Finally, the second proposition alone becomes true if we replace *X* by the idea of objects which have neither the attribute *a* nor *b*. Hence the two propositions are independent. But it is easy to see that the two propositions cannot become false together since the truth of the second follows from the falsity of the first. For if it is false that every *X* is an *A*, then it is certainly also false that every *X* is an [*A*]*b*. Finally, an example of a pair of independent propositions that can become false together are the two propositions considered in no. 2: 'every *X* is *Y*' and 'every *Y* is *X*'. Both of these propositions become false if we replace *X* and *Y* by a pair of mutually exclusive ideas.

6. If a pair of propositions *A* and *B* are independent, then their negations, the propositions *Neg.A* and *Neg.B*, are also either independent or incompatible, all with respect to the same ideas. For, if *A* and *B* are independent and of the kind that cannot be false together (no. 5), then propositions *Neg.A* and *Neg.B* are incompatible. But if they can both be false, then *Neg.A* and *Neg.B* stand in the relation of independence, for then there are ideas which will make them both true, and there will also be ideas which will make only one of them true. The latter result will be brought about by ideas which make only one of the two propositions, *A* or *B* true, but the other one false.

7. Propositions which are equivalent to independent propositions are themselves independent, all with respect to the same ideas. The proof is similar to § 156, no. 4.

§ 159. *Special Kinds of Incompatibility*

1. We have seen that there are several sub-varieties to the relation of compatibility; they were the subject of the preceding discussion

(§§ 155–158). Similar distinctions can be made with the relation of *incompatibility*. If we say of several propositions *A*, *B*, *C*, *D*, . . . merely that they stand in the relation of incompatibility to each other with respect to ideas *i*, *j*, . . ., then all we are saying is that there are no ideas whose substitution for *i*, *j*, . . . will make all of the propositions *A*, *B*, *C*, *D*, . . . true together. But by saying that *A*, *B*, *C*, *D*, . . . are incompatible with each other, we are not claiming that there may not be some of them, e.g. *A*, *B*, . . ., or *B*, *C*, . . . which are made true through certain common ideas, without, respectively, *C*, *D*, . . . or *A*, *B*, . . . becoming true also.

In § 155 we considered classes of compatible propositions *A*, *B*, *C*, *D*, . . . *M*, *N*, *O*, . . . and asked the question whether there are not some of them, *A*, *B*, *C*, . . ., which are of such a nature that every class of ideas whose substitution for the variables *i*, *j*, . . . makes all of them true, will also make one or several others *M*, *N*, *O*, . . . true. Let us now ask whether among several incompatible propositions *A*, *B*, *C*, *D*, . . . and *M*, *N*, *O*, . . . there may not be some *A*, *B*, *C*, . . . which are of such a nature that every class of ideas whose substitution for *i*, *j*, . . . makes all of them true, will make certain others, *M*, *N*, *O*, . . . *false*. If this is the case, then the relation of the propositions *M*, *N*, *O*, . . . to the propositions *A*, *B*, *C*, . . . is the exact opposite of the relation which we have previously called *deducibility*. I wish to call it the relation of *exclusion*; I shall say that one or several propositions *M*, *N*, *O*, . . . are *excluded* by certain others *A*, *B*, *C*, . . . with respect to variable ideas *i*, *j*, . . . if every class of ideas whose substitution for *i*, *j*, . . . makes all of *A*, *B*, *C*, . . . true, makes all of *M*, *N*, *O*, . . . false. *A*, *B*, *C*, . . . I call *excluding* propositions, *M*, *N*, *O*, . . . *excluded* propositions. Such a relation of exclusion holds, for example, between the propositions '*A* is *B*' and '*B* is *C*' on one hand and the proposition 'No *C* is *A*' on the other, if *A*, *B*, *C* are envisaged as the only variable ideas. For every class of ideas which make the first two propositions true will make the third one false. Hence I call the first two propositions excluding, and the last proposition excluded by them.

2. If propositions *A*, *B*, *C*, . . . exclude certain others *M*, *N*, *O*,... with respect to certain ideas *i*, *j*, . . ., then the negations of the latter, i.e. the propositions *Neg.M*, *Neg.N*, *Neg.O*, . . . must be deducible from the class of propositions *A*, *B*, *C*, . . . with respect to the same ideas. For every class of ideas whose substitution for *i*, *j*, . . . will make all of *A*, *B*, *C*, *D*, . . . true, must make all of *M*, *N*, *O*, . . . false, hence must make all of the propositions *Neg.M*, *Neg.N*, *Neg.O*, . . . true.

3. It can also be the case that the relation of exclusion holds mutually and with respect to the same ideas *i*, *j*, . . . between propositions *A*, *B*, *C*, . . . and *M*, *N*, *O*. . . . In this case every class of ideas which make all of *A*, *B*, *C*, . . . true will make all of *M*, *N*, *O*, . . . false, and every class of ideas which will make all of *M*, *N*, *O*, . . . true will make all of *A*, *B*, *C*, . . . false. We can properly call this relation between propositions *A*, *B*, *C*, . . . and *M*, *N*, *O*, . . . a relation of ***mutual exclusion***. The two propositions '*A* is as old as *C*', and '*B* is three times as old as *C*' stand in this relation to the following two propositions '*A* and *B* together are seven times as old as *C*' and '*B* is as old as *A* and *C* together', where *A*, *B*, *C* are envisaged as the only variable ideas. For, if the first two propositions are true, then the last two are false, and conversely.

4. When we say that propositions *A*, *B*, *C*, . . . and *M*, *N*, *O*, . . . are mutually exclusive with respect to ideas *i*, *j*, . . . then all that is claimed is that as soon as all of *A*, *B*, *C*, . . . are true, all of *M*, *N*, *O*, . . . are false, and as soon as all of *M*, *N*, *O*, . . . are true, all of *A*, *B*, *C*, . . . are false; in other words that propositions *Neg.M*, *Neg.N*, *Neg.O*, . . . are deducible from propositions *M*, *N*, *O*. . . . In such a case, nothing is said whether this double deducibility is itself unilateral or mutual. Let us consider the latter case; let us assume that the relation of deducibility between propositions *A*, *B*, *C*, . . . and *Neg.M*, *Neg.N*, *Neg.O*, . . ., and between the propositions *M*, *N*, *O*, . . . and *Neg.A*, *Neg.B*, *Neg.C*, . . . is mutual, i.e. that every class of ideas whose substitution for *i*, *j*, . . . makes all of *A*, *B*, *C*, . . . true or false will also make all of *M*, *N*, *O*, . . . false or true respectively. In this case we say that propositions *A*, *B*, *C*, . . . on the one hand, and *M*, *N*, *O*, . . . on the other stand in the relation of ***contradiction***, or that they contradict each other. We also say that propositions *A*, *B*, *C*, . . . taken together are the contradictory of *M*, *N*, *O*, . . ., and that the latter are the contradictory of the former. But if the relation of deducibility holds only in one direction, then we say that propositions *A*, *B*, *C*, . . . on one hand and *M*, *N*, *O*, . . . on the other merely stand in the relation of contrariety. The two propositions

'every *X* is *Y*' (*A*)
'the idea of a non-*Y* has a referent' (*B*)

stand in the relation of contradiction to the following two:

'it is false that every non-*Y* is a non-*X*' (*M*)
'the idea of an *X* does not have a referent' (*N*)

where ideas X, Y, are envisaged as variable. It is obvious that in this case the falsity of M and N follows from the truth of propositions A and B. For, if it is true that every X is a Y, then the idea X must have a referent, and hence the proposition that it has no referent must be false. Furthermore, if the idea non-Y has a referent (as is claimed by proposition B), then it must also be true that every non-Y is a non-X; hence proposition M must be false. In a similar way the falsity of propositions A and B follows from the truth of propositions M and N, for, if the idea X does not have a referent, as is claimed by N, the proposition 'every X is Y' cannot be true. Also, if the idea X does not have a referent, the idea non-X has the extension of the widest idea, namely of the idea 'something in general'. Hence the proposition 'every non-Y is a non-X' can fall short of truth (as is claimed by M) only if the idea non-Y has no object at all; consequently proposition B is false. But we can also show that the truth of M and N can be deduced from the falsity of propositions A and B. From the falsity of B the truth of M follows at once; for, if the idea non-Y does not have a referent, there is no doubt that the proposition 'every non-Y is a non-X' lacks truth. And if non-Y does not have a referent, it follows that the idea Y has the extension of the widest idea, namely of 'something in general'. Hence proposition A: 'every X is a Y' can fall short of truth only because the idea X itself does not have any object, which is exactly what is claimed in proposition N; hence N is true. Finally, from the falsity of propositions M, N we can deduce the truth of propositions A and B. For, if M is false, then it is true that every non-Y is a non-X; hence the idea non-Y must have a referent, as is claimed by B. Furthermore, if proposition N is false, then the idea X must have a referent; hence from the implied proposition that every non-Y is a non-X we can safely draw the conclusion that every X is a Y, as A claims.

Many readers will appreciate a mathematical example of contradiction especially one which can serve as a pattern for many others. Let the sign $\neq$ indicate an inequality between given magnitudes, where it is left undetermined which of them is larger; then the following two sets of six propositions stand in the relation of contradiction, given that the ideas a, b, c, d, and m are envisaged as variable. A resolution of the equations which are contained in them or their negations shows this quite clearly.

$$a+b+c = 3m; \quad 2b+c = 3m; \quad 2c+b = 3m;$$
$$a+b+d \neq 3m; \quad 2a+d \neq 3m; \quad 2d+a \neq 3m;$$

Are the joint contradictories of

$$a+d = 2m; \quad 2b+d = 3m; \quad 2d+b = 3m;$$
$$b+c \neq 2m; \quad 2c+c \neq 3m; \quad 2a+c \neq 3m.$$

An example of propositions which merely stand in the relation of contrariety to each other is found in no. 3 above.

5. It follows directly from the definition of no. 4 that if propositions *A*, *B*, *C*, . . . stand in the relation of contradiction to propositions *M*, *N*, *O*, . . . then propositions *A*, *B*, *C*, . . . must be equivalent to *Neg.M*, *Neg.N*, *Neg.O* . . .; and propositions *M*, *N*, *O*, . . . equivalent to *Neg.A*, *Neg.B*, *Neg.C*, . . ., all with respect to the same variable ideas. Similarly, if propositions *A*, *B*, *C*, . . . are all true or all false, then propositions *M*, *N*, *O*, . . . are all false or all true.

6. Among propositions which stand in the relation of contradiction to each other there can be no proposition which is formally true or formally false with respect to the same variable ideas. For, according to our definition, it must be possible to make any of a set of contradictory propositions true as well as false.

7. Any arbitrary proposition *A* and its negation *Neg.A* stand in the relation of contradiction to each other with respect to any ideas whatsoever, provided only that *A* is neither formally true nor false. For, if the latter is not the case, then there are ideas which make *A* true, and others which make *A* false. But all ideas which make *A* true (or false) make *Neg.A* false (or true) and vice versa.

8. If propositions *A*, *B*, *C*, . . . stand in the relation of contradiction to propositions *M*, *N*, *O*, . . . then propositions *Neg.A*, *Neg.B*, *Neg.C*, . . . stand in the same relation to propositions *Neg.M*, *Neg.N*, *Neg.O*, all with respect to the same ideas.

9. If some propositions stand in a relation of contradiction to each other, then all propositions equivalent to them stand in the same relation to each other, with respect to the same ideas. This follows from § 156, no. 4.

10. If propositions *A*, *B*, *C*, *D*, . . . are contradictory to *M*, *N*, *O*, . . . and *M*, *N*, *O*, . . . contradictory to *R*, *S*, *T*, . . . all with respect to the same ideas, then propositions *A*, *B*, *C*, . . . and propositions *R*, *S*, *T*, . . . are equivalent to each other with respect to the same ideas. For the same ideas which make all of *A*, *B*, *C*, . . . true or false make all of *M*, *N*, *O*, . . . false or true and consequently all of *R*, *S*, *T*, . . . true or false.

11. Let propositions *A*, *B*, *C*, . . . be contradictory to propositions *M*, *N*, *O*, . . . and propositions *E*, *F*, *G*, . . . contradictory to *P*, *Q*, *R*, . . . with respect to the same ideas; furthermore let propositions *A*, *B*, *C*, *D*, . . . *E*, *F*, *G*, . . ., as well as *Neg.A*, *Neg.B*, *Neg.C*, . . .

Neg.E, *Neg.F*, *Neg.G*, . . . be compatible with each other with respect to the same ideas. In this case, the totality of propositions *A*, *B*, *C*, . . . *E*, *F*, *G*, . . . stand in the relation of contradiction to the totality of propositions *M*, *N*, *O*, . . . *P*, *Q*, *R*, . . ., again with respect to the same ideas. Since propositions *A*, *B*, *C*, . . . *E*, *F*, *G*, . . . are to be compatible with each other, there must be ideas whose substitution for *i*, *j*, . . . makes all of them true. But since *A*, *B*, *C*, . . . are to be contradictory to *M*, *N*, *O*, . . . and *E*, *F*, *G*, . . . contradictory to *P*, *Q*, *R*, . . . these same ideas must make all of *M*, *N*, *O*, . . . *P*, *Q*, *R*, . . . false. Furthermore since all of *Neg.A*, *Neg.B*, *Neg.C*, . . . *Neg.E*, *Neg.F*, *Neg.G*, . . . are to be compatible with each other, there are ideas which make all of them true; but in this case all of *M*, *N*, *O*, . . . *P*, *Q*, *R*, . . . must also become true. In the same way we can show that the truth or falsity of all of *M*, *N*, *O*, . . . *P*, *Q*, *R*, . . . entails the falsity or truth of all of *A*, *B*, *C*, . . . *E*, *F*, *G*. . . .

12. The twofold condition which was just made, namely that propositions *A*, *B*, *C*, . . . *E*, *F*, *G*, . . ., as well as their negations, *Neg.A*, *Neg.B*, *Neg.C*, . . . *Neg.E*, *Neg.F*, *Neg.G*, . . . should be compatible with each other, is by no means superfluous. For, if propositions *A*, *B*, *C*, . . . are not compatible with propositions *E*, *F*, *G*, . . . then it can obviously not be the case that the totality of these propositions, namely *A*, *B*, *C*, . . . *E*, *F*, *G*, . . . should stand in the relation of contradiction to a certain other class of propositions, since the definition of this relation requires that the above propositions should be compatible. But from the compatibility of certain propositions *A*, *B*, *C*, . . . *E*, *F*, *G*, . . . with each other, it does not follow that their negations are also compatible (cf. § 154, no. 14); hence we must explicitly stipulate this condition since, by definition, it is required for a relation of contradiction.

13. Given that the totality of propositions *A*, *B*, *C*, . . . *E*, *F*, *G*, . . . stands in relation of contradiction to the totality of propositions *M*, *N*, *O*, . . . *P*, *Q*, *R*, . . ., and that part of the first, *A*, *B*, *C*, . . ., stands in the relation of contradiction of part of the second, *M*, *N*, *O*, . . ., with respect to the same ideas. It does not follow from this that the remaining propositions, *E*, *F*, *G*, . . ., also stand in the relation of contradiction to the remainder, *P*, *Q*, *R*. . . . For to say that certain propositions stand in the relation of contradiction to certain others means no more than that they are equivalent to the negations of the others and conversely. But from the equivalence of two classes of propositions and parts of them we may not draw any conclusions concerning the equivalence of the remainders (cf. § 156, no. 8).

14. Given that the relation of contradiction holds between a class of several propositions *A*, *B*, *C*, *D*, . . . and a certain other class *M*, *N*, *O*, . . . rather than between an individual proposition and a class of them, then, if *A*, *B*, *C*, *D*, . . . are not equivalent to each other, there are ideas whose substitution for the variables *i*, *j*, . . . will not make all, but only part of the propositions *A*, *B*, *C*, *D*, . . . true or false. By definition, these ideas will make propositions *M*, *N*, *O*, . . . neither all of them false, nor all of them true. But if the class of propositions *M*, *N*, *O*, . . . were replaced by an individual proposition, then either truth or falsity would have to occur. Hence it follows that several propositions, *A*, *B*, *C*, *D*, . . . can stand in the relation of contradiction to an individual proposition only if every substitution for ideas *i*, *j*, . . . will make either all of them true or all of them false, i.e. if they are equivalent to each other. But the examples of no. 4 show that a whole *class* of propositions can stand in the relation of contradiction to another class; the examples show that in this case it is not necessary that propositions occurring in a class must be equivalent to each other.

15. If several propositions *A*, *B*, *C*, . . . which are not equivalent to each other stand in the relation of contradiction to several others *M*, *N*, *O*, . . ., then there must be some among them, as well as among *M*, *N*, *O*, . . ., which are compatible with each other, all with respect to the same variable ideas. For, if propositions *A*, *B*, *C*, . . . are not equivalent to each other, then there are ideas which make some, but not all of them true; but these same ideas must also make some but not all of *M*, *N*, *O*, . . . false; hence some of the propositions *M*, *N*, *O*, . . . can become true at the same time as some of *A*, *B*, *C*, . . . i.e. they are compatible.

16. If two propositions *A* and *M* are not formally true with respect to certain ideas *i*, *j*, . . . and equivalent to each other with respect to the same ideas, then each of them stands in the relation of contradiction to the negation of the other, *A* to *Neg.M*, and *M* to *Neg.A*, with respect to the same ideas. For each idea which makes *A* true or false makes *Neg.M* false or true, and conversely.

17. But if a whole class of propositions *A*, *B*, *C*, . . . is equivalent to a class of others *M*, *N*, *O*, . . ., then we cannot conclude that a relation of contradiction holds between propositions *A*, *B*, *C*, . . ., on one hand and *Neg.M*, *Neg.N*, *Neg.O*, . . . on the other. For from the equivalence of those propositions it follows merely that every class of ideas which makes all of *A*, *B*, *C*, . . . true, will make all of *Neg.M*, *Neg.N*, *Neg.O*, . . . false, and that every class of ideas which makes all of *M*, *N*, *O*, . . . true will make all of *Neg.A*, *Neg.B*,

Neg.C, . . . false. But it does not follow, that every class of ideas which makes all of *A*, *B*, *C*, . . . false, will make all of *Neg.M*, *Neg.N*, *Neg.O*, . . . true and conversely, as is required by the relation of contradiction. Thus the two propositions $x+y = a$, and $x-y = b$ are obviously equivalent, with respect to ideas *a*, *b*, *x*, *y*, to propositions $x = \frac{a+b}{2}$, and $y = \frac{a-b}{2}$. But we cannot say that the first pair of propositions contradict the negations of the second pair, the propositions $x \neq \frac{a+b}{2}$, and $y \neq \frac{a-b}{2}$, since the falsity of the former does not follow from the truth of the latter.

18. If an individual proposition *M* is deducible from the individual proposition *A* with respect to certain ideas, and if the proposition *Neg.M* is deducible from the individual proposition *Neg.A* with respect to the same ideas, then proposition *A* contradicts *Neg.M*, and *Neg.A* contradicts *M* with respect to the same ideas. For every class of ideas which makes proposition *A* true or false will make proposition *Neg.M* false or true, and every class of propositions which makes the proposition *Neg.M* true or false will make proposition *A* false or true.

19. But the same does not hold for several propositions, i.e. if an individual proposition *M*, or several propositions *M*, *N*, . . . are deducible from several propositions *A*, *B*, *C*, . . . and if the propositions *Neg.A*, *Neg.B*, *Neg.C*, . . . are deducible from *Neg.M*, or *Neg.M*, *Neg.N*, . . ., it does not follow that propositions *A*, *B*, *C*, . . . contradict *Neg.M*, or *Neg.M*, *Neg.N*, . . . respectively. The falsity or truth of propositions *Neg.M*, or *Neg.M*, *Neg.N*, . . . does indeed follow from the truth or falsity of propositions *A*, *B*, *C*, . . .; but from the truth or falsity of proposition *Neg.M*, or the several propositions *Neg.M*, *Neg.N*, . . . it merely follows that not all of *A*, *B*, *C*, . . . can be true or false, but not that all must be false or true.

20. Of any two individual propositions which contradict each other, it holds that one of them must always be true, and the other false. For, the one, *A*, must be either true or false; but if *A* is true, then *M* must be false. Also, if a whole class of propositions *A*, *B*, *C*, . . . contradict an individual proposition *M*, then either all of *A*, *B*, *C*, . . . are true (if *M* is false) or all of *A*, *B*, *C*, . . . are false (if *M* is true). But if several propositions, *A*, *B*, *C*, . . . contradict several others, *M*, *N*, *O*, . . ., we may not conclude that the one class consists only of true propositions, and the other only of false ones; rather, it may be the case that both classes contain some true, and some false, propositions.

21. Propositions which are equivalent to contrary propositions are themselves contrary. The proof is analogous to § 156, no. 4.

22. Of a pair of contraries, whether they are individual propositions, such as *A* and *M*, or classes of propositions, such as *A*, *B*, *C*, . . . and *M*, *N*, *O*, . . . all may be false.

23. From any individual proposition *A*, as well as from any class of propositions *A*, *B*, *C*, . . . which are incompatible with an individual proposition *M*, we can deduce propositions which are contradictory to the latter, provided only that *M* is not formally false. (All with respect to the same ideas.) For, if proposition *M* is not formally false, then there are ideas whose substitution for the variables *i*, *j*, . . . will make it true. But since proposition *A*, or the class of propositions *A*, *B*, *C*, . . . exclude it, there are also ideas which make *M* false. Hence *M* is neither formally true nor formally false, and *Neg.M* is a proposition which contradicts it (no. 7). But, according to no. 4, *Neg.M* is deducible from *A* or from *A*, *B*, *C*, . . . respectively.

24. If certain propositions *M*, *N*, . . . which are not formally true are deducible from an individual proposition *A*, then the contradictory of *A* must be deducible from the contradictory of the propositions *M*, *N*, . . . all with respect to the same ideas. For, since *M*, *N*, . . . are deducible from *A*, all ideas which make *A* true must also make *M*, *N*, . . . true; hence all ideas which make *M*, *N*, . . . false and thus make their contradictory true (and there must be some such) must make proposition *A* false and hence its contradictory true. Hence the latter is deducible from the former.

25. No proposition which is not formally false is compatible with both of a pair of contradictory propositions. Rather, if it is not compatible with one of them, then the other is compatible with it, even deducible from it, all with respect to the same ideas. For, if proposition *X* is not compatible with one of a pair of contradictory propositions, *A* and *M*, e.g. with *A*, then every idea which makes *X* true (and there must be such) makes *A* false, and hence *M* true. Therefore *M* is not only compatible with *X* but deducible from it.

26. Given that two individual propositions, or two classes of propositions, exclude each other; whether they are contraries or even contradictories, there may be an individual proposition, or even a whole class of propositions, which are not only compatible with each member of the pair, but perhaps even deducible from them, all with respect to the same variable parts. Thus, in the example of no. 4, the propositions $a = m$, $b = m$ are not only compatible with the contradictory classes of propositions given there, but they are even deducible from each of them. To give another example, the proposition

'Caius deserves blame' is deducible from the proposition 'Caius is miserly' provided that the idea 'Caius' is considered variable. But it is nevertheless compatible with the contradictory of the latter, namely 'it is false that Caius is miserly' as well as with a proposition which is merely a contrary of it, namely 'Caius is profligate', from which it is even deducible.

27. If propositions *A*, *B*, *C*, . . . and *M*, *N*, *O*, . . . are contradictories of each other, and *A*, *B*, *C*, . . . are compatible with certain propositions *X*, *Y*, *Z*, . . ., then *M*, *N*, *O*, . . . are not deducible from the latter, and if *M*, *N*, *O*, . . . are not deducible from them, then at least one of the propositions *A*, *B*, *C*, . . . is compatible with them, all with respect to the same variable parts. For, if propositions *M*, *N*, *O*, . . . were *deducible* from *X*, *Y*, *Z*, . . ., then every class of ideas which made all of *X*, *Y*, *Z*, . . . true would also have to make all of *M*, *N*, *O*, . . . true and hence, because of the contradictoriness, all of *A*, *B*, *C*, . . . false. Consequently *X*, *Y*, *Z*, . . . and *A*, *B*, *C*, . . . would be incompatible. Hence, if *A*, *B*, *C*, . . . and *X*, *Y*, *Z*, . . . are compatible, then *M*, *N*, *O*, . . . cannot be deducible from the latter. Conversely if *M*, *N*, *O*, . . . are *not deducible* from *X*, *Y*, *Z*, . . ., then there are ideas which make all of *X*, *Y*, *Z*, . . . true without making all of *M*, *N*, *O*, . . . true, but these ideas do also not make all of *A*, *B*, *C*, . . . false. Hence there are ideas which make *X*, *Y*, *Z*, as well as one or several of *A*, *B*, *C*, . . . true. Hence, at least one of the propositions *A*, *B*, *C*, . . . is compatible with *X*, *Y*, *Z*. . . .

28. Every proposition of the form (I) '*X* has *y*' stands in the relation of contradiction with respect to ideas *X* and *y* to the following proposition (II) 'the class of the two propositions 'the idea *X* does not have a referent' and 'the idea of an *X* which does not have *y* has a referent' is not a class of false propositions'. Obviously, *X* and *y* can be replaced by ideas which make proposition (I) true, and by others, which make it false. But, as often as it is true the idea *X* has a referent, and the idea of an *X* which does not have *y* lacks reference, both propositions mentioned in (II) are false; hence the proposition which denies this is itself false. On the other hand, whenever (I) is false, either the idea *X* must be empty, or there must be some *X* which do not have attribute *y*, i.e. the idea of an *X* which is not *y* must have a referent. Hence it is true that not both of the propositions mentioned under (II) are false, i.e. proposition (II) itself is true. According to no. 18 it follows from this that propositions (I) and (II) stand in the relation of contradiction.

29. The proposition '*X* has *y*' and '*X* has the attribute non-*y*' are not contradictories but merely *contraries* with respect to variable

ideas *X* and *y*. If we replace *X* and *y* by other ideas, the two propositions will indeed never become true, but they will sometimes both be false. This happens whenever we replace *X* by an idea which applies to several objects, and *y* by an attributive idea which holds of some, but not of all, *X*.

In a note B. criticizes Aristotle for giving the impression (*De Interpretatione*, 17^b3 ff.) that the relation of contrariety occurred only with propositions of the form 'all *A* are *B*' and 'no *A* is *B*'. "Also the scholastics and even . . . Kant (*Logik*, § 49, A 183) . . . maintained that the distinction between contradictory and contrary judgments lies in this, that in a pair of contradictories one is a universal affirmative, and the other a particular negative, while with pairs of contraries one is a universal affirmative and the other a universal negative. But it is obvious that this definition does not apply to contrary propositions of the kind 'this is red', 'this is blue', etc.; this was already noted by others. . . . But even with respect to contradictories this definition is too narrow. If we call two propositions 'some *A* are not *B*' and 'all *A* are *B*' contradictories, we are justified in considering all other propositions which are equivalent to them also as contradictories."

In the remainder of the note B. discusses the views of various contemporary logicians on the subject of incompatibility.

§ 160. *Relations among Propositions which Result from the Number of Truths and Falsehoods in a Class of Propositions*

In this section B. considers distinctions between classes of propositions which result from a differing number of true propositions in them.

1. All propositions in a given class may be simply true or formally true (§ 147).

2. Sometimes we know of a given class of propositions that not all of them are false, or true, without knowing of any particular proposition that is true or false. "If we know of a number of propositions that not all of them are false, I wish to call them *complementary* . . . propositions . . . because, taken together, they exhaust the domain of assumptions which we can make with respect to the attributes of a certain object. . . . This relation of complementation can hold for particular propositions *M, N, O*, . . . which are given; in this case it may be called *material* complementation. But the relation may continue to hold, no matter what ideas are put in the place of certain variable ideas *i, j*, . . . where this substitution is either unrestricted or subject to certain conditions; in this case we speak of *formal* complementation."

3. If a given class of propositions contains precisely one true—or false—proposition we speak of *single member complementation* or *disjunction*. Disjunctions are also either material or formal. "If certain propositions *M, N, O*, . . . stand in the relation of formal disjunction to each other with respect to variable ideas *i, j*, . . . then every class of ideas whose substitution for *i, j*, . . . makes one of these propositions, e.g. *M*, true, must make all others false. Thus we can also say that each of the propositions *M, N, O*, . . . stands in the relation of exclusion to all the others, so long as it is not formally false (cf. § 159)."

4. If a class of complementary propositions *M*, *N*, *O*, . . . does not contain any formally true or false propositions, and if, upon any exchange of ideas *i*, *j*, . . ., there will be precisely one true proposition among them, then we cannot delete any of them without destroying the relation of complementarity. "For, if we left one of them out, there would be some ideas which make none of the remaining propositions true, i.e. precisely those ideas which made the discarded proposition true." Propositions of this sort are called *precisely complementary*.

5. Sometimes we can find out about a class of propositions *M*, *N*, *O*, . . . that they contain more than one true—or false—proposition, but we do not know how many. This relation is called multi-member or redundant complementation. This relation, too, can be either material or formal.

6. B. next considers the case where the class of propositions *M*, *N*, *O*, . . . contains a certain known number of true—or false—propositions in such a way that this number remains constant if certain ideas *i*, *j*, . . . are arbitrarily altered.

7. Some classes of complementary propositions are such that certain substitutions for their variable ideas will make all of them true simultaneously. This is the case if, and only if, they are compatible; hence they are called classes of complementary and compatible propositions.

8. So far, no restrictions were placed upon substitutions for variable ideas. Certain interesting relations result if variable ideas *i*, *j*, . . . are to be replaced only by ideas which make certain other propositions *A*, *B*, *C*, . . . true. The various relations of complementarity which result in propositions *M*, *N*, *O*, . . . under this condition will be given the names introduced in no. 2–7, except that they will be called conditional, in contrast to the above relations which may be called unconditional. "Hence if every class of ideas whose substitution for *i*, *j*, . . . makes all of *A*, *B*, *C*, . . . true also produces one or several truths in the class of propositions *M*, *N*, *O*, . . ., then I shall call propositions *M*, *N*, *O*, . . . *complementary under the condition that A*, *B*, *C*, . . ." Other conditional relations of complementarity are introduced analogously.

9. Given that ideas *A*, *B*, *C*, *D*, . . . are complementary with respect to the idea of something in general, i.e. given that taken together they exhaust the domain of the idea 'something in general', then the propositions 'the idea [*X*]*a* has a referent', 'the idea [*X*]*b* has a referent', 'the idea [*X*]*c* has a referent', etc. or, to put it in the more usual way, 'some *X* are *A*', 'some *X* are *B*', 'some *X* are *C*', etc., will have at least one true one among them.

In a note B. points out that of the above relations normally only subcontrarity is discussed; it occurs as a special case under no. 7 above.

§ 161. *The Relation of Relative Satisfiability, or Probability, of a Proposition with Respect to other Propositions*

1. In § 147 we already considered the concept of the satisfiability of a proposition; it is the concept of an attribute which can apply to an individual proposition if we envisage certain of its ideas as variable and consider the truth values of the propositions which result if arbitrary ideas are substituted for the given ones. One of the relations

among several propositions is very reminiscent of this concept of satisfiability: given that in an individual proposition *A* or several propositions *A*, *B*, *C*, *D*, . . . certain ideas *i*, *j*, . . . are variable, and given that *A*, *B*, *C*, *D*, . . . are compatible with respect to these ideas, it will often be important to find the proportion between cases in which *A*, *B*, *C*, *D*, . . . all become true and in which a certain other proposition *M* becomes true. For, if we accept propositions *A*, *B*, *C*, *D*, . . ., then this proportion tells us whether we should also accept *M*. If *M* becomes true in more than half of the cases in which *A*, *B*, *C*, *D*, . . . are true, the truth of *A*, *B*, *C*, *D*, . . . entitles us to accept *M* as well, otherwise not. I wish to call this relation between the indicated classes the relative satisfiability of proposition *M* with respect to propositions *A*, *B*, *C*, *D*, . . . or the probability which accrues to *M* from the premises *A*, *B*, *C*, *D*. . . . I call this relation relative satisfiability because of the similarity it has to the relation which I have called satisfiability of a proposition in § 147. When we ask for the (degree of) satisfiability of a proposition, we want to know the relation between the number of all propositions which can be generated by varying certain ideas, and the number of true propositions contained in that set. In the case of relative satisfiability of a proposition *M* with respect to certain others *A*, *B*, *C*, *D*, . . . we want to know the relation between the cases where *A*, *B*, *C*, *D*, . . . become true, and where *A*, *B*, *C*, *D*, . . . together with *M* become true. I call this relation ***probability*** [*Wahrscheinlichkeit*] since it seems to me that there is an increasingly wide-spread usage where by 'probability' we mean nothing but such a relation between given propositions, where it is not presupposed that these propositions should be represented or believed by a thinking being.

2. Being a relation between two sets, the relative satisfiability or probability of a proposition has a certain magnitude which, if it is determinable at all, can be represented by a fraction whose denominator and numerator are related as these two sets.

3. Since the number of cases in which both *A*, *B*, *C*, *D*, . . . and *M* become true can never be larger, but often smaller, than the number of cases in which propositions *A*, *B*, *C*, *D*, . . . alone become true, the degree of probability can never be larger than 1, and it reaches this magnitude only when all ideas which make all of *A*, *B*, *C*, *D*, . . . true also make *M* true, i.e. when *M* is deducible from *A*, *B*, *C*, *D*. . . . In this special case we tend to say that proposition *M* is *certain* with respect to propositions *A*, *B*, *C*, *D*. . . . On the other hand if there is not a single class of ideas whose substitution for *i*, *j*, . . . makes both *A*, *B*, *C*, *D*, . . . and *M* true, i.e. if *M* is incompatible with *A*, *B*, *C*,

D, . . ., then the degree of probability of *M* relative to *A*, *B*, *C*, *D*, . . . equals zero.

4. If premises *A*, *B*, *C*, . . . are equivalent to premises *A′*, *B′*, *C′*, . . . then the probability of *M* relative to *A*, *B*, *C*, . . . is the same as its probability relative to *A′*, *B′*, *C′*, . . . all with respect to the same variable ideas *i*, *j*. . . . Furthermore, if proposition *M* is equivalent to *M′*, then the probability of *M* relative to certain premises *A*, *B*, *C*, . . . is the same as the probability of *M′* relative to the same premises, all with respect to the same ideas *i*, *j*. . . . For the same ideas which make *A*, *B*, *C*, . . . true, also make *A′*, *B′*, *C′*, . . . true; and the same ideas which make *M* true also make *M′* true.

5. If the probability of a proposition *M* relative to *A*, *B*, *C*, *D*, . . . and with respect to ideas *i*, *j*, . . . equals μ then the probability of its negation, namely *Neg.M* equals $1-\mu$. Every class of ideas whose substitution for *i*, *j*, . . . makes all of *A*, *B*, *C*, *D*, . . . true also makes either *M* or *Neg.M* true; hence the number of cases in which either *M* or *Neg.M* are true must equal the number of cases in which *A*, *B*, *C*, *D*, . . . are true.

6. If the degree of probability of proposition *M* is the same as the degree of probability of its negation, i.e. if it equals one half, then we say that it is merely *doubtful*. Degrees of probability which are still lower are sometimes called improbabilities; we say that an improbability is greater the smaller the fraction in question (cf. no. 2).

7. If there is any idea which makes both *A*, *B*, *C*, *D*, . . . and *M* true, then there must also be an infinite number of such ideas, since every equivalent idea has the same result. Therefore the sets which were mentioned in no. 1, if they exist at all, must always both be infinite. Consequently the relation between them can never be found *directly*, i.e. by counting them; hence we must seek to determine it through considerations of another sort. In order to develop a general method which would accomplish this, we must note, first of all, that it is possible to determine that two or more propositions *k*, *k′*, *k″*, *k‴*, . . . all have the *same* degree of probability with respect to premises *A*, *B*, *C*, *D*, . . . without knowing what this degree of probability is. This will be the case whenever all of the indicated propositions stand in one and the same relation to the given propositions *A*, *B*, *C*, *D*, . . ., such that there is no difference between them other than that which arises from an exchange of variable ideas between them. For example, let the probability of a proposition *M* be relative to a premise *A*, namely that Caius has drawn a ball from an urn in which, among others, there is a ball numbered 1 and another numbered 2. I claim that, if no other premises are given, the

following two propositions: 'Caius has drawn ball no. 1' and 'Caius has drawn ball no. 2' both have exactly the same degree of probability provided that the ideas 'no. 1' and 'no. 2' are among those which are envisaged as variable in the context of this enquiry. For if we compare these propositions with the given premise *A*, we notice that both of them have exactly the same relation to *A*, since the only difference between them is that one of them has the idea 'no. 1', where the other has the idea 'no. 2'. Both of these two ideas are contained in proposition *A* in one and the same way (namely without any ranking, merely as a sum). Given, therefore, that among the ideas which we envisage as variable in this enquiry there are also the ideas 'no. 1' and 'no. 2', it is obvious that the same ideas which make both *A* and one of these propositions true will also make both *A* and the other one true. Given, for example, that the ball which was drawn by Caius was actually ball no. 3, it follows that both propositions become true if, and only if, we replace the variable ideas 'no. 1' or 'no. 2' by the idea 'no. 3' or an equivalent one. We must therefore necessarily consider the degree of probability of these two propositions as equal.

If we are asked to determine the probability of a proposition *M* from premises *A*, *B*, *C*, *D*, . . ., with variables *i*, *j*, . . ., we should first try to find a number of propositions *k*, *k'*, *k''*, . . ., all of which have the same degree of probability under the given assumptions *A*, *B*, *C*, *D*, . . ., and which are such that every class of ideas whose substitution for *i*, *j*, . . . makes all of *A*, *B*, *C*, . . . true will make one and only one of the propositions *k*, *k'*, *k''*, . . . true. In other words these propositions should stand in the relation of *single member complementation* to *A*, *B*, *C*, *D*, . . . (§ 160). If we succeed in this we may say that the cases in which *k*, *k'*, *k''*, . . . become true divide the entire infinite set of cases in which the assumptions *A*, *B*, *C*, . . . become true into as many equal parts. In other words, the set of cases in which *k* becomes true is exactly equal in size to that in which *k'* becomes true, etc., and the totality of these sets is exactly equal to the set of cases in which *A*, *B*, *C*, . . . become true. Let it furthermore be true that proposition *M*, whose probability we wish to determine, stands to the just-indicated propositions *k*, *k'*, *k''*, . . . in the following relation: none of these propositions leaves the truth of *M* undetermined; rather, from every one of them we can deduce either *M* or *Neg.M*. In this case we merely have to count the number of propositions *k*, *k'*, *k''*, . . ., and determine how many of them make *M* true in order to find the relation between the set of cases in which the assumptions *A*, *B*, *C*, . . . are true, and the set of cases in which both, these assumptions and M, become true. If the total number of propositions $k, k', k'', \ldots = k$; and if the number of these propositions from which M is deducible $= m$, then it is clear that the total infinite set of cases in which the assumptions $A, B, C, \ldots$ become true can be divided into k equal parts, and that m of these parts represent the infinite set of cases in which both $A, B, C, \ldots$ as well as M become true. Hence the desired degree of probability of proposition M is m/k.

Consider the following example: let the assumption be that Caius has drawn one ball from an urn in which there are 90 black and 10 white balls. We are to determine the degree of probability, given this assumption, of the proposition that Caius has drawn a black ball, where only 'Caius', 'ball', 'black', and 'white' are envisaged as variables. Let us, to begin with, designate the 100 balls in the urn as no. 1, no. 2, . . ., no. 100, so that we can mentally distinguish them. Now let us form the following 100 propositions: 'Caius has drawn ball no. 1', 'Caius has drawn ball no. 2', etc. until we come to the proposition 'Caius has drawn ball no. 100'. It is obvious that all of these propositions have the same degree of probability, provided that we add the just-introduced ideas 'no. 1', 'no. 2', etc. to the variable ideas. This must be permissible since these ideas do not appear in the originally given propositions at all, whose relation to each other can therefore not be changed by the introduction of these additional variables. That all of these propositions have the same probability follows from the circumstance that their relation to the given assumptions is in each case the same, so that every one of them becomes true if we replace the number which occurs in it by the number of the particular ball which was in fact drawn by Caius. Furthermore it follows from the nature of these propositions that every set of ideas whose substitution for the original variables, namely, 'Caius', etc., make all of the given assumptions true, will also make precisely one of these 100 propositions true. From this it follows that each of these propositions contains 1/100 of the total set of cases in which the assumption becomes true. As soon as we realize that, according to this assumption, only 90 of the balls in the urn are black, we see that the proposition that Caius has drawn a black ball is made true only by 90 of the above propositions, but fails in the 10 remaining ones, since only 90 of the indicated numbers can refer to black balls. Hence the probability of our proposition is 90/100 or 9/10.

8. It is likely that there are also assumptions $A, B, C, \ldots$ to which the just-described method of finding the degree of probability of a

proposition *M* cannot be applied, or where this probability is in itself indeterminate. For example, assume that you are told that the number of black and white balls is *unequal*, where it is left open which number is larger, and what their ratio is. From this information alone we certainly cannot determine the degree of probability of the proposition that Caius has drawn a black ball. We must therefore distinguish *determined* and *undetermined* probability.

In the remainder of § 161 B. states and proves a number of theorems of probability logic. We shall here merely give the theorems without their proofs.

9. Let proposition *M* have degree of probability μ under the assumptions *A*, *B*, *C*, . . . and with respect to ideas *i*, *j*, . . . and let proposition *R* be unilaterally deducible from *M* with respect to the same ideas. In this case the degree of probability of proposition *R* with respect to the same assumptions *A*, *B*, *C*, . . . is *never smaller* than μ.

10. Let *M* have degree of probability μ under the assumptions *A*, *B*, *C*, . . . and with respect to ideas *i*, *j*, . . ., and let *M* be unilaterally deducible with respect to the same ideas from some other proposition *L*. In this case the probability of the proposition *L* with respect to the same assumptions *A*, *B*, *C*, . . . can *never be larger* than μ.

11. Let proposition *M* have a degree of probability μ under the assumptions *A*, *B*, *C*, . . . and with respect to the ideas *i*, *j*. . . . Let proposition *N* have degree of probability ν under the assumptions *D*, *E*, *F*, . . . with respect to the same ideas *i*, *j*, . . ., and let *D*, *E*, *F*, be compatible with *A*, *B*, *C*, . . ., etc. Let *K*, *K′*, . . . be propositions of equal probability generated from *A*, *B*, *C*, . . ., and let *L*, *L′*, . . . be similarly generated from *D*, *E*, *F*, . . ., etc. If we can now generate propositions of the following kind: '*K* and *L* is true', '*K′* and *L′* is true', etc., and if these propositions are again all propositions of equal probability, then we can assert that the degree of probability that propositions *M*, *N*, . . . are all of them true under the combined assumption *A*, *B*, *C*, . . . *D*, *E*, *F*, . . . and with respect to the same ideas *i*, *j*, . . . is equal to the product $\mu \times \nu \times \ldots.$

12. If everything remains as in no. 11, and a proposition *R* is unilaterally deducible from *M*, *N*, . . . with respect to the same ideas *i*, *j*, . . ., then the probability of *R* with respect to the assumptions *A*, *B*, *C*, . . . *D*, *E*, *F*, . . . *can never be smaller* than the product $\mu \times \nu \times \ldots.$

13. Since the product $\mu \times \nu \times \ldots$ is smaller than each of its factors μ, ν, . . . and since it is the smaller, the larger the number of its factors, we can see that we can assume a smaller limit of probability for a conclusion the smaller the probability of each of its individual premises and the larger the number of probable premises. Premises which are certain do not decrease the probability of a conclusion; for if *M* is certain, then $\mu = 1$.

14. If everything is as in no. 11, and the proposition *R* has the probability ρ with respect to propositions *M*, *N*, . . . and the same ideas *i*, *j*, . . ., then the probability of *R* with respect to the assumptions *A*, *B*, *C*, . . . *D*, *E*, *F*, . . . and the ideas *i*, *j*, . . . is certainly not smaller than the product $\rho \times \mu \times \nu \times \ldots.$

15. If everything is as in no. 11, and propositions *M*, *N*, . . . are uni-

laterally deducible from R with respect to the same ideas $i, j, \ldots$, then the probability of R with respect to the assumptions $A, B, C, \ldots D, E, F, \ldots$ is certainly not larger than the product $\mu \times \nu \times \ldots$.

16. If everything is as in no. 11, then the probability of the assertion that there is a true proposition among $M, N, \ldots$ with respect to the assumptions $A, B, C, \ldots D, E, F, \ldots$ and ideas $i, j, \ldots$ is $1-(1-\mu)\times(1-\nu)\times\ldots$.

17. If everything is as in no. 11 and if we add the condition that propositions $M, N, \ldots$ must either all be true or all false, then the degree of probability for the assertion that they are all true with respect to the combined assumptions $A, B, C, \ldots D, E, F, \ldots$ and the ideas $i, j, \ldots$ is $\dfrac{\mu\times\nu\times\ldots}{\mu\times\nu\times\ldots+(1-\mu)(1-\nu)\ldots}$.

18. If the probability of one of the propositions $M, N, \ldots$ equals $1/2$, then this proposition will neither decrease nor increase the probability of the assertion that they are all true, for $\dfrac{\frac{1}{2}\nu}{\frac{1}{2}\nu+(1-\frac{1}{2})(1-\nu)}=\nu$.

19. Let μ be the probability of a proposition M with respect to the assumptions $A, B, C, D, \ldots$ and ideas $i, j, \ldots$; and let ν be the probability of a second proposition N with respect to the same assumptions and ideas, etc. If it is the case that these propositions are incompatible with respect to the same ideas, and if one of them is said to be true, then the degree of probability that M is the true proposition is $\dfrac{\mu}{\mu+\nu+\ldots}$.

20. Since $\dfrac{\mu}{\mu+\nu+\ldots}=1-\dfrac{\nu+\ldots}{\mu+\nu+\ldots}$, it follows that even a proposition which has a very low degree of probability can receive a very high one if circumstances arise where we have to choose between it and some other proposition which is even less probable. For, no matter how small μ is, if only $\nu+\ldots$ is still much smaller, $1-\dfrac{\nu+1}{\mu+\nu+\ldots}$ can be made as close to unity as we wish.

NOTE 1: B. defends his definition of 'probability' as a relation between propositions in themselves. He also defends his definition against the criticism that ordinary language would declare some propositions for improbable which would have to be called probable under his definition. B. assumes that the only way in which we can follow ordinary use is to employ a function which would assign a negative value to probabilities which are smaller than $1/2$ (under the present system), zero for $1/2$, and a positive value for probabilities which now get values larger than $1/2$. Also, the function should go to infinity in both directions. B. assumes that both common speech and common sense are subject to a conceptual error on this point, namely a confusion between probability and confidence. While probability should be defined as in the foregoing section, confidence can range from negative to positive values, with zero indicating indecision.

NOTE 2: The second note contains a critical discussion of the probability concepts of Aristotle, Wolff, Locke and others. B. then expresses his agreement with the definitions of Lacroix, Laplace "and other mathematicians who define probability as the relation between favourable cases and the total number of all possible cases." He thinks it necessary, however, to give a more careful definition of 'cases', 'possible cases' and 'favourable cases', since essential errors can occur in this connection. "We find that several very highly esteemed mathematicians have remarked that by possible cases we are to understand cases of equal possibility. But since it is *probability* rather than *possibility* which admits of degrees, we realize, that 'cases which have the same possibility' is only another way of putting the concept of cases which have the same probability; hence the definition of probability becomes circular." On the other hand, if cases of equal possibility are defined as cases for which there are equal but insufficient grounds (Huyghens), this is even more misleading, for if the reasons for the occurrence of each of two incompatible events are exactly equal, then it is certain that none of them will take place. "Hence if we are to have a rational expectation that a certain result will take place, for example that Caius will draw a certain ball from several balls in an urn, then we must presuppose that the relation between these balls and Caius is such that the reasons for drawing that particular ball are not exactly like the reasons for drawing some other ball, since otherwise he wouldn't draw any. In my opinion, the sense of this misleading expression was meant to be this: The given assumptions, on the basis of which the probabilities are to be calculated (in the present case the proposition that there are several balls in that urn, etc.), do not contain any reasons for drawing one ball rather than another. For the several propositions 'ball no. 1 will be drawn', 'ball no. 2 will be drawn', . . . all stand in the same relation to the given assumptions. This relation continues to be the same even if we add to the assumptions the proposition 'the balls in the urn do not all stand in the same relation to Caius, rather, there exists an inequality which will determine him to pick a certain ball.' Since this proposition does not contain the ideas 'no. 1', 'no. 2', . . ., it is obvious that the above propositions all stand in the same relation to it."

3. B. points out that the theorems which are proved in the preceding section are only the most simple and fundamental theorems of the probability calculus, but that they are more carefully phrased than it is usually done. For example, the theorem of no. 12 is usually put in this way: 'the probability of the conclusion is the product of the probabilities of its premises'. He points out that this product is merely the limit below which that probability cannot fall. For example if one of two balls in an urn is black and one is fragrant, the probability that Caius will draw the black ball is 1/2, and the probability that the black ball is the fragrant ball is also 1/2. Now we can construct a syllogism 'Caius will draw the black ball from this urn; the black ball is the fragrant ball; therefore Caius will draw the fragrant ball from this urn. If the probability of the conclusion is the product of the probabilities of the premises, the conclusion would seem to have the probability 1/4, while it has in fact a probability of 1/2. B. points out that 1/4 is merely a lower limit, but not the actual value of the probability of such a conclusion.

§ 162. *The Relation of Ground and Consequence* [Abfolge]

1. There is an important relation which holds among truths, namely that some of them are related to others as *grounds* to *conse-*

quences. I shall give a more detailed discussion of this relation in the next part. For example, the two truths that the three angles of a triangle are always equal to two right angles, and that every quadrangle can be divided into two triangles whose combined angles form the angles of the quadrangle, these two truths form the ground of the truth that the four angles of every quadrangle are equal to four right angles. Similarly, the truth that it is warmer in summer than in winter contains the ground of another truth, namely that the thermometer stands higher in summer than in winter, and the latter can be considered a consequence of the former. The above examples show that a truth which stands to certain other truths in the relation of consequence to grounds is often deducible from the latter, provided that we envisage certain ideas as variable. The proposition 'the thermometer stands higher in summer than in winter' is obviously deducible from the proposition 'it is warmer in summer than in winter', provided that only the ideas 'summer' and 'winter' are considered variable: no matter what ideas are put in their place, if they make the second proposition true, they will also make the first one true. But since propositions which are generated from others by the arbitrary exchange of ideas do not always have to be true, it is obvious that the relation of deducibility can also hold among false propositions in such a way that the truths which are generated by replacing certain variable ideas by others always stand in the ground–consequent relation to each other. This holds, for example, of the following two propositions: 'it is warmer in location *X* than in location *Y*' and 'the thermometer stands higher in location *X* than in location *Y*', provided that the ideas *X* and *Y* are envisaged as the only variable ones. It is obvious that both these propositions can become false if we can replace *X* and *Y* by arbitrary ideas. But as soon as we choose two ideas which make the first proposition true, the second one also becomes true, and, moreover, the first stands to the second as ground to consequence. It must be noted, however, that the latter does not always hold where there is a relation of deducibility. For example, the propositions which we just considered are mutually deducible from each other: the proposition 'the thermometer is higher in *X* than in *Y*' is deducible from the proposition 'it is warmer in *X* than it is in *Y*'; but conversely, the proposition 'it is warmer in *X* than it is in *Y*' is also deducible from the proposition 'the thermometer is higher in *X* than it is in *Y*'. Still, nobody will think for a moment that, when they are both true, the last proposition can be the ground and the other proposition its consequence. Nobody would wish to say that the real reason why it is warmer in summer than in winter lies in the fact that the thermometer stands higher in summer than in winter, rather, everybody looks upon the climbing of the thermometer as a consequence of the greater heat, and not vice versa. Hence not every relation of deducibility is such that it expresses a relation of ground and consequence whenever all its propositions are true. But there can be no doubt that a relation of deducibility which has this property will be important enough to warrant a special designation. I therefore want to call it a *formal* ground–consequent relation, while the relation that holds between *true propositions* will be called a *material* ground–consequent relation. Thus I say that propositions M, N, O, . . . are formal consequences of propositions A, B, C, . . . with respect to ideas i, j, . . ., if every class of ideas whose substitution for i, j, . . . makes all of A, B, C, . . . true, also makes all of M, N, O, . . . true, and if the truths generated from M, N, O, . . . are genuine consequences of the truths generated from A, B, C. . . .

2. The relation of ground and consequence also requires that a special division be made in the relation of *probability*. For, if propositions A, B, C, . . . are the assumptions on which M has probability μ, and if they are envisaged as parts of a larger class of propositions A, B, C, D, E, . . ., which are the ground of M, then the probability of M on the basis of A, B, C, . . . is called an *internal* probability. On the other hand if none of the propositions A, B, C, . . . belong to the indicated class of propositions, then that probability is called *external*. For example, a cloudy sky makes it internally probable that it will soon rain, while a drop in the barometer, or a prediction by a meteorologist, make it externally probable that it will soon rain.

§ 163. *Questions and Answers*

The discussion of § 145 is here continued. A distinction is made between answers in the wider sense (proffered answers), and correct answers. Answers that are too wide are distinguished from those that are too narrow.

CHAPTER 4

Several Types of Propositions which State Relations between other Propositions

§ 164. *Propositions which State a Relation of Compatibility*

§ 165. *Propositions which State a Relation of Incompatibility*

§ 166. *Propositions which State a Relation of Complementation*

§ 167. *Propositions which State a Relation of Probability*

§§ 164–167 have the purpose of bringing into normal form ('*A* has *b*') propositions which state relations between other propositions. For example, if propositions *M*, *N*, *O*, ... are deducible from *A*, *B*, *C*, ... with respect to *i*, *j*, ..., the following normal-form proposition will state that relation: 'Every class of ideas whose substitution for *i*, *j*, ... in propositions *A*, *B*, *C*, ... *M*, *N*, *O*, ... makes all of the propositions *A*, *B*, *C*, ... true—has— the attribute of also making all of the propositions *M*, *N*, *O*, ... true'. B. calls propositions of this form *arguments* [Schlüsse]. He points out that not all actual arguments state which ideas are to be considered as variable, but it is always implied that some of them are.

§ 168. *Propositions which State a Relation of Ground and Consequence*

The work of the preceding four sections is continued. After considering the notion of a partial ground, B. goes on to distinguish between the relation of ground and consequence on one hand, and cause and effect on the other. "I believe that these two words (cause and effect) in their proper sense refer only to objects which are real. We say of a real object α that it is the *cause* of truth *M*, if the proposition 'α is real' is a partial ground for truth *M*. We also say that the real object μ is an *effect* of object α if the proposition 'μ is real' is one of the consequences of the proposition 'α is real'. Thus we say that God is the cause of the existence of the world, and the world an effect of the activity of God, because the ground of the truth that there is a world lies in the truth that God exists. If this is correct, then the so-called causal propositions, i.e. propositions of the form '*X* is the cause of *Y*' or '*X* causes *Y*' or '*Y* is the effect of *X*', must be reckoned among propositions which state a relation of ground and consequence between certain other propositions. '*X* is the cause of *Y*' actually means 'the truth that *X* is the case is related to the truth that *Y* is the case as ground (partial ground) to consequence (partial consequence)'." B. goes on to define *condition* (or 'necessary condition') as a cause or partial cause in the following way: "if α is such a complete or partial cause whose existence (or the proposition 'α is real') is deducible with respect to some idea from the proposition *M* (or 'μ is real'), then we call α a *condition* of *M* or μ."

CHAPTER 5

Some Propositions whose Linguistic Expression Warrants Special Comment

§ 169. *Purpose of this Chapter*

The purpose is analysis of expressions which are fairly important and occur in several sciences.

§ 170. *Propositions Whose Linguistic Expression has the Form 'Nothing has attribute* b'

After giving some examples such as 'nothing is completely perfect' B. draws attention to the ambiguity of such sentences as 'Nothing is better than this medicine'; he then analyses expressions of the form 'Nothing has property *b*' in this way: 'the idea of an object which has property *b* does not have a referent'.

§ 171. *Propositions of the Form 'A certain* A *has* b'
cf. § 137

§ 172. *Propositions* whose *Linguistic Expression Contains the Word 'it' or 'one' or other Impersonal Forms*

This section is an analysis of the uses, in German, of the expressions '*es gibt* . . .' and '*man*'. Both these expressions are elucidated in terms of the reference of certain ideas. The same holds for such expressions as 'it is snowing right now', which is explicated as 'the idea of a snow-fall at the present time—has—reference'.

§ 173. *Propositions of the Form 'Some* (*or many*) A *are* B'

B. here distinguishes an absolute and relative sense of 'many *A* are *B*' and 'a few *A* are *B*'. "Thus three or four is in itself a small number of things, three or four billion a large one. On the other hand we call the number of stars that are visible to the naked eye small because we compare it to the number of all the stars there are, or the number of those that are visible through a telescope. In this case we find the number of *A* which are *B* small because we compare it to the number of all *A*; but sometimes we have other comparisons in mind. For example when we call the number of sick persons in a family considerable, we do not want to say that this number is in itself large (there may be only three or four); nor do we want to say that this number is a considerable proportion of the family as a whole (for no more than one-tenth of the family may be sick); all we want to say is that the relation between sick and healthy persons in this family is larger than normal." B. gives the following as the normal form for 'only a few *A* are *B*' in the absolute sense: 'the set of *A* which are *B* has the attribute of being a whole which is composed of only a small number of parts.' The normal forms for the other two senses are constructed accordingly. B. then goes on to criticize those logicians who have claimed that the expression 'some *A*' can be the subject of a proposition. "Let us consider, for example, the proposition 'some men are virtuous', and let us see what follows from the assumption that the idea 'some men' is the proper subject idea of this proposition, i.e. that 'some men' is the idea of the objects that the proposition is about. It cannot be denied that the idea 'some men' has subordinated under it, among other ideas also the ideas 'Nero', 'Caligula' and others, likewise the idea 'some vicious men'. From this it follows that Nero and Caligula, as well as all those men which fall under the idea 'some vicious men' belong to the objects with which the proposition 'some men are virtuous' deals; this is as certain as that 'some men' is the subject idea of that proposition. We would therefore have to accept the propositions 'Nero is virtuous', 'Caligula is virtuous' and 'Every vicious man is virtuous'. Actually we can consistently claim that 'some *A*' is the subject of 'some *A* are *B*' only if property *b* belongs to all *A*, hence if the proposition 'all *A* are *B*' is true. On the other hand, it might be claimed that the expression 'some men' in the proposition 'some men are virtuous' does not refer to all objects to which it is capable of referring, but only to certain of these objects, while it refers to *other* objects in the proposition 'some men are vicious'. But this amounts to an admission that the idea 'some men' is not the proper subject idea of these two propositions, for it is impossible that they should both have the same subject idea but different subjects."

§ 174. *Propositions of the Form* 'n A *are* B'

B. distinguishes two interpretations for this form of words, namely 'at least *n A* are *B*' and 'exactly *n A* are *B*'. The normal form for the first is 'the idea of a class of *n A* which are *B* has reference', and for the second 'the class of *A* which are *B* has the attribute that the number of its parts = *n*'.

§ 175. *Propositions of the Form* 'A *has* (*attribute*) b *to an equal, larger, or smaller degree than* B'

These propositions are analised as 'the relation between the magnitude of the attribute *b* in *A* to the magnitude of this attribute in *C* is a relation of equality or inequality, or a relation of a larger to a smaller, or a smaller to a larger, etc.'.

§ 176. *Propositions of the Form 'Only* A *is* B' *and* 'A *is nothing but a* B'

B. claims that the first of these has two senses, namely 'every *B* is *A*' and

'*A* and *B* are equivalent ideas'. The second of these senses he also expresses as '*A* and nothing else is *B*'. "When we hear these words we have in mind a thought which actually consists of two propositions, namely, '*A* is *B*' and 'what is not *A* is also not *B*'."

The locution '*A* is nothing but a *B*' is used when we have in mind a certain class of attributes of which only one is said to apply to *A*. Hence it can be put this way: 'of all the attributes which fall under the concept *c*, *A* has only attribute *b*'.

§ 177. *Propositions of the Form* 'A *is the case because* B *is the case*'

B. maintains that in its proper sense this locution signifies that *B* is the ground of *A*, but it is also sometimes used to indicate merely that *A* is deducible from *B*. Thus we say 'it will rain because the barometer is falling'; the relation here indicated is merely one of deducibility, and not ground and consequence.

§ 178. *Propositions of the Form* 'A, *as a* C, *is* B'

B. distinguishes three senses of this locution, namely, first, '*A* is *B*, since it is *C*'. Example: 'the earth as a complex body is destructible'. Secondly, 'the property *c* of *A* has (the property) *b*'. Example: 'Caius, as a musician, is incomparable', which is to be construed as 'the musicianship of Caius is incomparable'. Thirdly, it is sometimes the case that the interpretation of this phrase depends upon the particular terms involved, as for example in the sentence 'Titus, as judge, should not accept gifts'. "What is meant here is not that Titus should not accept gifts at all, since he is judge, nor could we want to assert of Titus' attribute of being a judge that it (this abstractum) should not accept gifts. [Rather, what is meant is that Titus should not accept gifts] which are offered in order to influence his judgments. But it is obvious that this interpretation depends upon the particular nature of the ideas designated by *A*, *B*, and *C*."

Locutions of the form '*A* as *A* is *B*' are also capable of being interpreted in any one of these three ways.

§ 179. *Propositions with 'if' and 'then'*

Every reasonably well developed language contains rather common modes of expression involving *if* and *then*; consider, for example, the following proposition: 'if Caius is a man, and all men are mortal, then Caius is mortal'. I have already stated (§ 164) that we use this form in order to express a relation of deducibility of a certain proposition from one or several other propositions. But I do not think that every time we use 'if' and 'then' we want to indicate the presence of certain ideas which are envisaged as variable and may be replaced by any other ideas without destroying the truth of the proposition.

This may indeed be the case with the proposition I gave as an example: everybody can see at once that this assertion remains true no matter what ideas are put in the place of the three ideas 'Caius', 'man', and 'mortal', and everybody feels that the sense of the assertion can only be to say that in each case where a substitution of ideas makes the antecedents true, the consequent will also express a truth. But since this locution is very convenient we often use it where we do not think of any variable ideas in the propositions in question, or at least where it is not necessary to consider such ideas. I believe that this occurs in the following proposition: 'If the digits of a given number are transposed in an arbitrary way, and if this new number is subtracted from the given number, then the remainder is always divisible by nine'. In this proposition there are no ideas at all that should be envisaged as variable, and the 'if–then' was used only for the sake of convenience. We could have expressed ourselves, for example, in the following way: 'the difference between two numbers which differ from each other only in the position of their numerals is always divisible by nine.' I believe that, in general, even when sentences connected by 'if–then' stand in the relation of deducibility, we do not have this relation in mind but a closely related one, unless we are faced with a familiar example of deducibility. Let me try, by giving several examples, to explain what this other relation is.

Consider, first of all, propositions which have the following form 'if *A* is *B*, then it is also *C*'; for example, 'if Caius is silent on this occasion, then he is ungrateful'. It may indeed be the case that the propositions which we here connect, namely 'Caius is silent on this occasion' and 'Caius is ungrateful' stand in a relation of deducibility. But if somebody forms this judgment, and thinks that he asserts a deducibility relation, he must also think that there is a certain idea in these propositions (for example the idea 'Caius') which can be varied in such a way that every substitution which makes the first proposition true also makes the second one true. But is this what we have in mind when we use those words? I do not think so; it seems to me that we have an entirely different thought in mind. The proposition 'if Caius is silent on this occasion, then he is ungrateful' is to indicate that everybody, if he were in Caius' position, would be ungrateful if he were silent on this occasion. Thus, the sentence we utter is actually the assertion that an idea has reference. It has the general form 'the idea of certain properties of *A*, of which it holds that every object which has these properties together with *b* must also have property *c*, has reference'. Sentences of the following form should be interpreted similarly: 'if *A* is *B*, then *C* is *D*'. For example, 'if

Caius is dead, then Sempronius is a beggar'. All we want to say here is that there are certain relations between Caius and Sempronius such that the following general proposition holds: it holds of any two persons who are related as Caius and Sempronius and one of whom dies (the one in the position of Caius) that the other (the one in the position of Sempronius) must become a beggar. One can easily see that this interpretation can also be applied to cases where there are several propositions in the antecedent or the consequent; I therefore think that we usually connect this sense with the locution 'if–then'; except in cases where the ideas *i*, *j*, . . ., which are to be varied in the related antecedents and consequence, are so clearly represented that they cannot be overlooked; in these cases the first interpretation applies.

NOTE: The following remark is more grammatical than logical: we use an *indicative* verb in connection with 'if–then' (if *A* is the case, then *M* is the case) if we want to leave it entirely undecided whether the compared propositions *A* and *M* are true or false; on the other hand we use the *subjunctive* ('if *A* were the case, then *M* would be the case), if we want to indicate that propositions *A* and *M* are false as stated.

§ 180. *Propositions of the Form* 'A *determines* B'

1. There is a closely related group of sentences in which we assert of an object *B* that it is *determined* by another, *A*; e.g. the cause determines its effect; the position of the barometer determines the atmospheric pressure; two sides of a triangle and the enclosed angle determine the whole triangle. That an object *A* determines another, *B*, can only have the sense that the attributes of *B* are deducible from the attributes of *A*, where it is presupposed that *A* as well as *B* can have several additional attributes, i.e. that the ideas *A* and *B*, under which we think these objects, leave several of these attributes undetermined; lastly, the relation of deducibility cannot hold between attributes as such, but only between propositions which state these attributes; hence the sense of the locution that *B* is determined by *A* turns out to be this: there are certain propositions that state attributes of *B* which stand in the relation of deducibility to certain other propositions that state attributes of *A*, all with respect to certain ideas which refer to attributes of *A*. As soon as these ideas are replaced by others in such a way that the statements about *A* become true, the statements about *B* must also become true. Thus

we say that a cause determines its effect, if we want to indicate that from certain propositions concerning the character of the cause we can deduce propositions which describe the character of the corresponding effect.

2. If we assert that *B* is not only partially, but wholly or completely determined by *A*, we mean to say that all propositions which concern attributes of *B* can be deduced from certain propositions concerning attributes of *A*, all with respect to ideas which refer only to attributes of *A*. Thus we say that the centre of a circle, the plane in which it is to lie, and its radius fully determine a circle, since all its attributes can be deduced from these three parts.

3. The determination of an object or objects by several objects is to be construed accordingly.

§ 181. *Propositions with 'Either–or' and Related Propositions*

According to B., sentences of this form can indicate either a material or a formal disjunction (cf. § 160), and may be taken in the exclusive as well as in the inclusive sense.

§ 182. *Propositions which Contain the Concept of Necessity, Possibility or Contingency*

Extremely important are propositions whose linguistic expression indicates either a *necessity*, a *possibility* or a mere *contingency*. But since the sense connected with these words is not always the same, these sentences themselves must be interpreted in various ways.

1. I believe that the words 'necessity', 'possibility', and 'contingency', as well as the related words 'must', and 'can', when taken in their stricter sense, are used only in a certain connection with the concepts of *existence* [*Sein*] or actuality. I think that every 'must', when it is taken in the stricter sense, is a 'must–exist', or a 'has to exist'; and every 'can', in the stricter sense, is a 'can–exist' or the possibility of being actual. We call something necessary or contingent in the stricter sense only if it is actual, and possible only if it can become actual. In particular, concerning the concept of necessity, we say that the actuality of a certain object *A* is necessary, or has necessity, or that it must exist, only if there is a purely conceptual truth of the form '*A′* exists (or has actuality)', where *A′* is an idea which refers to *A*. Thus we say that God is necessary, because the proposition that God is actual is a purely conceptual truth. Conversely if the proposition '*A′* is *not* actual', rather than the proposi-

tion '*A'* is actual' is a purely conceptual truth, we say that the object *A*, which falls under the idea *A'*, is impossible. Thus we say, for example, that an omnipotent creature is impossible, since the proposition that there is no such creature is a purely conceptual truth. We call the actuality of an object possible if it is not impossible. Thus it is possible that a man makes mistakes, since there is no purely conceptual truth which states the non-existence of a fallible man. If an object exists without being necessary, we call it contingent. Thus it is merely a contingent matter that there are, for example, fallible men.

2. According to these definitions, every object which can be represented by a pure concept that refers exclusively to that object is always either necessary, namely when it has actuality, or impossible, namely when it does not have actuality. For, if there is a pure concept *A'* which represents exclusively this object, then the proposition '*A'* is actual' is a purely conceptual proposition. Hence, if it is true, the object is necessary, but if it is false, then (since *A'* is a singular idea) the purely conceptual proposition '*A'* is not actual' is true, and the object is impossible. Hence, if an object is actual but not necessary, i.e. merely contingent, it must be an object which cannot be represented except by mixed ideas or intuitions (cf. § 73).

3. If neither the object *A* nor the object *M* are in themselves necessary, but it appears that the proposition '*M* is actual' is deducible from the proposition '*A* is actual' with respect to some ideas contained in *M* and *A*, then we say that *M* is necessary relative to presupposition *A*. We call this necessity of *M* a relative, or external necessity; by contrast, the necessity discussed under no. 1 is called internal. Thus we say that punishment is necessary only in a relative way, namely on the assumption that somebody has sinned, since the proposition 'somebody will be punished' is deducible from the proposition 'somebody has sinned'. Conversely if not the proposition '*M* is actual', but the proposition '*M* is not actual' is deducible from the proposition '*A* is actual', then we say that *M* is impossible relative to *A*. We call this impossibility of *M* relative or external, and contrast it with the internal impossibility of no. 1. If object *M* is not impossible not only internally, but also in relation to some other object *A*, then we say that *M* is possible with respect to object *A*. Finally if object *M*, which is not necessary relative to *A*, is nonetheless actual, we say that its actuality is externally contingent or contingent relative to *A*.

4. There is a second sense of the words 'necessary', 'possible' and 'contingent', which I call wider or improper. In this sense they are

not (as in nos. 1–3) applied to the actuality of things, but to truths in themselves. Whenever a proposition '*A* has *b*' is a purely conceptual truth, it is customary to say that the attribute *b* belongs necessarily to object *A* regardless whether or not this object, and hence that attribute, is actual. Thus we say that every equation of odd degree necessarily has a real root, although neither equations nor their roots are something that exists. If it is not the proposition '*A* has *b*' but rather the proposition '*A* does not have *b*, which is a purely conceptual truth, we say that attribute *b* is impossible for object *A*, even if nobody ever thought of looking for this attribute or that object among actual things. Thus we say that it is impossible that a simple idea is imaginary, since there is a purely conceptual truth which states that simple ideas are never imaginary. If it is not impossible that the attribute *b* belongs to *A*, hence if there is no purely conceptual truth of the sort 'no *A* has *b*', then we say that it is possible that *A* has *B*. Finally, if the proposition '*A* has *b*' is true, but is not a purely conceptual truth, i.e. if the attribute *b* belongs to objects *A*, without belonging to them necessarily, then we say that it belongs to them contingently. If it is neither the case that attribute *b* belongs to *A* nor that *p* belongs to *M*, each taken in itself, but if the proposition '*M* has *p*' is deducible from the proposition '*A* has *b*', with respect to certain ideas included in *A* and *M*, then we say that attribute *p* belongs to *M* with relative necessity, namely under the assumption that all *A* have attribute *b*. In a similar way the concepts of relative possibility and contingency are defined.

5. The words 'possible' and 'can', though not the words 'necessary' and 'contingent', are often taken in a third sense. We say that something is possible, or can be, if we want to indicate that *we* do not know any reason for its impossibility or, what comes to the same thing, that no purely conceptual truth which asserts the opposite is known to us. It obviously does not follow from the fact that we do not know such a truth that no such a truth exists; hence we must not confound anything that we call possible in this sense of the word with what is called possible in the proper sense (no. 1) or in the derived sense (no. 4), as is only too often the case. In order to avoid this confusion, we could use the word 'possible' with the qualification 'apparently', or we could use the word 'problematical'; I can think of no useful word which replaces the word 'can' in this connection. And since the words 'apparently possible' and 'problematical' are inconvenient, we have no option but to continue to permit the use of 'possible' and 'can' in all three senses, but we must make sure that every time these words are used their sense is clearly distin-

guished. It is not surprising that the word 'possible', in the sense of 'apparently possible', is often applied to (future or past) events. For example, if we are told 'it is possible that we are going to have rain today', this can only be understood as an apparent possibility, and nothing follows from it except that the speaker does not know anything from which it follows that it is impossible that it will rain today. If a historian writes 'it is possible that Attila was murdered', he only wants to say that he does not know of any fact which shows the falsity of Agnellus' surmise. Sometimes we even use the words 'can' and 'possible' in this third sense when we speak about purely conceptual truths. To the question whether the two-hundredth decimal place of π could be an even number, many a person will reply that this is possible; meaning thereby no more than that he does not presently know anything which proves the falsity of this assumption.

6. These definitions should make it obvious what I consider to be the sense of sentences which state a necessity, possibility, or contingency. Thus I believe, for example, that the assertion 'God exists with necessity' has no other sense than 'the proposition that God exists is a purely conceptual truth'. 'Every effect must necessarily have its cause' I take to mean nothing but 'it is purely conceptual truth that every effect has its cause', etc.

NOTE: B. contends that only an actuality which follows from purely conceptual truths may be called necessary; had the definition stated that everything which follows from a truth is necessary, it would have been a consequence that whatever is actual is necessarily actual, which B. takes to be clearly false. He goes on to say that the concept of possibility, according to his definition, is more complex than the concept of impossibility; but that it is not unusual than a simpler concept has a more complex sign in the language (cf. § 60).

The sense which was discussed under no. 4 above was said to be "improper" because the words in question are introduced only for the sake of emphasis and can be eliminated without loss of meaning. "Instead of saying of an assertion that it is actually and in fact true we could simply say that it is true; and instead of calling it necessary or saying that it follows necessarily from this or that other truth, or that it is merely possible, we could simply say, without qualification, that it is (or is not) a merely conceptual truth, or that it stands in the relation of deducibility to another truth, etc."

B. then considers a further division of necessity which depends upon the nature of the proposition relative to which some other proposition is necessary: "let some proposition '*M* is actual' be derived from the consideration of some other truths; depending upon the character of these other truths several further kinds of necessity, as well as possibility and impossibility, can be distinguished. Thus we speak of *metaphysical*, *physical*, *psychological* necessity, etc., depending on whether the most important truth which must be considered in order to realize that object M is actual, is a metaphysical, physical, psychological or some other truth. Also, two special kinds of possibility and impossibility are usually distinguished. An impossibility that could be detected at once, or which lies in the concepts (the expressions) themselves was called an absurdity or a contradiction in terms, or *contradictio in adiecto*, or in *ipsis terminis*. Conversely a possibility which holds in relation to all objects, whether they are known or unknown, i.e., whose non-existence cannot be deduced from any of them was called completely, absolutely, or in every respect possible. We can retain all of the above distinctions, even if they are of no particular importance, and add them to those made earlier." B. rejects, however, the concept of a *moral* possibility, impossibility and necessity if, what is meant by this, is that an act is morally permissible, forbidden, or required. His objection is based on the fact that even a morally impermissible act is still possible.

"It is quite customary to divide the possible into the *logically* and the *really* possible, where the former is defined as that which can be *thought* or *represented*, while the latter is said to be that which can exist, where it is often added that real possibility is dependent upon logical possibility. I do not wish to raise the question whether it is worthy of a special investigation and terminology that a thing can be thought or represented, and if the words 'thinkable' or 'conceivable' do not suffice for this purpose it is not really objectionable to use the word 'logically possible' in this sense. It is however a mistake to regard the conceivability or thinkability of a think as a requirement for the possibility of its *being* and to claim, for this reason, that real possibility is dependent upon logical possibility. So far as I can see, neither of these is required for the other; rather, something can be thinkable or conceivable without being possible (such as a round square) and, conversely, something can be possible, even absolutely possible, without being humanly thinkable or conceivable. Moreover if the possibility of being actual always rested on the possibility of a conception of this being, then nothing would in truth be possible. For, since every thinking is also a kind of being actual, something that exists (since thoughts have actuality in the being which has them at the time when it has them), it follows that the possibility of anything A presupposes the possibility of its idea, and since this idea in turn is something that exists, it requires the possibility of the idea of that idea and so on to infinity."

§ 183. *Propositions which Contain Time Determinations*

1. B. points out that a proposition, which is about something actual can be true only if it contains some time determination. In ordinary speech these time determinations are usually expressed through the verb. But the inflection of verbs generally allows only a distinction of past, present, and future. Where more precise determinations are required, special words must be introduced and they must be associated with the subject of the proposition (cf. § 127).

2. Ideas of a time can occur as determinations of the subject of an idea, but they can also occur in many other ways, e.g. they can themselves become the subjects of propositions. For example, the

judgment, 'cherries bloom earlier than grapes' should be interpreted as 'the relation between the time of year when cherries bloom, and the time of year when grapes bloom is the relation of an earlier to a later time'. Another type of propositions that belong into this class are those asserting *continuation*, *beginning*, or *end* of a certain state. . . . We say that an object or, to be more precise, a state, *continues* through time *t*, if the proposition that this state is actual at moment *x* continues to be true no matter what moment contained in the time span *t* we put into the place of *x*. If we say, furthermore, that a certain state continues *at this very moment*, i.e. at the present time, this means that it continues within a time span in which the present moment (i.e. the moment in which we have this thought) lies. From this it can be seen what components are likely to be contained in a proposition which asserts continuation, such as 'the sun continues to shine', namely 'there is a time span, in which the present moment lies, and every moment of which, when put into the place of *x* in the proposition 'the sun shines at moment *x*' makes this proposition true'. On the other hand, if we say that a certain state *A* begins (or ends) at moment *a*, what we say is that the proposition 'state *A* is actual at moment *x* whenever *x* is replaced by a moment which is later (or earlier) than *a*, so long as it does not exceed a certain span' is true, but that this proposition is false whenever *x* is replaced by a moment which is earlier (or later) than *a*. It is now easy to see how propositions are to be interpreted which claim that a certain state lasted for a *certain time span*, namely from moment *a* to moment *b*.

3. There is another important kind of propositions with time determinations, namely those asserting that something comes about (gradually) or becomes. But this can happen in several ways, so that I wish to distinguish four separate kinds of such propositions: first of all, we can simply say that something comes about, without indicating either the object from which, nor the object through which, it was generated. But we can also add a determination of the object through which, and the object from which, something was generated, or both. The following propositions are examples for the four cases: 'A storm forms', 'Ice turns into water', 'Error and ignorance produce moral defects', 'A good person becomes evil when he associates with evil persons'. The general forms of these four kinds of propositions are '*M* comes about', '*M* is generated from *A*' or '*A* turns into *M*', '*M* comes about because of *P*', and '*A* turns into *M* because of *P*'. Obviously, every object of which we can truly say that it is coming about or that it is in the process of becoming, must

belong to the class of those things which can become actual. But it must not possess this actuality at the time when we claim it to be coming about, but only at a later date. For, what is already there is not in the process of becoming, and what never will be cannot be said to be in the process of coming about. Although we can say of something which is in the process of becoming that it will have actuality at a future time, it is not the case that 'becoming' and 'future actuality' are one and the same concept; rather, to say something is in the process of becoming is to say more than merely that it will have existence at some future time, even if there is a sense in which one can say of everything which will exist at a future time that it is already in the process of becoming. The difference between becoming and mere future actuality does not rest upon the shorter or longer time which has to pass until the object is actual, as if we said that an object is in the process of becoming only if we expect it to acquire its existence shortly. This distinction would not only be vacillating and arbitrary but also at variance with common usage; for we often say of things which will become actual only after centuries that they are already in the process of becoming, while in some other case we would not want to say of a thing that it was forming as little as an hour before it came to exist. Thus, to say truly that something is in the process of becoming, some change must take place and this change must be connected with the object of which we say that it is in the process of becoming, i.e. it must be through this change that the object comes about, or the cause of the (future) existence of that object must lie in that change. But a change cannot be thought without a something which changes; a change consists in the fact that the (internal) structure of this something does not remain the same through any finite part, no matter how small, of the time through which the change lasts. In order to be able to say that an object *A* changes throughout time *t* there must be no finite segment of that time span within which *A* retains exactly the same inner structure. If we want to say, furthermore, that this change takes place *now*, i.e. at the present moment, then the present moment must lie within that time span. . . . Hence if we assert the proposition '*M* is coming about' (is presently in the process of becoming), what we say is that a certain change, which is the cause that *M* will exist at some future time, is just now taking place. Hence our proposition belongs to the class of assertions of reference, which were considered in § 137; to show its components as clearly as possible, it should be formulated in this way: 'the idea of a change which is just now taking place and which contains the cause of the future existence of object *M*—has—

reference. If we say '*M* is being generated from *A*' or '*A* turns into *M*', we merely want to say that the object which is presently *A* (falls under the idea *A*) is undergoing a change which is the cause that it will be *M* in the future (will fall under the idea *M*). Hence the immediate parts of this proposition become apparent if we express it thus: 'object *A*—has—the attribute of undergoing a change whose effect is going to be that it will be *M* at a future time'. On the other hand, if we want to say that *M* is generated through *P*, we only want to indicate that *P* is a cause, at least a partial cause, of the change whose effect is *M*. Hence our proposition can be displayed in the following way: 'object *P*—has—the attribute of being a cause (partial cause) of the fact that a change takes place whose effect (at a future time) is *M*'. If all this is correct, then propositions of the form '*M* is generated from *A* through *P*' can be expressed thus: 'object *P*—has—the attribute of being the cause of the fact that the same object which is now *A* undergoes a change through which it will be *M* at a future time'.

In a note, B. mentions that most logicians have placed time determinations in the copula. A notable exception is Wolff, who placed them in the subject.

§ 184. *Expressions which Must be Interpreted as a Class of several Propositions*

B. here considers expressions of the general form 'not *X*, but *Y* has (attribute) *b*', for example 'not the body, but the soul of man is immortal'. Expressions of this sort are to be interpreted as consisting of two propositions, namely '*X* does not have *b*' and '*Y* has *b*'. "The particular connection between these propositions is to indicate that they should be compared. Similar considerations hold with respect to related expressions such as '*X* has *b*, but not *a*'. . . . 'All *A*, including those that are *X*, have *b*'. We indicate with these words, first, that we take all *A* to have attribute *b*, and, secondly, that we assume that our interlocutors will not find this credible, especially that they would wish to except those *A* which are also *X*, an exception we do not wish to admit; etc."

APPENDIX

Earlier Treatments of the Subject Matter of this Part

§§ 185–194. *See Table of Contents*

B. examines the claim that logic is concerned only with the form and not with the content of judgments, and rejects all previous interpretations of that notion. However, he accepts this view, if it is taken in a certain sense. His views are the following: "I cannot accept the view that logic is concerned only with the form of judgments, given the previously indicated definitions of the concept of form; but perhaps there is another sense in which it can be justified. This must indeed be expected, since otherwise this view could hardly have come to be so widely accepted. Let us then find this sense. Nobody can deny that the distinctions that are made in logic should be of such a kind that every special logical concept should include a whole class of propositions, i.e. not only a single, but several different propositions. From this it follows immediately that the attributes which determine the logical category into which a certain proposition belongs can only concern matters which several propositions have in common. Now if such attributes are called the common *form* of these propositions, i.e. their *Gestalt*, then we can justly claim that all distinctions made in logic concern only their form, i.e. only matters which several, or even infinitely many, propositions have in common (§ 12). I do not think, however, that these remarks suffice to determine whether a given distinction belongs into logic. For, according to this definition, there will be infinitely many distinctions which concern only the forms of propositions; but logic is supposed to distinguish only forms that have a particular use, i.e. it is to introduce us only to those kinds of propositions that require a special scientific investigation."

The remainder of this appendix is concerned with a critical discussion of the Kantian table of judgments. B. points out that infinite judgments do not differ from negative ones and, concerning the category of relation, he remarks that no proof is given that the indicated three kinds of proposition are the only ones that can occur, and that there is no reason why only categorical propositions but not hypothetical and disjunctive ones are said to have subject and predicate ideas. "If somebody assumes that there is a subject as well as a predicate idea in a proposition like 'Caius should forgive Titus' which means the same as 'Caius—has—the duty to forgive Titus' (propositions of this kind are generally called categorical), then he should also not hesitate to claim subjects and predicates for hypothetical judgments such as: 'since *A* is the case, *B* is also the case' which merely means 'the relation between propositions *A* and *B* is a relation of ground and consequence'; the same holds for disjunctive judgments: 'either *A* is the case or *B* is the case' which expresses only that the idea of a true proposition among *A* and *B* is a singular idea. If we can say of the first proposition that Caius is its object or is the object of which it treats, then we can say of the second one that its object is the relation between propositions *A* and *B*, and of the third that its object is the idea of a true proposition among *A* and *B*. Just as this is the reason why we call the subject of the first proposition the man Caius, so we should call the subject of the second proposition the relation between propositions *A* and *B*, and that of the third one the idea of a true proposition among *A* and *B*. The fact that in the first case the subject is a real thing (Caius) but in the other two it is not, should not make any difference. For we also judge about things which do not have reality, such as mere ideas; and nobody would deny the name of a categorical proposition to a proposition like the following: 'the concept of a triangle includes the concept of a figure.' And just as in the first proposition we say of a subject that it has a certain duty, so we say in the second one that the given relation is a relation of ground and consequence, and in the third that the given idea has the attribute of being a singular idea. And if we call the assertion of an obligation a predicate, why should we not call the assertion of the other two attributes predicates as well? Hence I maintain that we can show subject and predicate in every proposition, if we only understand them correctly."

Later B. criticizes the general assumption that modality is determined by the relation between a proposition and the faculty of judgment. It is his view that this assumption should not result in the division of judgments into problematic, assertoric and apodictic; rather the question here is whether a proposition was merely entertained, or a judgment actually formed. In the latter case the degree of confidence with which they are accepted must be taken into account.

Finally, B. considers various theorems concerning hypothetical and disjunctive judgments. In particular, he is concerned with the question if hypothetical and disjunctive judgments can be changed into one another or into categorical judgments. Some logicians have denied this, for example Kant, Jacob, and Krug; others have affirmed it, for example, Herbart and Maaß. The latter view is based on the assumption that the linguistic expression of our judgments is quite arbitrary, and that we can express a judgment in various different forms. "This holds especially of the form generally called hypothetical, which we often use to express judgments that are really not hypothetical while on the other hand a truly hypothetical judgment can also be expressed without the use of 'if' and 'then'. This has already been said several times (cf. § 179). If somebody says 'if a triangle is equilateral, then it is also equiangular', he does not utter a hypothetical judgment, although he uses 'if' and 'then'. On the other hand if somebody says 'proposition *B* is deducible from proposition *A*' then he forms a hypothetical judgment although he does not use 'if' and 'then'. On the other hand, I do not hesitate to affirm that from a given hypothetical or disjunctive judgment many other judgments can be deduced which do not belong to the class of the given judgment, though some of them are even equivalent to it. For example, we can certainly deduce from the hypothetical proposition 'if *A* is the case, then *B* is the case' the disjunctive proposition 'either *B* or *Neg.A* is true' and, if we add to this the following propositions '*A* is not formally false', then the two last propositions together (they can easily be combined to a single proposition) allow us to deduce the hypothetical proposition in turn. . . . [On the other hand] from the judgment 'if *A* is *B*, then it is also *C*' I cannot always deduce the categorical judgment 'an *A* which is a *B* is also *C*'. The reason is that the hypothetical judgment can be true even if no *A* is a *B*; but in this case we cannot call the categorical judgment true since it does not have any object. Thus the judgment 'if the soul were corporeal, then it would also be composite' is a correct hypothetical judgment if we consider the idea 'soul' as variable. On the other hand, the categorical proposition 'a soul which is corporeal is also composite' does not express a truth. It would be even less correct to call a categorical judgment of this form *equivalent* to the hypothetical judgment, as Kiesewetter did."

PART III

Of True Propositions

§ 195. *Contents and Purpose of this Part*

B. takes pains to point out that he is still concerned only with *objective* truths, i.e. truths in themselves.

§ 196. *Some Attributes which all Truths have in Common*

Before we proceed to the investigation of those attributes which differentiate some classes of truths from others, it is reasonable to ask whether there are any attributes which *all* truths have in common. It seems to me that aside from those attributes which truths share with propositions in general, and which I have already mentioned in the previous part, only a few things need to be mentioned here.

1. First of all, let me point out that true propositions, just as propositions in general, do not have real existence. Hence it is not proper to say that certain truths (i.e. purely conceptual truths) have eternal existence, as if we wanted to say of others that they are perishable, or that something has stopped being true or will become true only in the future. How these phrases should be interpreted was already explained in § 25.

2. I have already pointed out that all *true* propositions must have an object with which they deal, though perhaps this does not hold for propositions in general. Hence all true propositions must contain an idea which refers to this object, so that the so-called subject idea (or basis) must in all true propositions be a proper referring idea (cf. § 66). The same was felt by Wolff when he wrote (*Logik*, Frankfurt and Leipzig 1740, § 533): "In a true proposition, the notion of the subject ought to be possible". After the discussions of the preceding part it should not be difficult to refute any apparent counter example.

a. To the objection that there must be truths whose subject is not a mere idea, but a complete proposition, we reply that this could be the case at best with truths that make an assertion about a whole proposition. If the object about which we make a statement is itself a proposition, then (it might be thought) the subject of our judgment is not a mere idea, but a complete proposition, namely the proposition of which we then speak. But we will give up this opinion as soon as we realize the following: in order to say correctly that a certain proposition deals with a certain object it is necessary, not that this object itself, but an *idea* of it be a component of our proposition. Thus, for example, when we make a judgment about this house, the proposition must contain an idea referring to this house; similarly, every truth which deals with a complete proposition must necessarily contain an idea of this proposition as a part, and this idea must be the subject idea. Hence there is no doubt that in truths of this kind, too, the basis is an idea, indeed a referring idea.

b. The matter is similar with those truths which seem to be concerned with so-called *imaginary objects* such as 'there is no round square'. We already know from § 137 that propositions of this kind mean merely this: 'the idea of a round square does not have a referent.' Hence its subject is again a referring idea, for only the idea 'round square' is non-referring; the idea of this idea (which is the subject of the proposition) is a referring idea. Its referent is that first idea.

c. Furthermore, a critic could point to propositions whose linguistic expressions have the form 'nothing—has—(attribute) *b*'. But I have already interpreted these linguistic expressions in § 170 in a way which shows that whenever they express a truth they do not lack a referring subject.

d. Finally, one could call attention to a very large number of propositions which are considered to be true, but where we do not care whether their subject ideas have a referent; in these cases, the existence of such an object depends upon quite accidental extraneous circumstances. For example 'a golden mountain would be bald'; 'whoever shows up at this place at such and such an hour must be suspected and should be taken into custody', and others. Closer inspection shows that expressions of this kind, whenever it becomes doubtful whether their subjects really have a referent, should only be understood as conditional assertions, roughly in the following way: 'if a mountain were made of gold, then it would be bald' and 'if somebody were to show up . . ., etc.'. After the interpretation of § 179, it will not be necessary for me to produce further proofs that expressions formed in this way from 'if' and 'then' contain a proper referring idea.

3. Just as the subject of every true propositions must be a proper referring idea, so the predicate part must be a proper attributive idea. It is presupposed that the copula in every proposition is the concept of having (§ 127) and that in negative propositions the concept of negation never applies to the copula, but to the predicate part (§ 136).

Hence it is obvious that if we want to say of something that an object has it, it must be a certain attribute of that object.

4. It is similarly obvious, given this assumption, that the *concretum* which belongs to the predicate idea of a true proposition must always be a referring idea which stands to the subject in the relation of inclusion (§ 95), hence which must be either equivalent to or higher than it. For, if the proposition '*A* has *b*' is to be true, then the idea *B*, or 'something which has *b*', must include every object that falls under the idea *A*.

5. As indicated in § 154, there is a theorem regarding all truths, to the effect that they must all be consistent with each other, no matter what ideas in them are envisaged as variable.

6. Every truth can be envisaged as deducible from innumerably many others, and from every truth innumerably many others can be considered deducible with respect to any ideas whatever.

7. Every individual truth, as well as every class of several truths, can be considered part of a single further truth; i.e. if *A*, *B*, *C*, *D*, . . . are truths, then the proposition that *A*, *B*, *C*, *D*, . . . are truths is a new truth, and is a single truth in which each of *A*, *B*, *C*, *D*, . . . appear as parts.

§ 197. *There are Analytic as well as Synthetic Truths*

To show that there are analytic truths B. cites the example 'every *A* which has (the attribute) *b* has the attribute *b*'. He points out that this proposition is analytic, because its predicate idea is a constituent of its subject idea, but he hastens to add that one cannot prove the existence of *synthetic* truths simply by citing an example of the form '*A* has *b*', where *b* is not part of *A*. Because, according to the definition of § 148, such a proposition can still be analytic if it contains any idea at all that can be arbitrarily altered without changing the proposition into a false one. B. considers the following example: 'this triangle has the attribute that the sum of its angles equals two right angles'. One of the constituent ideas of the subject of this proposition is the idea 'this'. The proposition will remain true under any substitution for 'this', so long as this substitution does not render the subject non-referring. Hence, the proposition is analytic, although its predicate is not contained in the subject. It is analytic because its subject contains a part (the idea 'this') such that every substitution for this part generates an idea which, if it refers at all, has a referent of which the predicate holds. "But without doubt there must also be propositions which do not have this peculiarity and which are not analytic truths merely because their subject ideas contain a part which can be arbitrarily changed while the predicate idea is kept constant. Propositions whose subject ideas are simple are of this sort, likewise propositions whose subject ideas are of the form 'something that has the attribute *x*', where the idea *x* is simple. For, if the entire idea which forms the subject of a given proposition is simple, then the only variable thing in that idea can be the idea itself. Hence, by arbitrary alteration of the subject idea, such a proposition can be made to deal with any object whatever . . . but then this proposition will not remain true if we keep its predicate idea constant and alter only its subject idea, provided only that the predicate is not the most general attribute of something in general." Hence, if such a proposition should be analytic, it must be due to the fact that its predicate idea contains some part that can be arbitrarily altered *salva veritate*. This would be the case with a proposition of the following form: 'every A has the attribute of falling under one of the ideas X or non-X' or 'every A has the attribute of falling under one of the ideas $[B]x$ or $[B]$non-x'; where the ideas X or x alone are considered variable. "For whatever is put in the place of X or x, the first proposition remains true so long as A has a referent and the second remains true if, in addition, B is an idea under which all A fall. But nobody will doubt that there must be predicate ideas which do not have this peculiarity. For if a predicate idea is simple, or if its composition is of the following sort: 'each of the attributes $x+y+z$ . . .', where the ideas x, y, z, . . . are simple, we can obviously neither consider the entire predicate idea, nor any of its parts as variable. For if the proposition 'A has x', or the proposition 'A has each of the attributes $(x+y+z+ \ldots)$' is to remain true, then neither the idea x in the first case, nor any of the ideas x, y, z, . . . in the second case, must be arbitrarily altered, since this proposition could certainly become false if any of these ideas is replaced by a non-referring one. After these preliminaries it can easily be proved that there must be truths of the following form: 'A (or something which has a) has attribute b (or each of the attributes $b'+b''+\ldots$)' where the signs 'A' and 'b', or in any case the signs 'a' and 'b''', 'b'''', . . . designate simple and distinct ideas, and where the attribute b, or any of the b', b'', . . . are not so general that they hold of anything whatever. For, to begin with, it is certain that there are simple ideas which have a referent, since every pure intuition is an idea which is simple, but still represents only a single object. It is even more obvious that there are attributive ideas which are simple and still do not hold of all objects. An example is the concept of reality; it is moreover clear that if all simple attributive ideas represented attributes which hold of all objects whatever, then there would be no composite attributive ideas which represent an attribute that holds only of some, but not of all objects. Finally, in order for an attribute to belong to all objects of a certain kind, it is not necessary that it should be represented as part of the idea of these objects. Hence among the objects that can be represented by the simple idea A, or the idea which is generated from a simple attributive idea a of the form 'something which has a', there will be some of which we can assert an attribute whose idea is altogether different from A or a. An example is the proposition 'A has reality' which is a truth of this sort, whenever A is a pure intuition. It is obvious that a proposition of this form cannot contain a single idea which can be arbitrarily altered without changing its truth. Hence we may call it a synthetic truth". B. then contends that there must also be synthetic truths of the form 'A has non-b'. "Who would doubt that there are very many attributes b that can be represented by a simple concept which can be denied of a certain object A which is represented by another simple idea, e.g. by an intuition."

Still another form of synthetic truth is the following 'the idea [something]

$(a+b+c\ldots)$ has a referent'. For if any of the a, b, c, ... is considered variable, it may be replaced by one which is inconsistent with the rest, which will make the proposition false. Likewise, the predicate idea can not be arbitrarily replaced. Similar considerations hold for propositions of the forms 'the idea [something] $(a+b)$ is imaginary'.

"Whoever admits the above will not hesitate to allow not only that there are analytic as well as synthetic truths, but that these truths are also found in both classes of propositions, namely intuitive as well as purely conceptual propositions. Thus the proposition 'every triangle is a figure' is an example of an analytic truth which is also purely conceptual, while the proposition 'this triangle is a figure' is an analytic truth, but intuitive. The example '*A* has actuality' where *A* is a pure intuition, is a synthetic intuitive truth, while all propositions of the form '[something] $(a+b+c+\ldots)$ has a referent' are synthetic conceptual truths."

In a note, B. criticizes Kant mainly because he never attempted to show that there are synthetic truths; he merely adduced all manner of examples from different sciences, especially mathematics and pure physics, and attempted to show by analysis of their constituent concepts that the idea of the predicate did not lie in that of the subject. "It seems to me, however, that the matter cannot be satisfactorily settled in this way. If it is not shown, first of all, that there are attributes which hold of all objects of a certain class, but are not found in their concept, then one cannot but retain a suspicion that in every proposition of the form '*A* has *b*' the concept of the predicate *b* is contained in the concept of the subject *A* even though analysis has so far not disclosed it . . .".

§ 198. *The Concept of the Relation of Ground and Consequence between Truths*

In my opinion, the most important of all relations that can hold between truths obtains when some of them form a ground and others their consequence. I have considered this relation on several previous occasions (especially in § 162) but now I must discuss it at greater length. To begin with, let us properly determine the *concept* of this relation. Consider the following three truths:

> 'nobody should prefer his own advantage to a greater advantage for others'.
>
> 'if somebody destroys the essential means of livelihood of other persons in order to provide himself with unnecessary sensual pleasure, then he prefers his own advantage to the greater advantage for others'.
>
> 'nobody should destroy the essential means of livelihood of other persons merely in order to obtain an unnecessary sensual pleasure'.

If we compare these truths with each other we notice that the first two stand in a very special relation to the third: the presence of this relationship is shown by a certain effect, namely that the last truth is seen with the greatest clarity if we have first taken cognizance of the other two, and have become conscious of them.

But it seems to me that further consideration shows that the nature of this relation is not exhaustively described if we consider only this *effect*, namely that knowledge of the last truth can be obtained from the first two. This relation also holds between truths which obviously do not stand in the same relation as those given above, for example between the following three:

> 'if the thermometer registers higher, then it is warmer'
> 'in summer, the thermometer tends to register higher than in winter'
> 'in summer it tends to be warmer than in winter'.

It is quite proper to say that we *recognize* the latter as soon as we have knowledge of the first two truths and have become conscious of them. Nonetheless, we cannot but feel that the relation between the truths of the first example is quite different from the relation in this case. If we want to express the peculiarity of the first relation in words we feel almost forced to call it a relation between *ground* and *consequence*, and to say that, in the first example, the first two truths were the ground of the last, and the last the consequence of the first two. But we cannot say the same for the second example. We sometimes say that the truth that it is warmer in summer than in winter is grounded on the truth that the thermometer registers higher in summer than in winter, but we must admit that in saying this we speak only of the recognition of these truths, and that we merely want to indicate that the recognition of one of these truths *causes* the recognition of the other. However, we do not want to say that the first truth is in itself the *ground* of the second; on the contrary, we want to assert that the exact opposite relation holds in this case. Now everybody must admit that the above example is not an isolated case; hence I conclude that there are truths which stand to each other in a relation which we can best describe as the relation of *ground* to *consequence*. The example shows that sometimes it is not just one, but a whole class of truths, which stand to one or more truths in the relation of ground to consequence; it is permissible in such a case to call the individual truths which make up these classes, *partial grounds* and *partial consequences*. . . . It happens all too often that the words 'ground' and 'consequence' makes us think of merely subjective grounds and consequences in the order of knowing, i.e. of truths which, as premises, produce a cognition, or which follow from another cognition. By contrast, I will sometimes call the grounds and consequences in the present sense *objective* grounds and consequences, in order to indicate that this relation holds among truths *in themselves* independent of our recognition.

NOTE: I have been strengthened in the view that there is a special relation of ground and consequence between truths by noticing that some of the most penetrating thinkers have been of the same opinion. It is known that Aristotle already (*Analytica Posteriora*, 78ª 22ff.) and for many centuries after him the scholastics have distinguished two kinds of proofs, namely those which show only the ὅτι, i.e. which show *that* something is the case, and those which show the διότι, i.e. *why* something is the case. Now since truths which show the διότι (the why) of another clearly stand in the relation of ground to the latter, we may conclude that the Stagirite was not unfamiliar with this relation between truths, even if he did not give it a special name. Nothing clearer can be desired than Leibniz' explanation of this relation (*Nouv. Ess.*, IV, XVII, 3): "The reason is the known truth whose connection with another less known one makes us give our assent to the latter. But particularly and pre-eminently we call it reason, if it is the cause not only of our judgment, but also of the truth itself, which we also call *reason a priori*; and the *cause* in things corresponds to the reason in truths. This is why cause indeed is often called reason, and particularly final cause. Finally, the faculty which perceives this connection of truths, or the faculty of reasoning, is also called *reason*." Thus he speaks of a connection (*liaison*) between truths, not as cognitions (*jugements*), but as truths in themselves (*de la vérité meme*). He attributes to them grounds (*raisons*) upon which they rest; and since the connection between these grounds and their consequences is called a connection between truths (*une liaison des vérités*), it is obviously assumed that these grounds are also truths. . . . It cannot be denied that this connection between truths in themselves has not often been discussed; but this is not surprising. First of all those who had framed the concept of a truth in itself at all did not spend a lot of time over it; secondly, the relation between ground and consequence has great similarity with a pair of other relations, namely the relation between truths that are deducible from each other, and the relation that holds between real things when one is the cause of the other. Hence it is not surprising that our relation was confused with one or the other of these and that this confusion did not become apparent since it did not lead to any obvious contradiction. The schoolmen already distorted the matter by using words '*ratio* (ground), *causa* (cause) and *principium* (origin) as synonymous, and defining the concept of these words as that which *determines* another (*id, quod determinat*); then they went on to distinguish two kinds of the latter, the *principium cognoscendi*, which determines a cognition, and the *principium essendi*, which determines a thing outside of our knowledge. German logicians tend to call the former the *logical ground* or *ground of knowledge*, or simply *ground*, and the latter the *real ground* or *cause*. . . .

§ 199. *Whether the Deduction Rule can be Considered a Partial Ground of a True Conclusion*

In the example which we used to illustrate the concept of ground and consequence in the preceding section, the truth which was the consequence stood in the relation of deducibility to the two truths which we considered to be its grounds. For each relation of this kind there is a certain *rule* which describes it, i.e. a proposition which indicates what attributes the premises *A*, *B*, *C*, *D*, . . . and the conclusions *M*, *N*, *O*, . . . which follow from them, must have. And if the rule according to which we want to deduce certain propositions *M*, *N*, *O*, . . . from others *A*, *B*, *C*, *D*, . . . is invalid, then it is obvious that we cannot claim that the propositions *M*, *N*, *O*, . . . are truths which *follow* from the truths *A*, *B*, *C*, *D*. . . . In reflecting on this one could get the idea that the rule, according to which propositions *M*, *N*, *O*, . . . are deducible from propositions *A*, *B*, *C*, *D*, . . . ought to be envisaged as a *truth* which must be added to the truths *A*, *B*, *C*, *D*, . . . in order to give the *complete* ground of the truths *M*, *N*, *O*. . . . On this view the two truths 'Socrates was an Athenian' and 'Socrates was a philosopher' do not constitute the complete ground for the third truth, which is deducible from them, namely 'Socrates was an Athenian and a philosopher'; rather, the rule, which is employed in this deduction must be added, namely the truth 'if the two propositions '*A* has *b*', and '*A* has *c*' are true, then the proposition '*A* has $(b+c)$' is also true'. I am quite sympathetic with this view but the following consideration forms a strong reason against it.

Assume that the complete ground of the truths *M*, *N*, *O*, . . . includes, besides truths *A*, *B*, *C*, *D*, . . ., from which they are deducible, also the rule which allows the deduction of propositions *M*, *N*, *O*, . . . from propositions *A*, *B*, *C*, *D*. . . . This amounts to saying that propositions *M*, *N*, *O*, . . . are true only because this rule of inference is valid and because propositions *A*, *B*, *C*, *D*, . . . are true; hence it is a tantamount to the following argument:

'If propositions *A*, *B*, *C*, *D*, . . . are true, then propositions *M*, *N*, *O*, . . . are also true; but propositions *A*, *B*, *C*, *D*, . . . are true, therefore propositions *M*, *N*, *O*, . . . are true'. But since every deduction has a rule, this one does too. If we abbreviate the first of the above propositions by *X*, and the second by *Y*, then the rule can be put in this way:

'If propositions *X* and *Y* are true, then propositions *M*, *N*, *O*, . . . are also true'. Now if it was required in the first place that the complete ground of the truths *M*, *N*, *O*, . . . must include, besides the truths *A*, *B*, *C*, *D*, . . . also the rule of their deduction, then it must be required, for the same reason, that the second rule of deduction also be added to the ground on the same footing as the first, since we can also say of the second rule that the truths *M*, *N*, *O*, . . . would not follow if it were invalid. We can see at once that this type of argument can be repeated *ad infinitum*, and that, if it were legitimate to add *one* rule of deduction to the ground of the truths *M*, *N*, *O*, . . ., an *infinite* number of them could be claimed to belong to this ground, which seems absurd.

§ 200. *Whether the Relation of Ground and Consequence is a Species of Deducibility*

Let us consider again the example of § 199, and let us assume that the deduction rule which allows the deduction of the consequence does not belong to its ground: then it is likely that the two stated truths, taken together, form the *complete* ground of the third. This would be an example of several truths which stand to each other both in relation of deducibility, and of ground and consequence. Considered as a relation of deducibility, the first two propositions form the premises, when viewed as a relation of ground and consequence, they are the grounds; and the truth which forms the conclusion under the first aspect is, under the second aspect, the consequence. The question is now whether these two relations always coincide in this way, or what distinction there is between them.

From what has been said in § 198, it is obvious that the relation of deducibility is not exactly like the relation of ground and consequence; it is also obvious that the difference between them does not merely lie in the fact that the former holds between propositions in general, the latter only between true propositions. (The propositions exemplifying deducibility without ground–consequence relation were also true.) But even if we do not assume that there must be a relation of ground and consequence wherever there is a relation of deducibility, the converse could still hold. It could be held that propositions that are related as ground and consequence must also be *deducible* from each other. In this case the relation of ground and consequence would have to be envisaged as a special kind of deducibility; the former concept would be subordinate to the latter. Though this seems quite probable to me, I do not know any proof that would justify such a view. I call propositions deducible from others, if I find in them ideas, which can be exchanged for other arbitrary ideas with the result that whenever some of these propositions become true, the others become true also. Hence, if it is asserted that the relation of ground and consequence is subordinate to the relation of deducibility it would have to be shown that a relation of ground and consequence can hold between certain propositions *A*, *B*, *C*, *D*, . . . on one hand and *M*, *N*, *O*, . . . on the other, only if there are certain ideas in them which can be exchanged for others in such a way that propositions *M*, *N*, *O*, . . . become true whenever propositions *A*, *B*, *C*, *D*, . . . are true. But how can we prove that this is, and always must be, the case?

On this view it is understood that the same relation of ground and consequence which holds between propositions *A*, *B*, *C*, *D*, . . . and *M*, *N*, *O*, . . . also holds between all those truths that result from them by alteration of the variable ideas *i*, *j*. . . . From this it follows that for every class of truths *A*, *B*, *C*, *D*, . . . *M*, *N*, *O*, . . ., which are related as ground and consequence, there is an infinite number of other classes of propositions which stand in the same relation, provided that infinitely many truths can be generated from the propositions *A*, *B*, *C*, *D*, . . . by alteration of the ideas in them which are considered variable. But is it even probable that for every class of truths from which another class follows as from their ground, there exists an infinite number of other classes of truths from which other truths follow in the same way, such that the particular nature of the ideas of which these classes of truths are composed has no influence upon their relation of ground and consequence? The following example seems to prove the opposite.

If we don't want to deny in general that there are relations of ground and consequence, we will be inclined to admit that there is a certain practical truth of the form 'you should do *A*' (or 'you should will *A*'), of such a nature that *all* other practical truths, e.g. that we should not lie, etc., can be derived from it as a consequence, provided that certain theoretical propositions of the form '*X* is necessary in order that *A* can take place' are added. But even that first truth (the so-called supreme moral law) seems to have a ground. For, if *A* is impossible, then there can be no duty to will it. Hence, the duty to do *A* is grounded either in part or completely in the truth that *A* is possible. But if we assume that this truth, or a class of several truths of which it is a part, is the complete ground of the supreme moral law, it is obvious that the latter does not follow from them as from its ground according to the usual rules of deduction. For, by assumption, there cannot be a single truth containing the concept of obligation among the truths from which the proposition 'you should do *A*' flows as a consequence; (if such a truth asserted an obligation, it would be practical). Hence we can see that none of the ordinary rules of deduction justifies the inclusion of the concept of obligation into the conclusion.

§ 201. *Whether the Concepts of Ground and Consequence Include the Concepts of Cause and Effect*

1. If the relation of ground and consequence is not a species of deducibility, we cannot hope to explain the former from the latter; hence we must look for another related concept. A concept closely akin to the concept of ground and consequence is doubtlessly the concept of causality, i.e. of the relation which holds between causes and effects. Perhaps we can use one of them to define the other.

2. The words 'cause' and 'effect' in their proper sense designate objects that are *real*. Only something which is real (something which has existence) can be called an effect, and something can be a cause only if it can bring about real things, and hence is itself real. From this it follows that grounds and consequences should not be considered kinds of causes and effects and that the relation of ground and consequence is not subordinate to the relation of causality; grounds and consequences are truths, not something which has reality, such as causes and effects.

3. This does not show, however, that the concepts of cause and effect are not part of the concepts of ground and consequence. The best surmise is that those truths which assert the existence and attributes of a cause are a *ground*, and those which deal with the existence and the attributes of the effect are to be considered *consequences*. The truth 'God is' could be considered the ground of the truth 'there is a world', since the existence of God is the cause, and the existence of the world the effect. But in this manner the relation of ground and consequence could hold only between truths which refer to something real, which state either the existence or the attributes of real objects. However, I think that this relation also holds

between truths with an entirely different content. Thus, mathematical truths can be related as ground and consequence, although they do not deal with objects that have reality. The truth that in an equilateral triangle all angles are equal is a consequence of the truth that a scalene triangle has two equal angles.

4. Hence we shall have to give up the idea of defining the relation of ground and consequence via the relation of causality. Rather, we shall have to derive the concepts of cause and effect from those of ground and consequence in the manner of § 168.

§ 202. *The Parts of the Concepts of Ground and Consequence*

B. points out that both of these concepts are *concreta*, hence that they contain the concept of a something as well as a concept of a certain attribute which determines them to be ground and consequence respectively. He points out that they are correlative terms, so that a definition of one of them will at once lead to the definition of the other. He takes the concept of ground to be the simpler of the two, indeed surmises that its associated *abstractum*, namely the idea of the particular attribute which makes something a ground, is simple and hence not definable.

§ 203. *That only Truths are Related as Ground and Consequence*

B. reiterates his view that only truths are related as ground and consequence and that other uses of these expressions can be explained in one of three ways, namely, (a) if we say of real things that one is the ground of another, we have in mind the relation of cause and effect; (b) if we speak of a ground–consequence relation in the case of propositions, regardless whether true or false, we have in mind the relation of deducibility; (c) if we assert that such a relation holds between mere ideas, we either have in mind actual propositions (as when we speak of a false idea) or we have in mind ideas as they actually occur in the mind of a thinking being; in the latter case the idea is a real object, and the case reduces to one of cause and effect.

§ 204. *Whether Something can be Ground or Consequence of Itself*

After pointing out that propositions and classes of propositions may be said to be deducible from themselves, B. denies that they can be their own grounds. In a note B. criticizes the view that so-called axioms are their own grounds. He thought that this was done only to satisfy the maxim that there is a ground for everything. He thinks that the grounds for axioms lie in true propositions which describe the concepts of which these axioms consist.

§ 205. *Whether Ground and Consequence are Always only Single Truths, or Classes of several Truths*

B. thinks that the total ground of a given proposition can sometimes consist of only a single truth, but he does not believe that the total consequence of any one proposition or class of propositions will ever consist of one truth only. Hence the complete consequence of a given ground will always be a class of propositions.

§ 206. *Whether a Ground can have several Consequences or a Consequence several Grounds*

It follows from the preceding section that a given ground can have only one total consequence, namely a certain class of propositions. On the other hand, it might seem as if different grounds can have the same consequence, but examples which seem to establish this only show that these consequences have some parts in common. "The complete consequence of certain truths *A*, *B*, *C*, *D*, . . . includes, among others, also the truth that each of the propositions *A*, *B*, *C*, *D*, . . . is true. But this is a consequence (a partial consequence) only of the class of truths *A*, *B*, *C*, *D*, . . ., and of no other. Thus it follows that each ground has precisely one consequence which is peculiar to it at least in some of its parts."

§ 207. *Whether the Consequence of a Part can be Envisaged as the Consequence of the Whole*

From what has been said in § 206 it follows that a consequence of part of the truths *A*, *B*, *C*, *D*, . . . is not a consequence of the total class of these truths.

§ 208. *Whether one and the same Truth, or a Whole Class of Truths, can be both Ground and Consequence, though in Different Respects*

B. points out that a given truth or class of truths can be the consequence of some, and the ground of other truths.

§ 209. *Whether Truths or a Whole Class of Truths can be both Ground and Consequence in one and the same Respect*

While there can be a relation of equivalence between propositions, B. doubts that the ground–consequence relation can hold in both directions between two propositions or classes of propositions. For example, if *M*, *N*, *O*, . . . are the complete consequence of *A*, *B*, *C*, *D*, . . . then they will include the proposition that *A*, *B*, *C*, *D*, . . . are all true. Hence that proposition would have to be a partial ground of the truths *A*, *B*, *C*, *D*, . . ., which is absurd.

"We sometimes speak of the mutual effects of things on each other . . .

(according to my own definition, an object can be the cause of another only if the truth which asserts its existence is the ground of the truth which asserts the existence of the other; thus it appears as if there are truths each of which is the ground of the other). But it soon turns out that, strictly speaking, not the objects which stand in this mutual relation are the cause, but only certain of their forces, and then it is not these things which have the effect but only certain of their changes. . . . For example, if a pair of spheres hit each other, they will change their shape by flattening somewhat; but the flattening of one is not the cause of the flattening of the other (in which case the cause would also be an effect); rather, the flattening of both of them is caused by the quantity of motion with which the two spheres approach."

B. then considers a further apparent counter-example, namely contraposition. From a proposition of the form 'what is *A* is also *B*', we can derive 'what is non-*B* is non-*A*', where the latter is a consequence of the former. In the same way we can drive the proposition 'what is not non-*A* is not non-*B*' from the second of the above propositions. But the ideas 'not non-*A*' and 'not non-*B*' are equivalent to the ideas *A* and *B*. Hence the third proposition can be translated into the first, and we may claim that the first proposition is a consequence of the second. But this would be a mistake. The first and third propositions are indeed equivalent, but they are not the same, and while the third proposition is indeed a consequence of the second, the first is not, since it is simpler, "and who would want to claim that a simpler truth is a consequence of a more complex one?"

§ 210. *Whether a Class of several Grounds can be the Joint Ground of the Class of their several Consequences*

B. denies this: "Who does not feel that the connection between ground and consequence, cause and effect, is much more *intimate* as that a mere conceptual combination of grounds and consequences should result in *one* ground and *one* consequence? . . . Would anyone wish to say that the class of the two truths 'God is perfect' and 'A scalene triangle has two equal angles' is the (joint) ground of the two truths that 'this is the best of all worlds' and 'in an equilateral triangle all angles are equal'? Do we not rather say that these are two grounds and two consequences?"

§ 211. *Whether the Parts of a Ground or of a Consequence are Rank Ordered*

In the relation of deducibility, there is no rank ordering among the premises; similarly, since a ground or consequence are a class of truths, "hence a mere *set*" (cf. § 84), there is no such ordering among grounds and consequences.

§ 212. *Whether the Partial Grounds of a Truth can be Grounds or Consequences of each other*

It was permissible, in the case of the relation of deducibility, to include among premises propositions which were themselves deducible from other premises. The same may also hold with respect to the ground–consequence relation. "There are truths which stand to each other as ground to consequence (at least as partial ground and partial consequence), and which can be combined as parts to bring about a new consequence." B. gives the following example, "the proposition that every truth is incorporeal is a truth from which it follows that this proposition itself is incorporeal. Hence if we are asked for the ground of the following truth 'the proposition that every truth is incorporeal is itself incorporeal', we cannot but reply that this ground lies in the two following truths: 'every truth is incorporeal' and 'the proposition that every truth is incorporeal is itself true'. Obviously, the second of these two truths is itself a consequence (partial consequence) of the first."

§ 213. *Whether a Consequence of a Consequence can be Considered a Consequence of the Ground*

B. begins this section by pointing out that the relation of deducibility is clearly transitive, and that common usage would also consider the relation of ground and consequence to be transitive. Against this he urges the following consideration. "Assume that the several real things *a*, *b*, *c*, . . . together are the complete cause of the real things *m*, *n*, *o*, . . ., and that the latter contain the complete cause of the real thing *r*: then the complete ground of the truth '*r* is the case' must lie in the truths '*m* is the case', '*n* is the case', '*o* is the case', . . . and the complete ground of these must lie in the truths '*a* is the case', '*b* is the case', '*c* is the case'. . . . But if the ground of a ground is always ground of the consequence, then the complete ground of the truth '*r* is the case' must also lie in the truths '*a* is the case', '*b* is the case', '*c* is the case'. . . . Consequently, the existence of things *a*, *b*, *c*, . . . is a complete cause for the existence of thing *r*. Hence both the things *m*, *n*, *o*, . . . as well as things *a*, *b*, *c*, . . . are a complete cause of *r*. Hence one and the same object has several complete causes for its existence. Should we admit this? Are we not, rather, justified in asserting that different causes (providing they are not partial causes, but complete causes) have different effects?"

Another argument against the transitivity of the ground–consequence relation is that often the ground for a single truth *M* lies in several truths *A*, *B*, *C*, *D*, . . . such that each of the latter has its separate ground. In such a case we cannot say that the ground of the ground is the ground of the consequence, because there is no (individual) ground of the ground, and according to § 210 the sum of the grounds is not the ground for the sum of their individual consequences.

§ 214. *Whether every Truth can be Envisaged, not only as Ground, but also as Consequence of Others*

In the case of deducibility, there is no doubt that every proposition can be considered both a premise of, as well as a conclusion from, other propositions. The question is now whether a similar situation holds with respect to the relation of ground and consequence, i.e., whether every truth can be considered ground as well as consequence of other truths.

The first part of this question, namely whether for every truth there are others that follow from it, has already been decided above. I have repeatedly assumed that the proposition '*A* is a truth' can be considered a true consequence of the truth *A*; if this is true, then it is obvious that for every given truth at least one other truth can be found which is a consequence of it.

The only contentious part of our question is whether every truth can also be envisaged as a *consequence* (at least partial consequence), i.e. whether for every truth we can find another truth or class of them which can be considered its ground. It is my surmise that this must be answered in the negative; I think that there are and must be some truths which have no further grounds for their truth, but I must confess that I have not been able to find a satisfactory proof for this.

What led me to this surmise are mainly several examples which seem to me to belong to the kinds of truths which have no ground. One of them seems to be the proposition that there is something, or (as I think it should be put) that the idea of something has a referent. Every other truth which could be thought to be its ground, must itself be a consequence of it or of others that are its consequences etc.

Once it is admitted that there are truths that have no further ground they will be considered important enough to give them a special name. Since they are only ground, but not consequences, I will call them *basic truths*, all others *derived truths*.

NOTE: It is somewhat surprising that the existence of basic truths in the just-mentioned sense has so far rarely been recognized. Leibniz speaks several times of *verités primitives* (e.g. in *Nouv. Ess.*, I, II [?]; II, XXI, 3; IV, II, 1) and he seems to understand exactly the same thing that I call basic truths. But most logicians substitute for this the concept of a truth which is *self-evident*, i.e. one which we recognize without any proof, where some logicians add the remark that they are truths for which no proof can be given at all. But that this is not a definition of a relation that holds between truths in themselves is obvious. Others, who extended the range of the well-known law of sufficient reason somewhat too far, define basic truths as truths which contain in themselves a sufficient reason for their truth.

The note concludes with a criticism of several contemporary logicians.

§ 215. *Whether there are several Basic Truths*

B. assumes that the question must be answered in the affirmative, since one truth would not be sufficient to have all others as its consequence and consequence of consequence, etc.

§ 216. *Whether the Ascension from Consequence to Ground must Find an End for every Given Truth*

B. calls the ascertainment of the ground of a given truth, and the ground of this ground, etc., etc., the *ascension from consequence to grounds*. In the case of certain truths, namely basic truths, this ascension cannot be carried out, and for consequences of basic truths, and their consequences, it can go back only as far as these basic truths themselves. But from what has been said, it does not follow that the ascension to the grounds must come to an end no matter what truth we started with. "For even if there are truths which do not depend upon further grounds, there can still be other truths which have a ground, and the truths which are their ground again have a ground and this goes on *ad infinitum*. If I am not mistaken, an example of such a truth is every proposition which describes one of the contingent states or attributes of a created substance. Since every such state has a cause at least part of which lies in the preceding state, we find here a sequence of causes which goes to infinity. But if there is an infinite sequence of causes, there must also be infinite sequences of grounds, since the real thing M can be a cause of the real thing N only if the truth that M is the case is a ground or partial ground of the truth that N is the case."

§ 217. *What the Author Means by Supporting Truths*

Partial grounds, complete grounds, and grounds of grounds, etc., of a truth are called the supporting truths of that truth.

§ 218. *No Truth can be a Supporting Truth of Itself*

Otherwise we would have the absurdity that the truths through which we recognize a given truth A include A among their number.

§ 219. *Whether one and the same Truth can Appear as Supporting Truth several Times*

The question is answered in the affirmative.

§ 220. *A Graphic Representation of the Relation of Ground and Consequence among Truths*

B. offers the following diagram.

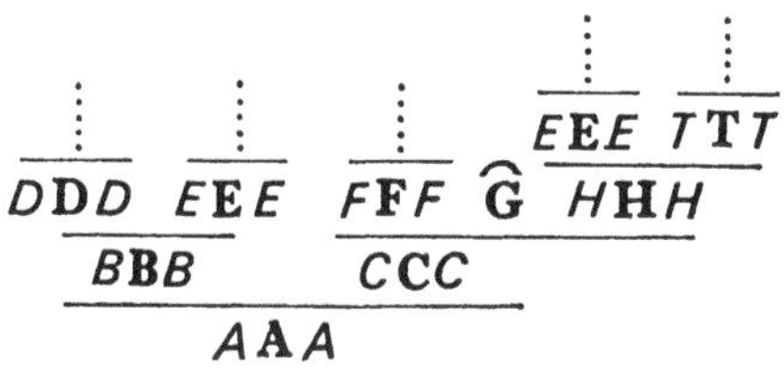

A is to represent the truth with which we begin in the ascension to the grounds. The immediate and proper ground of *A* is to be the junction of the truths *B* and *C*; the arc above *G* is to indicate that *G* is a basic truth. According to § 205, every truth is accompanied by several others which follow from the same ground. But since these are not always required for the derivation of further consequences, they are indicated by italic letters. The horizontal lines above letters and the dots forming a vertical line are to indicate that the business of ascension has not ended here. Since one and the same truth can occur several times as a supporting truth, it has to be entered several times. The following alternative representation avoids this (*E* is written only once).

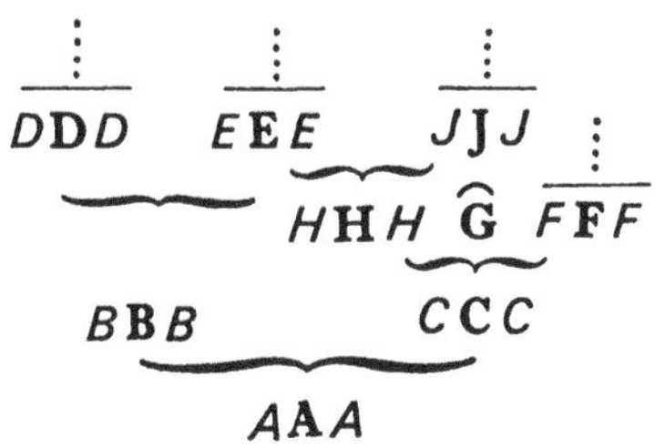

"But it is obvious that in cases where one and the same truth is repeated several times the representation becomes very confusing. Let others find a better method."

§ 221. *Some Criteria which Allow us to Determine whether certain Truths Stand in a Relation of Dependence to each other*

Nothing would be more desirable for the purposes of logic than to find several general criteria which allow us to find out whether given truths stand in a relation of ground and consequence, or at least in a relation of dependency to each other, i.e., whether one of them can be called a ground or partial ground of the other, or at least a supporting truth. "But I can only give a few surmises."

1. First of all, it seems to me that the distinction made in § 133 between conceptual and intuitive propositions allows an important application at this point. It seems that truths which contain only pure concepts (conceptual truths) have the peculiarity that they depend only on other conceptual truths, never upon intuitive propositions. However, it is the case that intuitive truths (experiences) are often helpful in the *discovery* of a purely conceptual truth; but the objective ground of such a truth cannot lie in them; if they have any ground at all, it must lie in other conceptual truths.

2. Furthermore, I believe that a purely conceptual truth upon which a second one depends must never be more complex than the latter, though it does not have to be simpler. Propositions which form the objective ground of a purely conceptual truth must separately contain no more simple parts than the truth which depends upon them. Consider the common argument form:

Whatever has *a* has *b*
Whatever has *b* has *c*

Whatever has *a* has *c*

In my opinion, such an argument does not represent a relation of ground and consequence if either the concept *b* is more complex than *a* or *b* more complex than *c*. In the first case, the second, in the second case the first, of the two premises would be more complex than the conclusion. But I do not wish to generalize my assertion and extend it to intuitive propositions, since they appear to furnish some counter-examples. Let *J* be an intuition, i.e. an idea which, though simple, refers only to a single real object; in this case there is no doubt that there is an infinite number of truths which have the idea *J* as their subjects, and which have as their predicate an attribute which really holds of this object. Many of these attributes (in particular those whose idea is quite complex) will be grounded in others. Thus, for example, if *J* designates a cry which I have just heard, then the truth that this cry obliges me to find out whether there is some victim, has its ground in several other truths, for example in the truth that this cry has a similarity to a cry for help. But it is quite obvious that in such a case the complete ground of a truth such as '*J* is *B*' cannot lie in a single truth of the form '*J* is *A*', but only in the combination of the latter with still another truth of the form 'every *A* is a *B*'. No matter how simple the idea *a* (the *abstractum* which belongs to the *concretum A*), the proposition '*J* is *B*' is always somewhat simpler than the proposition '*A* is *B*', the latter of which is nonetheless a partial ground of the former.

3. If the views expressed above in nos. 1 and 2 are accepted, and if we also accept the hypothesis that there is only a finite number of simple concepts, then it is easy to give a strict proof of the assertion given in § 214, that there are truths which do not have a ground, though this proof will be restricted to conceptual propositions.

If the number of all simple concepts is finite, then the number of all pure conceptual truths which do not exceed a certain degree of complexity, for example, which do not contain more parts than a given proposition *M*, can also be only finite. But if somebody asserts that every pure conceptual truth must have a ground, then he must also assert that every truth *M* rests upon an infinite number of supporting truths; for if every ground requires another ground, *ad infinitum*, then the number of propositions upon which truth *M* depends is infinite. But if these supporting truths are all purely conceptual propositions and are to be of lower complexity than *M* itself, then this leads to a contradiction with what we have said above. This then shows that, at least for the domain of purely conceptual truths, there are genuine basic truths. If all this is accepted, I should like to go on to the following assertion: if we consider the class of all purely conceptual truths whose degree of complexity does not exceed a certain limit (e.g. all truths which do not contain more than 100 simple parts), then it follows from what has been said that every truth which is a member of this class and which is not itself a basic truth must find its complete ground in this class of propositions; this holds not only for immediate, but also for remote grounds up to the final ones which have no further grounds. We also know already that this class will contain truths *A*, *B*, *C*, . . . which have no ground at all, in addition some other truths *D*, *E*, *F*, . . . which have a ground, but which are not deducible from this ground (cf. § 200). Finally, all other truths *M*, *N*, . . . *Z*, which are also contained in this class are consequences which are deducible from their ground. But if we can change the order of these truths at will, then it must obviously be easy to arrange them in groups in such a way that, if only a number of them are accepted, the others are deducible from them. Now I wish to assert that if we only always observe the rule never to deduce a simpler truth from premises, any of which is more complex, then the number of propositions which we must accept outright (i.e. without deduction from others) will be the smallest if we arrange the propositions according to their objective connection. Hence, it seems to me that the relation of ground and consequence has the following peculiarity: if we order propositions according to this relation, then the smallest number of premises will allow us to deduce the largest number of conclusions, so long as they are more complex than their premises.

4. Another peculiarity of this relation seems to lie in the fact that each of the truths *A*, *B*, *C*, *D*, . . ., which together form the ground of truth *M*, is simpler than any of the propositions that are equivalent to it. Another characteristic is that every truth which must be considered a consequence of the several truths, *A*, *B*, *C*, . . ., is always simpler than any proposition that is equivalent to it. Thus it follows that all truths which are equivalent to a given individual truth *M*, and which are more complex than it, are always consequences of this truth alone, etc.

5. In general, it seems to me that the truths *A*, *B*, *C*, . . . which form the ground of truth *M*, which is also deducible from them, must always be the simplest class of truths from which *M* is deducible, provided that always the same ideas are considered variable, and that none of the premises are individually more complex than the conclusion.

6. Not only must propositions *A*, *B*, *C*, . . . be the simplest, but they must also be the most general propositions from which *M* is deducible. For example, I do not consider the proposition that Caius has duties toward God a consequence of the true proposition that Caius is a man and that all men have duties toward God. For these two propositions are not the most general from which the first is deducible, since not only all men, but also all rational and finite beings have duties toward God. It seems to me that the argument cited in no. 2 indicates a relation of ground and consequence only if *B* and *C* are equivalent ideas; for, if this is the case, the two propositions 'all *A* and *B*', and 'all *B* are *C*', are certainly the propositions of greatest scope from which the conclusion 'all *A* are *C*', is deducible.

7. If a proposition *M* stands to other propositions *A*, *B*, *C*, . . . in the relation of irredundant deducibility (cf. § 155, no. 26) with respect to ideas *i*, *j*, . . ., if, moreover, propositions *A*, *B*, *C*, . . . and *M* are the simplest propositions in their respective equivalence

classes, and if none of *A*, *B*, *C*, . . . is more complex than *M*, then we may assume that *M* stands to *A*, *B*, *C*, . . . in a true relation of ground and consequence, in such a way that whenever *i*, *j*, . . . are replaced by ideas which make proposition *A*, *B*, *C*, . . . not only true, but keep them from becoming redundant, then truth *M* is a proper consequence of truths *A*, *B*, *C*. . . . Hence I do not hesitate to claim the relation of ground and consequence for the following propositions.

Every *A* is *B*
Every *A* is *C*

Every *A* is *B* as well as *C*

The last proposition is deducible from the first two, but they are also deducible from it and they all fulfil the above stipulation of simplicity.

NOTE: A good deal of what I have said in this section is generally accepted, for if it is said that *a priori* truths cannot be proved from experience, it can only mean that the objective ground of a purely conceptual truth cannot lie in observational propositions. And if it is said that in a truly scientific exposition we must proceed from the more general to the more specific, then this can derive only from the notion that the more general and simpler truths are the ground of the more particular and more complex ones. This also seems to be the most suitable place to admit that I occasionally doubt whether the concept of ground and consequence, which I have above claimed to be simple, is not complex after all; it may turn out to be none other than the concept of an ordering of truths which allows us to deduce from the smallest number of simple premises the largest possible number of the remaining truths as conclusions.

§ 222. *What the Author Calls Conditions of a True Proposition and Connections between Truths*

Sometimes the grounds for a certain proposition are deducible from it, sometimes they are not. For example, the propositions 'every *A* is *B*', and 'every *A* is *C*' are the ground of 'every *A* is *B* as well as *C*', and are also deducible from the latter. On the other hand, in the argument form cited in no. 2 above, the premises, which may be grounds for the conclusion, are not deducible from it. Whenever a supporting truth *A* of *M* is deducible from the latter, it will be called a *condition* of it. In this sense the proposition '*A* is *B*' is a condition for the truth of the proposition '*A* is *B* as well as *C*'.

I have now given the reader a concept of all the more important relations which are more or less directly connected with the relation of ground and consequence. I should like to call these relations the *connections* which hold *amongst truths in themselves*, sometimes also the *objective* connections between truths. Hence, if I use later on the expression that we must investigate the objective connection between certain truths, I shall always mean that we must investigate whether these truths stand in the relation of ground and consequence to each other, which of them is the (complete or partial, immediate or remote) ground of another, which is supporting truth or condition, which is (immediate or remote, complete or partial) consequence, which must be considered basic truths, etc.

NOTE: This objective connection between truths as well as the concept of objective truths themselves is presently almost completely forgotten. However, there is a lot of discussion of a certain *organic* connection; but it seems to me that it is usually discussed in the context of cognition, rather than truths in themselves, and that it is claimed to be a requirement of scientific exposition; hence it will be discussed later.

PART IV

Of Arguments

§ 223. *Contents and Purpose of this Part*

Part of the purpose of logic is to state the most general rules according to which conclusions can be drived from premises. But since from every (consistent) proposition an infinite number of others can be deduced, "it would be unreasonable to demand a listing of all propositions that can be deduced from one or several given propositions. Moreover, according to the very wide sense in which I have taken the word deducibility (§ 155) the validity or invalidity of some deductions can be assessed only if we have knowledge of matters outside logic. Thus from the proposition 'this is a triangle' we may deduce the proposition 'this is a figure the sum of whose angles equals two right angles' (with respect to the idea 'this'), and from the proposition 'Caius is a man', we can deduce the proposition 'Caius has an immortal soul' (with respect to the idea 'Caius'). For whenever we replace the indicated idea by some other idea, the conclusions become true whenever the premises are true. But to realize this, we must know two truths, namely that the sum of the angles in any triangle equals two right angles, and that the souls of all men are immortal. Since these are truths which are not at all concerned with logical objects, i.e. with the nature of concepts and propositions, or rules according to which we must proceed in scientific exposition nobody will demand that logic should teach deductions of that sort. Hence, what can be expected in this place is only a description of those modes of deduction whose correctness can be shown from logical concepts alone, or, what comes to the same thing, which can be expressed in the forms of truths, in which nothing is mentioned except concepts, propositions, and other logical objects. An example is the way in which the proposition that Caius has an immortal soul can be deduced from a union of two premises, namely that Caius is a man, and that all men have immortal souls, where the two ideas 'man' and 'soul' are considered variable in addition to the idea 'Caius'. To see the correctness of this deduction nothing is required save knowledge of the general truth that from two propositions of the forms '*A* is *B*' and '*B* is *C*' a third proposition of the form '*A* is *C*' is deducible. But we can see this without knowing anything at all about the nature of man, about death, etc." Completeness in this undertaking could be achieved only if all the different kinds of proposition discussed in Part II are taken as premises, both singly and in combinations of two and more. But the scope of this undertaking would be far too large, hence only the most important forms, and the most noteworthy combinations of them, especially combinations of two premises, will be considered.

In the written version of these arguments, only those ideas will be considered variable which are indicated by general signs (letters) "for example when I claim that from the two propositions 'whatever has *a*, has *b*', and 'whatever has *b* has *c*', we derive the proposition 'whatever has *a* has also *c*', it is to be understood that the relation of deducibility between these propositions obtains if all the ideas indicated by letters *a*, *b*, *c*, and no others, are considered variable. In this way my words will never represent propositions themselves which stand in the relation of deducibility to each other, but only the *forms* which these propositions must have; hence I will not represent arguments themselves but only their forms (the rules according to which they must be formed)."

§ 224. *Some General Rules for Finding Conclusions to Given Premises*

Before I begin the development of individual conclusions which can be deduced from a given combination of premises, it would seem reasonable to premise some generally useful rules on how conclusions to given premises can be found.

1. Given that we have already deduced certain conclusions *M*, *N*, *O*, . . . from certain propositions *A*, *B*, *C*, *D*, . . ., and given that we know that from one or more of *M*, *N*, *O*, . . . some new conclusions *R*, *S*, *T*, . . . can be deduced, always with respect to the same variable ideas *i*, *j*, . . ., then, according to § 155, no. 24, it will be permissible to consider the latter to be conclusions from the given premises. However, if we have found propositions *R*, *S*, *T*, . . . in this way, we will not know whether they stand in the relation of irredundant deducibility to the given premises *A*, *B*, *C*, *D*, . . ., even if we know that the arguments according to which *M*, *N*, *O*, . . . are derived from *A*, *B*, *C*, *D*, . . . and *R*, *S*, *T*, . . . from *M*, *N*, *O*, . . . are irredundant arguments (cf. § 155, no. 34).

2. Assume that we are to find conclusions to certain propositions *A*, *B*, *C*, *D*, . . .; assume furthermore, that we have on another occasion considered *A*, *B*, *C*, *D*, . . . together with some other propositions *E*, *F*, *G*, . . ., and that we then deduced from the total of these two sets the conclusions *M*, *N*, *O*, . . . with respect to ideas *i*, *j*. . . . In this case we may say that propositions *M*, *N*, *O*, . . . become true whenever the truth of *E*, *F*, *G*, . . . is added to the truth of propositions *A*, *B*, *C*, *D*. . . . Hence we may form the following hypothetical judgment as a conclusion which follows from propositions *A*, *B*, *C*, *D*, . . . alone: 'if *E*, *F*, *G*, . . . are true, then *M*, *N*, *O*, . . . are also true'. In this way we obtain the following argument:

A, *B*, *C*, *D*, . . .
____
If *E*, *F*, *G*, . . . are true then *M*, *N*, *O*, . . . are also true.

It is quite different from the argument:

$$\frac{A, B, C, D, E, F, G, \ldots}{M, N, O, \ldots}$$

The difference between them can be seen if we express both of them in a way in which the sense of every argument should be expressed, according to § 164. Then the second argument becomes 'every class of ideas whose substitution for *i*, *j*, . . . makes all the propositions *A*, *B*, *C*, *D*, *E*, *F*, *G*, . . . true also makes all the propositions *M*, *N*, *O*, . . . true'; but the other argument becomes 'every class of ideas whose substitution for *i*, *j*, . . . makes propositions *A*, *B*, *C*, *D*, . . . true also makes the proposition true that every class of ideas, whose substitution for *i*, *j*, . . . makes propositions *E*, *F*, *G*, . . . true also makes propositions *M*, *N*, *O*, . . . true'. In this way, from every given argument with *n* premises, we can derive other arguments which have only $n-1$, $n-2$ . . . even only one of those premises. Thus, from the two premises '*A* is *B*', '*B* is *C*', the conclusion '*A* is *C*' follows. Hence we will be justified to deduce, from the single premise '*A* is *B*' the conclusion 'if *B* is *C*, then *A* is also *C*'. In this case, if the argument, which allows us to deduce *M*, *N*, *O*, . . . from the class of propositions *A*, *B*, *C*, *D*, *E*, *F*, *G*, . . . is irredundant, then the new argument is also irredundant.

3. If we have found that certain propositions *H*, *J*, *K*, . . . allow the deduction of the negation of one or more given premises *A*, *B*, *C*, *D*, . . ., or even the negation of a conclusion *M* or *N* that is deducible from them, we may conclude that propositions *H*, *J*, *K*, . . . are never all true when *A*, *B*, *C*, *D*, . . . are. Hence, we obtain the conclusion 'the class of propositions *H*, *J*, *K*, . . . is not a class of nothing but true propositions'. But this argument does not allow us to ascertain whether this conclusion is irredundantly deducible from its premises. For, even if propositions *H*, *J*, *K*, . . . yielded by an irredundant argument the negation of one of *A*, *B*, *C*, *D*, . . ., or the negation of a conclusion which followed by an irredundant argument from *A*, *B*, *C*, *D*, . . ., it could be still the case that the falsity of one of the propositions *H*, *J*, *K*, . . ., hence the correctness of the proposition that the class of these propositions is not a class of nothing but true propositions, could be deducible only from a part of the proposition *A*, *B*, *C*, *D*. . . . Thus, from the two propositions:

All men are mortal (*A*)
Caius is a man (*B*)

We can irredundantly deduce:

Caius is mortal (*M*)

Similarly from the two premises 'All men are immortal' (*H*) and 'Caius is a man' (*I*) we can irredundantly derive a proposition from which the negation of the above conclusion can be deduced. Hence it is true that from the two first premises, *A* and *B*, we can derive the conclusion 'the propositions, 'all men are immortal' and 'Caius is a man' are not both true'. But this deduction is not an irredundant one, since the falsity of the proposition 'all men are immortal' follows from the first of these premises alone; hence the deduction of this conclusion does not require any more than that first premise.

§ 225. I. *Conclusions from One Proposition of the Form* 'A *has* b'

1. First of all I wish to investigate what conclusions can be obtained from a single proposition when we assume as many variables in it as any proposition can have. According to § 127 a proposition is present if two ideas *A* and *b* are connected by the concept of the word 'has'. Thus the expression '*A* has *b*' can represent a proposition.

2. It is obvious that every true proposition remains true if its subject idea *A* is replaced by an equivalent idea, e.g., by the following: 'anything having the attribute of falling under idea *A*'. Thus it follows that the generality of the investigation is not limited if we replace the form '*A* has *b*' by the form 'anything having *x* has *b*' or 'whatever has *x* has *b*'. I make this remark since I shall use the latter form in the sequel, it being more convenient for certain purposes.

3. It is similarly obvious that every true proposition '*A* has *b*' remains true if we replace *b* by the idea 'the attribute non-non-*b*'. Through this change the proposition becomes negative, if it was not already negative; hence it follows that all conclusions that can be derived from affirmative propositions can also be derived from negative ones.

4. If I was correct in § 196 in asserting that every true proposition has an object with which it deals and which is represented by its subject, then it follows that every determination of ideas *A* and *b* which makes '*A* has *b*' true will also make true the proposition 'the idea *A* has a referent'. Hence it will be permissible to consider the latter a conclusion which follows from the former.

5. If I was also justified in saying that the predicate part of every true proposition must be an attributive idea, then it follows from '*A* has *b*' that *b* is an attribute; other propositions that follow are 'the idea *b* has reference'; likewise 'the idea *B* has reference'.

6. Furthermore, if '*A* has *b*' is true, then the attribute *b* applies to

the objects that fall under A; this can be envisaged as a relation between A and b; we are thus led to the following conclusion: 'The relation between the A's and b is a relation between certain objects and an attribute that applies to them.' This conclusion is obviously equivalent to the given proposition and can be considered an objective consequence of it.

7. If every A has the attribute b then the idea of an A which has attribute b is a referring idea, but the idea of an A which does not have attribute b is a non-referring idea. Hence we can draw the following two conclusions; 'the idea $[A]b$ has a referent' and 'the idea $[A]nb$ does not have a referent'.

B. goes on to maintain that the following conclusions can also be drawn from the given proposition: 'the class of all B is not a part of the class of A'; 'the idea B is either equivalent to A or higher than A'; 'A and $[A]b$ are equivalent ideas'; finally 'either the idea B has the extension of the widest idea, i.e. of the idea of something in general, or the proposition holds that whatever does not have b does not have a'.

In a note, B. points out that the proposition under discussion is usually written as 'every A is B', or in the case of negation 'no A is a B'. Furthermore, the proposition 'the idea of an A which is b has a referent' is generally written as 'some A are B'. "The so-called particular propositions of all logicians, such as the proposition 'some A are B' are by no means to be taken literally, i.e. they are not to be understood as if there must be some, i.e. several, A's which are B in order for the proposition to become true; rather, it suffices for the truth of this proposition if only a single A is a B, i.e. if the idea of A which is B has any referent at all." In the sequel B. maintains that contra-position is not "strictly valid". "Providing that I was correct in asserting (§ 196) that it is required for the truth of a proposition that the subject idea has a referent, then I do not see how we can say that the proposition 'every non-B is non-A' is always true whenever the proposition 'every A is a B' is true. For, if we put in the place of B one of the ideas which have the largest extension (the idea of something in general is of this sort), then the proposition 'every A is B' becomes true no matter what A is. Now if this argument is valid then the words 'every non-B is a non-A' must also assert a truth. However, the idea non-B (not something, i.e. nothing) does obviously not have any extension at all; hence these words do not express a true proposition. The invalidity of this argument becomes even more obvious with negative propositions such as 'no A is a B', where we replace B by an idea which has no extension at all, for example the imaginary idea of a round square. Then this premise becomes true no matter what A is; hence we can claim the status of valid arguments for the most absurd combination of concepts. It is undoubtedly true that no object which is not in space is a round square. The ordinary argument from contra-position allows us to deduce the following proposition from this truth: 'every round square is spatial'. In the same way the following proposition is undeniably true: 'whatever is not a man is not a round square'; but by contra-position we arrive at 'every round square is a man'. Hence, if we want to save this form of argument, we must stipulate for the affirmative case that the idea non-*B* has reference, and for the negative case that the idea *B* has reference. This condition must be added either to the premise or to the conclusion, i.e. either we must introduce two premises in the following way:

> Every *A* is *B*
> The idea non-*B* has a referent
> ___
> Every non-*B* is also a non-*A*.

or we must choose a single premise, and use a disjunctive conclusion in the following way:

> Every *A* is *B*
> ___
> Either the idea non-*B* does not have a referent, or every non-*B* is also a non-*A*."

In the sequel B. criticizes Herbart, who maintained that the universal affirmative judgment does not imply the assumption that the subject term has a referent. A similar criticism is levelled against Fries: "Fries also seems to be of the opinion that from the truth of a proposition we cannot infer with certainty that its subject idea has a referent. He writes (*System der Logik*, Heidelberg 1811, p. 215): the assertoric modality is most clear in propositions which merely affirm an existence, e.g., 'eagles exist', 'there are no griffins'. But if I say 'all griffins are birds' then I merely consider the apodictic modality of a law of thought. If I want to convert to 'some birds are griffins', then the particular judgment, since it cannot be maintained without experience, claims the assertoric modality, and is false, since there are no griffins." "This remark is based upon a correct observation concerning language usage. The particular proposition 'some birds are griffins' does indeed have no sense other than that there are birds which have the attribute of being griffins, and hence that the idea of a bird which is a griffin has a referent. However, the general proposition 'all griffins are birds' is indeed often used merely to express the hypothetical judgment 'if a given object is what we think by a griffin, then it is a bird' (§ 196). The latter we can indeed say without asserting the existence of such a bird; hence it appears as if general propositions of the form '*A* has *b*' could be true although the idea *A* does not have a referent. But this illusion disappears as soon as we realize that using it in this sense is just a misuse of the form '*A* has *b*'. If the proposition '*A* has *b*' (all griffins are birds) is to retain its proper sense, then its truth demands undeniably that there should be griffins; otherwise how could logicians maintain that from every general proposition a particular one can be deduced, unless the proposition 'some birds are griffins' already lies in the proposition 'all griffins are birds'."

§§ 226–229. *Conclusions from the Combination of several Propositions of the Form* 'A *has* b'

In these sections B. considers arguments with premises which are chosen from the following list:

1. whatever has *a*, has *b*
2. whatever has *b*, has *a*
3. whatever has *a* does not have *b*
4. whatever has *b* does not have *a*
5. whatever does not have *a*, has *b*
6. whatever does not have *b*, has *a*
7. whatever does not have *a* does not have *b*
8. whatever does not have *b* does not have *a*

He then considers conclusions that may be drawn from subsets of two of them, among them a number of the customary syllogisms.

§ 230. II. *Conclusions from a Proposition which is the Negation of One of the above Forms*

After describing a number of obvious conclusions from the indicated premises, B. takes exception to the following arguments:

I It is false that all *A* are *B*
———
1. It is false that every non-*B* is a non-*A*
2. Some *A* are not *B*

II It is false that no *A* is *B*
———
1. It is false that no *B* is *A*
2. Some *A* are *B*

"In my opinion none of these arguments is without flaw, for, if we replace *A* by any non-referring idea, then the premises in both arguments become true, no matter what is meant by *B*, although the conclusions do not. It is certainly correct to assert that the proposition 'every round square is a sensitive being' does not have truth. But according to the argument, the following propositon should then also be false: 'whatever is not a sensitive being is also not a round square'. But must we not admit that this proposition is completely correct? On the other hand, according to the second half of the first argument, it should be true that certain round squares are not sensitive beings hence, that there are round squares which are not sensitive beings, consequently that there are round squares."

§ 231. *Combinations of Propositions of Forms I and II*

In this section, B. considers the negations of the eight propositions stated in § 229, and discusses conclusions that can be drawn from sets of propositions taken from the combined lists. This leads to a discussion of some syllogisms and also of arguments of the following sort:

Whatever has *a* has *b*
It is false that whatever does not have *a* also does not have *b*
———
Either the idea *A* has the extension of the idea of something in general, or the idea *B* is higher than *A*.

§ 232. *Combination of Negations with Negations*

Here B considers such arguments as:

It is false that all *A* are *B*
It is false that *B* are *A*

If the two ideas *A* and *B* have referents, then they stand in the relation of either exclusion or independence.

§ 233. III. *Conclusions from the Assertion that an Idea has References*

B. considers premises of the form 'the idea *A* has a referent' or 'the idea of an *A* which is *b*' has a referent. This leads to the consideration of syllogisms with particular premises, and several other argument forms.

§§ 234–238. *See Table of Contents*

In these sections B. explores the combination of various forms of propositions as premises.

§ 239. IV. *Conclusions from the Assertion that an Idea is a Singular Idea*

According to B. an assertion of the form 'the idea *A* is a singular idea' allows the conclusion 'of the two propositions '*A* has *x*' and '*A* does not have *x*' one is always true, no matter what idea is put in the place of *x*'.

Later, B. explores combinations of the assertion of the singularity of an idea with various other propositions mentioned earlier. The following is an example of the many argument forms considered in these sections:

The idea [something] $(a+b)$ is a singular idea
The idea [something] (non-$a+b$) is a singular idea

There are exactly two *B*

Example: *B* = a human, not born of man, who appeared on earth; *A* = a male

§§ 240–243 *See Table of Contents*

§ 244. V. *Conclusions from the Assertion that an Idea is General*

The assertion that an idea is general has the following form: 'the idea of a class each of whose parts falls under the idea *A* has a referent'. This is generally put as 'there are several *A*'. B. considers arguments such as the following:

There are several A
Whatever has a also has b
———————————
There are several B

§ 245. VI. *Conclusions from Determinations of the Scope of an Idea*

Here B. considers premises of the form 'the totality of A's has the attribute of numbering m'. He considers such arguments as:

The totality of A's is equal to m
Whatever has a has b
———————————
The totality of B's is not smaller than m.

§ 246. VII. *Conclusions from Propositions which Assert a Relation between Ideas*

B. considers conclusions from assertions to the effect that certain ideas are equivalent, subordinate one to another, etc.

§§ 247–252. *See Table of Contents*

§ 253. XIII. *Conclusions from Propositions which State a Relation of Probability*

B. gives a brief discussion of (incomplete) induction: Because we have seen b in a number of A's, we conclude that all A are b. "This argument contains two premises, namely a major 'A property which belongs to all A that have so far been observed probably belongs to all A in general.' (The probability depends in part on the absolute number of observed A, in part on the proportion between observed A and all A.) The minor premise states that b is such an attribute; in the conclusion that attribute b is assigned to A with a certain degree of probability."

A similarly brief discussion of arguments from analogy follows, and the section concludes with a note: "Several of the best recent logicians claim that both argument forms do not hold absolutely, but only on a further assumption, namely that nature, in spite of her variety, is subject to certain general laws. Some claim that we can persuade ourselves of the validity of this assumption only through experience, which teaches us the applicability of these forms of argument (cf. Tieftrunk, Kiesewetter (*Grundriß einer allgemeinen Logik*, part II, Berlin 1806, p. 182) Jakob (*Grundriß der allgemeinen Logik*, Halle 1791, § 460), Krug (*Logik*, Königsberg 1806, § 166 note), Klein (*Anschauungsund Denklehre*, Bamberg 1818, § 218), Bachmann (*System der Logik*, Leipzig 1828, §§ 224, 239) and others. Cf. also Senebier, *L'art d' observer*, Geneva 1775, IV, 7.) I am of the opinion that the grounds for these two forms of argument, e.g. the proposition that a certain attribute b

belongs to all *A* if it has been found in several, are absolutely valid, purely conceptual truths, just like the rules of inference for ordinary syllogisms or subalternation. We do not have to consult experience to find out if these rules hold; rather, we must presuppose their validity to have experience in the first place. Also, these rules must not be called mere assumptions (*praesumptiones*); at most we could apply this name to individual propositions *M* which are held true because of them. In any case, the proposition which is said to be a presupposition of those forms of argument, and which is said to be confirmed by experience, can easily be seen to be a purely conceptual truth: we can say not only of nature, but of every existing and even non-existing object whatever that there must be certain general laws which it obeys. And must we not make the same assumption with every other inference, e.g. syllogisms? The only difference is that the one kind (e.g. the syllogism *Barbara*) is much more irresistible than the other, which can be found only after some thought. Finally I should like to call attention to the fact that there is nothing in the nature of these two kinds of argument which would restrict their application to merely empirical objects. It is sometimes permissible and appropriate to form a judgment concerning the truth of a given purely conceptual proposition on the basis of an (incomplete) induction or argument by analogy."

APPENDIX

Earlier Treatments of the Subject Matter of this Part

§§ 254–261. *See Table of Contents*

§§ 262. *The Syllogism in the Received Logic*

The most important part of this appendix consist of a criticism of Aristotle's view on the syllogism. Early in the prior analytics, Aristotle has defined a syllogism (*Anal. Priora* 24b18): "A syllogism is a discourse in which, certain things being stated, something other than what is stated follows of necessity from their being so;" but this definition obviously fits every argument, not only with two, but also with three and more premises, no matter whether the form of argument is direct or indirect. On the other hand, Aristotle argues later on that all syllogisms have to belong to the three (Aristotelean) figures. B. makes the following objection against this view.

a. "First of all it is a one-sided view that the conclusion should have the form 'β does or does not have α'. This form results if it is presupposed that every proposition that is to be proved should have neither more nor less than two variable ideas (α and β), where one of these is the complete subject idea, the other the complete predicate idea. But this is not at all the case; often we have to prove propositions in which only a single idea is to be viewed as variable and frequently we have to prove others where three or more ideas are simultaneously considered variable. These variable ideas are distributed in many different ways: sometimes they are in the subject, sometimes in the predicate idea, sometimes they form these by themselves, and at other times in conjunction with other ideas. Thus the conclusion in the so-called hypothetical syllogism in *modo ponente*, as well as in *tollente* has only a single variable idea (of an entire proposition). Furthermore the disjunctive syllogism is an example of an argument whose conclusion can have a great number of variable ideas (of entire propositions), as many as desired. And all these ideas are found only in the subject idea of the proposition, so that the predicate idea can be an altogether invariable concept. . . .

b. We must also criticize Aristotle for assuming in his proof that a proposition which leads to the conclusion that α is or is not a β, must necessarily contain α. The three propositions:

'Caius plays the flute'
'Titus plays the organ'
'Caius and Titus are two of distinct persons'

allow us to draw the conclusion 'among Titus and Caius there is a flute and an organ player'. Nonetheless, neither of the premises has either a subject or a predicate idea which is the same as the subject or predicate idea of the conclusion.

c. It is also false that from a single proposition nothing follows, for all immediate inferences flow from a single premise.

d. The above example disproves the contention that a conclusion which follows from a pair of premises of the form 'γ has α' and 'δ has α' cannot refer to β. For, if $\beta = \gamma + \delta$, then the two indicated premises allow the deduction of the conclusion 'β has α'.

e. Hence we must also deny that a conclusion which is to prove such and such of so and so must be compounded of propositions (namely premises), which are concerned with such and such and so and so (to say that a proposition is concerned with α means that α is either its subject or its predicate idea). It is quite possible to deduce conclusions which have certain variable ideas that do not appear in any of the premises at all. Thus, from the proposition 'there is an *A*' (i.e. *A* has a referent) we can deduce the conclusion 'the proposition 'some *A* are *B*' and 'some *A* are not *B*' are not both false' (i.e. the ideas *A* which is *B*, and *A* which is not *B* are not both without referent, no matter what *B* is), etc."

Subsequently B. considers various other proofs to the effect that the syllogistic form of argument is the only correct one. He points out that a criticism of the arguments which are to establish this view cannot establish the falsity of this contention. He reasons:

The proofs can be deficient, but the assertion itself, that the syllogistic form of argument is the only one, could nevertheless be true. Hence it is necessary to show more explicitly than it has already been done that there are many forms of argument which differ essentially from syllogisms. But first of all I must clarify the sense of this assertion. If we assume that every argument, whether syllogistic or not, can be put in the form of the following proposition, 'every class of ideas whose substitution for *i*, *j*, . . . makes propositions *A*, *B*, *C*, *D*, . . . true also makes propositions *M*, *N*, *O*, . . . true', and if we are furthermore willing to call every proposition which has this form an *argument*, then there is no doubt that there are arguments with any desired number of premises as well as conclusions. For, any number of propositions *A*, *B*, *C*, *D*, . . ., so long as they are consistent with respect to ideas *i*, *j*, . . . can function as premises, and any number of propositions *M*, *N*, *O*, . . ., so long as they become true whenever the former are true (and a great number of these can always be found) can be envisaged as conclusions. Among the propositions *M*, *N*, *O*, . . . there may also be some which do not require the truth of all of the propositions *A*, *B*, *C*, *D*, . . ., but only of one or several of them. In this sense we can give the name of an argument to the following combination of propositions:

All *A* are *B*
All *C* are *D*
———————————
Some *A* are *B*, and some *C* are *D*

But this was not intended, when it was asserted that every true argument consists essentially of only three propositions. They obviously had in mind only arguments with a single conclusion, indeed arguments which have no more premises than are required for the derivation of this conclusion; in other words, what they had in mind was a kind of argument which we called (in § 155) *irredundant* arguments. But even if an argument is irredundant, nothing is easier than to indicate an example where the number of premises can be as large as desired. An argument of this kind is the well-known sorites:

$$\begin{array}{l} A \text{ is } B \\ B \text{ is } C \\ C \text{ is } D \\ \ldots\ldots \\ L \text{ is } M \\ \hline A \text{ is } M \end{array}$$

It could be objected to this example that the conclusion can be derived only by a repeated application of the syllogistic argument form. However, it is not necessary that this argument form should be interposed in order to arrive at the indicated conclusion; hence this example actually suffices to prove that there are other than syllogistic argument forms. But if these other argument forms were all of the nature of the sorites, i.e. if we could in every case arrive at the conclusion by a series of ordinary syllogisms, then the theory of non-syllogistic argument forms would at least be dispensible. However, I should like to claim that this is not the case; there are argument forms which tell us how to derive certain propositions from others which cannot be derived from them by any syllogisms, no matter how often they are repeated. To be convinced of this important truth, nothing is required but a review of the various arguments which were discussed in this part, and to try to derive the indicated conclusions from their premises through one or several syllogisms. I wish to emphasize only a few examples.

a. Given the two premises 'every *A* is *B*' and 'every *B* is *A*'; in § 226, no. 5, we deduced from them the conclusion 'every object which falls under one of the ideas *A* or *B* also falls under both of them.' This conclusion can never be derived by ordinary syllogistic methods; rather, the latter lead us to the identical proposition 'every *A* is *A*'.

b. From *n* proposition of the form:

There is only one *A*
There is only one *B*, etc.

and the further proposition:

> No object of one of the ideas *A*, *B*, *C*, . . . is an object of two of them,

we deduced the conclusion (cf. § 243, no. 2) 'the number of *A*, *B*, *C*, . . . is *n*'. Through what single syllogism or combination of several of them could this conclusion be derived?

c. According to § 252, no. 9, we can derive the conclusion '*A* is true' from the premises:

> Either *A* or *B*
> Either *A* or *C*
> Either *A* or *B* or *C*

But is this done through a syllogism, etc.?

5. If anybody wanted to justify the ordinary doctrine, he might reply that if the indicated arguments are to be expressed completely, they must be presented in syllogistic form; that each of these arguments must be prefaced by a hypothetical major premise, which expresses a rule to the effect that the conclusion follows if the set of indicated premises is added as a single minor premise. There can be no doubt that in this way every desired argument can be turned into a so-called *hypothetical syllogism*. For let the number and form of the premises and conclusions *A*, *B*, *C*, *D*, . . . and *M*, *N*, *O*, . . . be whatever it will, we can always form the following argument, which I am glad to acknowledge as a syllogism:

> *Major Premise:* Whenever propositions *A*, *B*, *C*, *D*, . . . are true, then propositions *M*, *N*, *O*, *P*, . . . are also true.
> *Minor Premises:* Propositions *A*, *B*, *C*, *D*, . . . are true (as they stand)
> *Conclusion:* Therefore, propositions *M*, *N*, *O*, *P*, . . . are also true (as they stand).

We must note that the major premise in this argument is itself a complete argument; if this argument has any usefulness, and if its validity can be judged merely from logical concepts, then it must itself be discussed in logic. Hence nothing was gained by this expedient.

§§ 263–268. *See Table of Contents*

BOOK THREE

THEORY OF KNOWLEDGE

§ 269. *Purpose, Contents, and Divisions of this Book*

According to § 20 a treatise on logic should concern itself with the conditions of human knowledge. In this book *B*. discusses the theory of knowledge to the extent necessary to show the correctness of the rules set forth in the subsequent 'Heuretic'. Since every cognition [Er Kenntnis] is a judgment, and since every judgment consists of (subjective) ideas, this book will treat of both, and of the truth of our judgments as well as of the confidence we may put in them.

PART I

Of Ideas

§§ 270–279. *See Table of Contents*

In §§ 270–279 B. considers subjective ideas (as opposed to ideas in themselves), and points out that a subjective idea is the appearance, in a mind, of an idea in itself (§ 270). To every subjective idea there corresponds an idea in itself (§ 271), and every subjective idea is something real, in the sense that it is a certain attribute of a living thing (§ 272). Every subjective idea is in time; but I can have several different ideas at the same time; several different beings can represent the same objective idea, but these representations would be different subjective ideas; finally, one and the same person can represent the same objective idea on different occasions; each of these occasions would then count as a separate subjective idea (§ 273). The problem then arises whether in a judgment like 'equiangular triangles are equilateral' the concept of equality is represented twice at the same time, so that the same objective idea would occur doubly. B. surmises that the idea of equality occurs only once in this case "just as one and the same point can lie on several different lines" (§ 274).

Subjective ideas can take on various degrees of vividness (§ 275). All distinctions among objective ideas can be carried over to subjective ideas (§ 276). Thus we have simple as well as complex subjective ideas (§ 277), and we have intuitions as well as concepts (§ 278); (cf. §§ 72 and 73). An immediate cause (or partial cause) of an intuition is called a sensible object. Ideas of sensible objects are called sensible ideas; "things which are not the object of a possible intuition are called supersensible things" (examples: immortality, piety) and their ideas are called supersensible ideas.

§ 280. *Clear and Obscure Ideas*

1. Sometimes the notion of the clearness or obscureness of an idea is taken in a quantitative sense; we do this, for example, when we say that we can think of one thing more clearly than another. But there is something vacillating about this usage: whether we should call an idea clear will depend upon the idea with which we compare it.

2. "We say of many ideas that we are conscious of them; could it be that the proper difference between clear and obscure ideas is connected with this? Is it reasonable to say that an idea is *clear* when we are conscious of it, and *obscure* when this is not the case? To decide this matter we must first determine more precisely what we are to understand by this *consciousness*, which is said to accompany every clear idea. If we judge of an idea that we have it, then there is no question that we have it, nor that we can be said to be conscious of it: if somebody says that he has a certain idea how can we deny that he is conscious of it? But we cannot turn this into a definition of the concept of a clear idea; we cannot require that for an idea to be clear,

we must have formed the judgment that we have it; this definition would be much too narrow, and whether an idea should be called clear or obscure would depend upon a quite accidental characteristic, which has nothing to do with its inner attributes." B. suggests that this difficulty could be avoided if we define an idea as clear if it has those characteristics which an idea must have to enable one to judge that one has it. "Indeed, this seems to be what we want to express by the form of words 'to be conscious of an idea', or 'to *know* that we have an idea'. For, if we attribute to somebody knowledge of something, namely knowledge of truth *A*, then we do not want to indicate that he forms judgment *A* at the very same moment at which we attribute to him this knowledge; rather it is sufficient that he should have formed this judgment at some other time, and that he requires only a certain external stimulation in order to repeat it. Hence it is not the actual formation of judgment *A*, but only a certain capacity to form it, which we call knowledge of truth *A*." Similarly, we might say of somebody that he is conscious of idea *a* not only at the moment when he forms the judgment 'I have *a*', but also at all other times at which *a* was present, and when he could have formed this judgment, if it had not been for absence of some stimulating circumstance. B. rejects this suggestion on the grounds that, since such a judgment does not have to be formed, it should not be necessary to mention it as part of the concept of a clear idea. "It must be possible to express the nature of a clear idea by using a characteristic which is found, not merely in some of them, but in all of them, and which not merely *may* be present, but *must* be present".

3. "That we may find this characteristic let us see what is necessary in order to form the judgment that we have a certain idea *a*. There is no occurrence, in this judgment, of the idea *a* itself, but only of an *idea of* this idea. Hence if somebody simply has an idea *a*, but not an idea of this idea, then he could not form the judgment that he has idea *a*. However, it can happen, conversely, that somebody not only has a certain idea *a*, but also an idea of it, without using the latter in order to form the judgment that he has *a*. But if it has come to the point that we not only have an idea *a*, but also an idea of it, then the only thing required for the formation of the judgment that we have *a* are certain changes in external circumstances; no changes in *a* will be necessary. Could we then not define a clear idea as an idea which we represent to ourselves? It is understood that the form of words 'to represent something' is to be taken in the narrower sense of § 101; hence we require for a clear idea that it be represented by an idea which refers *exclusively to it*. But this can still occur in two ways: the idea which is to become clear can be represented either by an intuition, or by some complex idea which happens to refer to it because of its particular composition." B. rejects the latter suggestion; an idea which I had one hundred days ago at exactly 1 a.m. does not become any clearer if I form the idea 'idea which I had one hundred days ago at exactly 1 a.m.' "This leads to the following definition: An idea is *clear* if we represent it to ourselves by way of an *intuition*. It will be called *obscure* whenever this is not the case."

4. B. here answers the following objection: would it not be reasonable to call an idea clear if it is merely *capable* of being intuited, without such an intuition being actually present. He says "that we are entitled to demand more than the mere possibility of intuition from a clear idea." He maintains

that "according to the law of reaction" something will happen to the original idea when it is intuited "just as a pain becomes more vivid when we represent it in intuition".

5. The proffered distinction between clear and obscure ideas meets the purposes of logic because (a) there really are ideas of both sorts (there must be obscure ideas for "a being all of whose ideas are clear would have to have an infinite number of them") and (b) "it is certainly of importance whether one of our ideas is clear or obscure, i.e. whether we have arrived at an idea of it since, so long as we do not have such an idea, we are rarely, if ever, in a position to say anything about that idea which applies exclusively to it."

In a note, B. criticizes Leibniz' definition of clearness:

> "A cognition is clear if I have it in such a way that I can re-identify the thing represented" (I. C. Gerhardt, ed., *Philos. Schriften*, vol. IV, 422) and "An idea is clear when it suffices to recognize and distinguish the thing; without this the idea is obscure" (*Nouv. Ess.* II, XXIX, 2).

He points out that the form of words 'object of an idea' can have four senses, namely (a) the corresponding objective idea, which is the *matter* of a subjective idea; (b) that which is represented by the idea (this is Bolzano's sense, cf. § 49); (c) that to which in a given judgment the predicate-idea is actually applied; finally, (d) the external real object which causes a certain idea in us. B. contends that Leibniz did not use 'object' in the sense of (a). If he used it in the sense of (b), the following objections can be made: not every clear idea has an object; there are non-referring ideas which are clear. For example, I can have a very clear idea of 'a man with a golden tongue' even if there is no such thing. Secondly, assume that I form the idea of a celestial body which has the form of a ring. What could be meant by saying that if this idea is clear, then I can differentiate their object from the objects of other ideas? "To say that somebody can distinguish the object of a certain idea from the objects of other ideas means that he is in a position to name certain attributes which hold of the one or several objects of the first idea, but not of the objects of the other ideas. But what happens if that idea is the widest of all ideas, the idea of something in general? Since this idea comprises every object, there can be no characteristic by which we can distinguish its objects from those objects that do not fall under it. Hence, according to that definition, it would be impossible to form a clear idea of something in general. Let us then try the third sense (c) and let us understand by the object of an idea *A* not that thing which is actually represented by that idea, but that to which *we* relate it in a judgment of the form 'this is *A*'. I admit that if we take 'object' in this sense, then the concept which emerges is of great importance for logic: it is the concept of an idea which suffices to distinguish the object to which we apply it from many or all other objects. But I would maintain that not the designation 'clear' but some entirely different designation, e.g. 'complete' or 'exhaustive', etc. would be suitable to name this concept (cf. §§ 110, 112). By the clarity of an idea we usually mean a certain relation between that idea and our consciousness; but whether a given idea *A* suffices to distinguish the object to which we apply it depends not upon the way in which this idea is represented in our consciousness, but on its relation to that object. One and the same idea would have to be called clear if we apply it to one object, and obscure if we apply it to another; i.e. clearness and obscurity of an idea would depend upon to what object we wish to relate it. . . . Finally, if we take the expression 'object of an idea' in its fourth sense, then an obscure idea is one through which the object which has caused it can barely be distinguished from other objects, i.e. an idea from which it is almost impossible to infer the properties of the object which caused it. I believe that the case which is here described could easily occur with ideas which are very clear before the mind and which could not be called obscure. Thus, for example, the visual idea of a distant object does perhaps not allow to distinguish this object from others; nonetheless, I can pay close attention to this idea and can be intimately conscious that I have it, and what it is like; for example, I may be aware that I have something very shiny before me, etc. I do not deny that ideas of this kind are frequently called obscure . . . a more suitable designation would be 'deficient, and not suitable to determine the object by which they are caused'."

B. goes on to criticize various opinions on clearness and obscurity, among them Hegel's view that "it must remain undetermined what an obscure concept is; for otherwise it would no longer be obscure, but clear."

In a further note, B. maintains that some philosophers, notably Descartes, Leibniz, and Wolff, have claimed that we can have ideas of which we are not conscious, while Locke and others denied it. The reason for the denial was, generally, that an unconscious idea would be a contradiction in terms. B. defines 'unconscious idea' as 'idea of which we do not have an intuition' (the latter could be expressed through the words '*this* idea'). In this sense there are many unconscious ideas.

§ 281. *Distinct and Confused Ideas*

B. claims that a complex idea can be clear even if its parts are not clear, i.e. we may have an idea of that idea, without having ideas of its parts. On the other hand, it is sometimes the case that we can form "a true judgment of the following form: 'the idea A consists of the ideas a, b, c, . . . in such and such a combination'. This is a bit of information which can be of great service, especially in scientific investigation. When we are in a position to make such a judgment about an idea, we call it *distinct* [*deutlich*]; compound ideas whose components we do not know are called *vague* or *confused* [*undeutlich* or *verworren*]."

In the sequel, degrees of distinctness are distinguished in the following manner: if an idea is distinct, but its immediate parts are not distinct, then a is said to have distinctness of degree one; if an idea is distinct and its immediate components are also distinct, then its distinctness is said to be of degree two, etc. Finally, if an idea is distinct, and all of its components, immediate, and remote, are also distinct, and its simple components are known to be simple, then this idea is called completely distinct.

§ 282. *Whether each of our Ideas has a Beginning and an End*

B. here maintains that the presence of an idea depends not only upon the presence of certain substances, but also upon the relation between substances. Since these relations change, ideas come and go. Hence it is false that our ideas have always been in us. It also follows that every subjective idea has a beginning and an end in time.

§ 283. *Whether every Idea Leaves a Certain Trace after it has Disappeared*

B. answers this question in the affirmative, his reason being that recollection must be explained in terms of traces of earlier ideas.

§ 284. *Effects which Result from the Traces of our Ideas*

In this section, B. argues that the recollection of an idea *A* is the same as the judgment that we have once had *A*. But to form this judgment, we must have an idea of the idea *A* (an idea which refers exclusively to *A*, i.e. an intuition of *A*). It follows that every recollected idea is clear in the sense of § 280. He goes on to ask what, on a given occasion, causes a certain recollection to occur, and develops an associationist theory which differs only in minor detail from others then current.

§ 285. *Signs for our Ideas*

1. "As we saw in § 283, our mind has the peculiarity that ideas that were once connected with each other also mutually stimulate each other. We use this peculiarity in the following way: certain ideas which can be easily stimulated are used as means in order to produce others which are more difficult to generate, but which are associated with the former. An object which is used for this purpose, i.e. one through whose idea we want to stimulate in a thinking being some other, associated idea, is called a *sign*. The objective idea whose corresponding subjective idea is supposed to be stimulated by the idea of the sign is called the *signified* idea, also the signification [*Bedeutung*] of the sign. If the signified idea has an object, then this is often called the signified object, or again, the signification of the sign. Sometimes the words 'sense' [*Sinn*] and 'meaning' [*Verstand*] are used synonymously with the word 'signification'. But a certain distinction could be made between them in the following way. We could call the *signification* of the sign that particular idea which it is intended to produce, and succeeds in producing; while we could confine the words 'sense' and 'meaning' to those ideas which are merely intended to be produced by a certain sign in a given case. Obviously, somebody who does not understand a sign can take it in a sense or meaning which is quite different from its real signification. To use certain signs in order to produce ideas in others is to *say something* in the widest sense of these words. To inspect given signs in order to find out which ideas their author wanted to generate is called *reading* in the widest sense. To grasp what ideas their author wanted to produce is to *understand* them. To assume that they have such and such a sense, while they have another, is to *misunderstand* them. Finally, a sentence in which it is asserted that certain signs have such and such a sense is called an *interpretation* of these signs . . .". B. continues to define further semiotic con-

cepts: a sign which is used by all men in the same sense is called a universally accepted sign. The class of all signs which a given person is wont to employ, especially those which he uses to reveal his thoughts to others, is called the *language* of that man. A sign is called a *natural sign* if the nature of man, i.e. attributes which we all have in common, determines which idea it designates. All other signs are called *contingent*; if a contingent sign was created by a decision, it is called *conventional*. Thus, wringing one's hands is probably a natural sign, wagging a finger a contingent sign, and the way in which the Romans indicated that a gladiator was to die, is merely conventional. A sign is usually called an *expression* if it is an external object, and the signified thing is something internal, e.g. a thought, a sensation, etc. Thus, clapping one's hands is generally taken to be a (natural) expression of delight. Most signs are visible or audible objects.

2. Many complex ideas are signified by presenting the signs for their component ideas in a certain order; for example, propositions are so signified. When we speak, we produce sounds in a certain order, when we use visible signs, we produce a string of signs which has a certain direction, which determines in what order they are to be looked at.

3. A *single* sign is a sign which is not a class of others, thus a word is generally a single sign.

4. A sign which is used to signify more than one idea is called ambiguous.

5. Sometimes we choose a certain sign to signify an idea *B* because that sign was first used to signify a certain different idea *A*. "In this case the signification *B* is said to be *derived from A*. A signification which is not derivative is said to be *primary*. Thus the signification of the word 'movement', when it is used for a mental change, is obviously derivative from another, which indicates change of location; the latter is likely to be primary."

6. A sign is *indeterminate* if we do not know what idea the speaker wanted to designate by it. If there are two or more significations which are equally likely, the sign is called *vacillating*. If we cannot guess at any idea which could be signified, we call it altogether incomprehensible.

7. When a sign is used in the way in which it was meant, it is used in its *proper* signification, otherwise in an *improper* or borrowed sense.

8. If a sign is used improperly, but its signification is obviously and closely related to its proper signification, we call it metonymical, otherwise *metaphorical*.

9. In addition to its signification, every sign also produces other ideas. These are called *accompanying ideas*. "If the accompanying ideas which come with a certain sign have a detrimental influence upon our judgments, I call them *harmful*, and the sign itself . . . I call impure. On the other hand, a sign which does not have such harmful accompanying ideas I call a *pure* expression."

10. There can be signs of signs, e.g. a written language may signify the sounds in the spoken language. Signs of signs are called indirect, others *immediate*. A sign which was originally indirect can eventually become immediate.

11. Persons who have once learned a language generally accompany all their thoughts with signs, sometimes deliberately, sometimes not. This could be called *speaking to oneself*.

NOTES 1–5: After these historical and critical notes, B. concerns himself with the questions whether thought without language is possible, and how the origin of language is to be explained:

NOTE 6: It can be seen from No. 11 that and why (in my opinion) everybody who has once learned a language also uses it even if he does not actually intend to communicate his thoughts to others. I should like to add that all of our thinking, if we did not use the medium of language, would become very uncertain; particularly, it would be difficult to handle complex ideas which do not have a sensible object, unless we had definite signs for them. However, I do not dare assert that without this medium no thinking, especially no judging, could take place at all.... For one thing, beings of a higher order than we are (at least God) do not need signs in order to think; but it is also the case that creatures less perfect than ourselves, namely animals, do not need signs for their ideas in order to form judgments. Hence from what human characteristic does it follow that we of all beings require signs in order to judge? Hoffbauer (*Die Psychologie nach ihren Hauptanwendungen*, Halle 1808, § 126) has pointed out correctly that the use of signs presupposes thinking without signs. For in order for the word 'rose' to become a sign of the idea *rose*, it is not enough that both ideas should have occurred simultaneously in my soul and should thereby have become associated. For there are thousands of associated ideas none of which is the sign of the other. A sign requires intention: either I must utter the word 'rose' with the intention of producing the idea of a rose in myself or in somebody else, or I must recognize that somebody else is acting with such an intention. But for such intentional actions, as well as for their recognition, judgments are required: thus it follows that even prior to the first sign which we invent, or which we recognize as invented by others, judgments must have taken place.

7. Treatises on logic frequently investigate the problem of the origin of language. I should also like to make some suggestions on this subject. It seems certain that man did not invent language in order to speak to himself, but to speak to others; hence it depends upon the social connections which he has formed with other men. Man is so constituted that most of his thoughts, sensations, wishes, and intentions, even without his knowledge and intent, are with more or less accuracy reflected and recognizable in his facial expressions gestures, sounds, etc. From the fact that he is fixing his eyes upon a certain object, we can see that he is thinking about this object; and it is even more obvious whether somebody experiences pleasure or pain, desire or aversion, courage or fear, etc. We can see these

things in each other, and the outward appearance of another person is for a sign [*Kennzeichen*] of his inner life; this does not require the degree of sophistication which can be acquired only through language, but even deaf mutes, children, and idiots can achieve it; even animals, devoid of language and reason, dogs, apes, and others can see from our facial expressions whether we are angry or afraid, etc. A man, unless nature has ravaged him, will acquire more capacities as he grows older: he will not only see what others experience inside, but he will also know *by what signs* he sees it and, consequently, what others can see in him. It seems that even among the animals some reach this degree of understanding, though not with clear consciousness. How else can we explain that they sometimes behave as if they did something without really doing it? In any case, there is no doubt that children and mutes have this ability. But when man had come to the point of knowing that such and such facial expressions, gestures, sounds, movements, etc. made others believe that he had such and such ideas in him, then he was in a position to use these external changes as *signs* in the narrower sense of the word, i.e. as objects from whose perception others are to infer the presence of certain ideas inside him. This means that a certain language had been invested; it did not consist of anything but certain voluntary motions of the body which [normally] follow automatically upon certain changes in the soul, and which were not now used merely to *represent* these changes. Men cried out when they wanted to represent pain, and stomped their feet when they wanted to represent anger, etc. If we call this oldest language a language of gestures, we should not forget that sounds as well as looks and facial expressions often were connected with the gestures, at least those sounds which were imitations of certain involuntary expressions of our inner life.

Most objects which were found in the environment of man, especially those which attracted his attention to a higher degree, namely animals, made certain noises from time to time: the sheep bleated, the lion roared, the spring murmured. From a mere desire to imitate, if for no other reason, man would attempt to copy these various tones and sounds. When he succeeded to some extent, he found that the idea of the entire object was renewed in his mind as often as he produced the sound which was appropriate to this object, or whenever he heard this sound. It is not astonishing that, later on, the production of these sounds was used in order to cause in others the ideas of the corresponding objects; thus these sounds now became signs of those objects. But the sound of a thing is not the only thing which can be copied. It was probably soon realized that other peculiarities of things, e.g. their way of moving, their posture, etc. can sometimes be imitated with the result that this imitation produces an idea of the thing itself. Hence these imitations, too, will have been accepted as signs, and a limping person will have been signified by limping about, a blind person by groping to and fro with closed eyes, etc. Hence a second, much more complete, language was invented, namely the language of imitation or similarity; it is characteristic of this language that reference to an object is made by producing something resembling it. Just as the language of gestures was not confined to gestures in the narrower sense, so the language of imitation did not use merely gestures and sounds, or mere motions of the body, but probably also other objects which were at hand; for example, it is likely that drawings were made. The combination of these two kinds of language—it is understood that the first was not given up after the second was invented—made it possible not only to signify most inner states (namely those whose high degree of vividness produced physical changes), but also most external objects, so long as they had a property which made them conspicuous and which could be somehow imitated. Things which were directly in front of them they probably indicated very early, simply by pointing with their hand or finger: a sign which, in a sense, is of a special kind. It probably derives from the fact that one has to extend his hand in order to grasp an object; consequently, the extension of the hand was initially merely a sign for the desire for this object. But once it was realized that others look at an object to which one extends his hand and that a certain idea was thus generated in them, they must have had the idea of using this extension of hands even if they did not want to express their desire to have the object, but merely wanted others to look at it and form its idea. In order to avoid misunderstanding they probably changed the gesture somewhat: the grasping of the object became a mere pointing with the finger. This example was probably not the only one which led men to the discovery that, in order to signify an object, it is not necessary to produce something that is similar to it; all that mattered was to make others understand that such and such an utterance indicated that they were to form an idea of a certain object. Once this discovery was made it became possible to invent new signs every day; the only thing required was to point to a certain object while uttering a certain sound or perform an otherwise insignificant action. In this way a third kind of language, the conventional language, came into being. This language was a sound or word language, i.e. most of its conventional signs were sounds or tones; this can be explained from more than one circumstance. Sounds are the signs which can be most easily produced. When hands and feet are bound or busy, the tongue is still free. Sounds can be perceived by day and night. Sounds are signs with which we can force our thoughts upon somebody who does not, at the moment, pay attention to us, or even does not *want* to pay attention to us. Sounds are signs which the author perceives in the same way as the others, a circumstance which makes learning easy. Sounds are signs through which, aside from their actual meaning, merely through the way in which they are produced, we can express our state of mind; we can indicate whether we speak seriously or in a jocular way, whether we feel well or ill, whether we like the listener or dislike him, etc. Sounds, finally, are signs whose modulation can influence the mood of the hearer, can make him glad or sad, etc. It is understandable that, given all these advantages of sounds, they were used almost exclusively, except when the addressee was absent, or deaf. Considering that the subject of the origin of human language cannot be exhausted in a note, but only hinted at, I break the discussion off at this point. Degerando's thoughts on the subject are similar (*Des signes et de l'art de penser*, 4 vols., Paris 1798, vol. I, p. 10 ff.).

§ 286. *Description of the most Important Ways in which Ideas Originate*

We must now consider in a few words the difficult problem of the origin of our ideas. To begin with, we must make a distinction between simple and complex ideas.

1. We can say only little about the mode of origin of simple ideas precisely because they are simple. All we can claim is that the mind must have a faculty for producing them. This faculty becomes active only under certain circumstances and stimuli. Especially the kind of simple ideas which we called *intuitions* differ sharply from the other simple ideas (i.e. pure, simple concepts) in that they are the next and immediate (hence inexplainable) effect of certain changes within us; consequently these changes are the objects represented by these ideas. Depending on whether these changes are themselves already ideas (or even judgments) we can distinguish two kinds of intuition. Intuitions which have as their object another idea present in our mind may be called inner intuitions; if the object is a change which is not itself an idea, the intuition may be called external, because the question of the origin of such intuitions leads us immediately to suppose that there is an external object which effected this change in

our mind. Experience shows that under certain circumstances the most divergent external objects can cause intuitions in us which are so similar that we can often detect no difference at all between them. Thus we see red not only when we are confronted with a red body, but sometimes also without one, namely when the eye is pathologically affected in certain ways.

2. Pure, simple concepts originate in a much different manner. Their origin betrays a much greater activity of the mind. There is no doubt that they are always occasioned by the presence of other ideas; it is likely that a new simple concept arises only when it proceeds from ideas presently in us in order to combine them to a new idea or, rather, to a complete judgment.

3. According to § 277 it does not suffice for the explanation of the origin of complex ideas to assume merely that the several simple ideas which are its parts become active at the same time. Rather, we must assume an activity of the mind which brings it about that these several ideas are combined into a whole which is itself an idea. This formation of ideas is called concatenation or *synthesis*.

4. Our use of the faculty of judgment can also lead to the formation of new ideas. "For example, the concept of *having* is likely to have originated on the occasion of our first judgment, as the copula of that judgment."

5. Concentrating on certain matters does not immediately produce new ideas, but since this concentration may make some ideas more vivid, and since vivid ideas produce more associations, it follows that the direction of attention has a certain influence upon the formation of ideas.

6. We can form new ideas, and renew old ones by "seeking them out". It might be objected that if we want to produce in us a certain idea *A*, we must already have *A*. "My reply is that this decision does not require the presence of the idea *A* itself, but only the presence of an idea of it". For example, I can decide to find out (or remember) what 'bread' means in Russian without yet having an idea of that name.

7. Abstraction does not lead to the formation of new ideas, since abstraction consists merely in the directing of attention to some parts of complex ideas. But abstraction can raise ideas already present to the level of clarity and distinctness.

8. It is true that all our complex ideas are synthesized by an activity of our own, but it does not follow from this that we are always conscious of the parts of such ideas, even if we are conscious of the ideas themselves (cf. § 281). It is common, especially with ideas that we have formed in early childhood and have frequently repeated since, that we can no longer identify their parts. This is because whenever we renew these ideas they fly past our mind with such a speed that our intuitive faculty does not have enough time to grasp the individual parts. I believe that this holds particularly for

the lower generic ideas under which we subsume external intuitions, i.e. the general ideas of colours, sounds, odours, etc., thus particularly ideas like 'red', 'blue', 'yellow', 'sweet', 'bitter', etc. I consider all generic ideas of sensation to be complex, though I am still quite unable to indicate the simple parts of even a single one of them. If any one of these ideas were simple, it would have to be either an intuition or a concept. But these ideas are not intuitions, since they are general ideas, i.e. since they have not one, but several objects. But I can also not agree that they are simple concepts if for no other reason than that for every one of them there is another which is incompatible with the first, but where the two domains are so closely adjacent that it is often doubtful whether a given intuition should be subsumed under one or the other. Thus red and yellow are so closely adjacent that we often do not know whether we should call a given intuition one or the other. The same holds for the colours blue and green, green and yellow, etc. A similar point can be made regarding sounds, odours, tastes, etc. This phenomenon is difficult to explain on the assumption that the indicated ideas are pure concepts; but it is easily understood if we assume that they are complex ideas which contain concepts of magnitude [*Größenbegriffe*]. Thus the only remaining question is whether these complex ideas are pure concepts, or whether they contain some intuitions as parts. I must confess that I have of late been inclined toward the first alternative. I came to this conclusion by considering that each intuition of which man is capable has as its proper object a *change* which presently occurs in the mind, hence something which is in existence only for a very short time. From this it follows that we can detect a certain reference to a point in time (a present or past *now*) in any idea which contains an intuition as a part. Thus the mixed nature of the idea 'Socrates' becomes apparent, because I think of a philosopher by that name about 2000 years before the present moment. But in the ideas 'red', 'sweet', 'fragrant' I do not detect even the most remote reference to a moment present or past. The more I think about it, the more probable it seems to me that the general ideas which are presently under discussion are, all of them, nothing but *concepts of certain laws that govern the changes in the mind which are the objects of our intuitions.* It seems obvious that the mental changes which take place when we see must be different from those that occur when we hear or smell, they must be of a different kind when we see something that we call red than when it is something we call yellow or blue, etc. But should it be so unlikely that our mind brings the peculiarity of each of these changes under concepts? It judges, for example, that a certain state repeats itself several times, or that this repetition takes place in equal or unequal or constantly increasing or decreasing, etc. intervals. If this is the case, would it not be understandable that it happens in a way which makes it impossible for us ever to become clearly aware of it, since the speed with which the individual ideas pass by is much too fast to allow us to perceive them individually?

9. The preceding and § 285 make it obvious that language and social intercourse have a very great influence upon the origin and formation of our ideas. We designate our ideas with special words, i.e. we associate idea and word through simultaneity; this has as a consequence that every recurrence of the idea makes the word come to mind, and conversely. Through social relations with other persons, especially if we hear them call several objects by the same name, we are caused to consider what the common attributes of these objects are, and we form the concept of a thing which has all these attributes as the presumed concept designated by that name.

10. It is clear from this that most of our ideas originate without our wanting to produce them. We can call them only *given* ideas; those which are produced by a conscious and deliberate activity may be called *factitious*. Since the creation of a compound idea is conceivable only if its component simple ideas are already at hand, we may call simple ideas *original*, and compound ideas *derivative*.

11. We have already seen that the simple ideas must be either pure intuitions or pure concepts. Now the external causes which play a role in the generation of both kinds have a much greater share in the production of intuitions (if they are external) than of concepts. From this it follows that intuitions point to [*hinweisen*] a certain object as their cause. For the sake of this difference it should be permissible to say of our simple concepts that they are *innate*, but of all other ideas, i.e. of simple intuitions, compound concepts and mixed ideas, that they are *acquired*. But this must not be interpreted as if all simple concepts are with us from birth, but only that the external world plays a much smaller role in their origin than is the case with external intuitions.

In a note, B. considers objections that might be raised to his contention that general ideas of colours, sounds, etc. are pure concepts. The objections are: 1. if these ideas were pure concepts, then it should be possible to teach them without any sensory stimulation. But it is well known that this cannot be done. 2. There is an enormous difference between these ideas and pure concepts such as 'possibility', 'necessity', 'moral goodness'. 3. If these are pure concepts, how is it to be explained that even the lowest animals can master them? B. replies: we could teach the meaning of 'red' without objects only if we knew the composition of these concepts. But even if the concept 'red' were taught in this way to a person born blind, his concept would be different from ours, since it could not produce the same associations or accompanying ideas, but at best vaguely related ones, the like sound of a trumpet. Here lies also the distinction between such ideas as 'red' and 'necessity'. The former are associated with a wealth of intuitions, and the latter are not. Concerning animals, he remarks that "animals, too, might have the capacity for concepts, only not the capacity to become conscious of them; in this case it would not be strange that animals approach and even surpass man in the minuteness of their sensory discrimination. But even without this assumption, is it more surprising that animals discriminate between colours, sounds and odours, than that they comprehend spatial relations, which man always represents through concepts?"

§§ 287–289. *See Table of Contents*

PART II

Of Judgments

§ 290. *The Concept of a Judgment*

Cf. §§ 19 and 34.

§ 291. *Some Attributes Common to all Judgments*

1. Every judgment is an appearance of some proposition in itself in the mind; the proposition is the content of the judgment.
2. Unlike propositions, judgments have real existence, being attributes of minds: "they have reality in the mind of the being that thinks them."
3. The parts of a judgment are subjective ideas. There are as many subjective ideas in a judgment as there are objective ideas in the corresponding proposition.
4. If several subjective ideas are to result in a judgment, they must be in the mind at the same time, or at least they must show some temporal overlap. But this simultaneity does not altogether explain that a judgment results. The ideas must have "a peculiar connection with each other. If I am asked what this connection is, I can answer only that it must be a sort of mutual influence of these ideas upon each other. But I am unable to determine what sort of mutual influence there must be in order for a judgment to result from the presence of several subjective ideas."
5. "Our will has a certain influence upon the formation of our judgments, since we can direct our attention upon some ideas and withdraw it from others. In this way we can completely change the course of our thoughts. But no matter how significant this influence of the will is, it never directly depends on our will alone whether or not we form a judgment. The formation of judgments follows a certain law of necessity, and depends only upon the nature of the ideas that are present in the mind at that time."

§§ 292–299 *See Table of Contents*

Except for § 293, these sections contain considerations analogous to those given in §§ 270–284 for ideas.

§ 293. *Strength or Vividness of Judgments, and Confidence in a Judgment*

Since each idea has a certain vividness, it is reasonable to attribute vividness also to judgments, in so far as they are combinations of several ideas; but from this vividness, which results from the vividness of its individual parts, we have to distinguish the strength or efficacy which results from the particular way in which those parts are combined to form a judgment. Since a judgment is not a mere sum of ideas, but consists in a certain efficacious combination of them, and since this combination can be more or less strongly felt, the efficacy of the judgment will depend upon this strength, even if content and vividness of the individual ideas do not change. I am going to call this degree of efficacy or strength of a judgment the degree of *confidence* with which we form it. With respect to the proposition which forms the content of the judgment, the degree of confidence, with which it is formed, is also called the degree of *approval* (*assensio*) which is attributed to the proposition. Sometimes it is called the degree of our certainty or acceptance. If two different persons each form a judgment with the same content, so that both contain the same proposition, and if the individual ideas of which these judgments consist all have the same vividness, it is still possible that there is a great difference in the degree of *confidence* with which each of them expresses his judgment. The judgment will show greater effects in the person who forms it with greater confidence than in the other person, and conversely where the efficacy of a judgment is more pronounced, given the same degree of vividness, confidence must be higher.

NOTE: It is necessary to distinguish the vividness of a judgment, i.e. its efficacy in so far as it derives from the strength of its constituent ideas, from the confidence, i.e. its efficacy which derives from the degree of strength with which the individual parts are united to a whole. This becomes particularly obvious when one of these two magnitudes almost disappears, while the other reaches a relatively high level. It turns out that in both cases the efficacy of the judgment is very small. Thus if we point out to a voluptuary that lust bears pain, he will hardly be moved; not because the ideas of which this truth consists do not have enough vividness for him, but because he does not form this judgment with sufficient confidence. It occurs equally often that persons who have not the least doubt concerning certain religious truths nonetheless show little effect of them in their behaviour, simply because the concepts of which these truths are composed have no vividness in their soul. . . .

Because of its importance for logic, the concept of confidence, in the sense here used, is discussed in almost every treatise of that science. However, I take exception to the fact that it is not sufficiently distinguished from another, related, concept. This other concept is the concept of *probability*, not in the sense of § 161, but in the sense in which it is employed in ordinary life. We usually call a proposition *probable* if it stands to propositions accepted by a certain being in the relation of probability in the sense of § 161, and if, in addition, the degree of this probability is greater than one-half. Since I shall speak about this concept only later, I shall indicate the difference between it and the concept of confidence later on.

§ 300. *Mediation of a Judgment by other Judgments*

1. We must now investigate the important question concerning the *origin* of our judgments. It is obvious that not all judgments originate in the same way. The judgment that the relation between diameter and circumference of circle is irrational certainly comes about in a different way than the judgment 'I feel a pain just now', or a similar one.

2. Many of our judgments arise because we have already formed other judgments. In this case we shall say that judgment *M* was *caused* or *mediated* by other judgments, namely *A*, *B*, *C*, *D*. . . . "Often, the action of the mind when it moves from judgments *A*, *B*, *C*, . . . to judgment *M* is called an inference, and *M* is called inferred or concluded or derived, while judgments *A*, *B*, *C*, *D*, . . . are called the premises, and frequently (though improperly) the grounds." Judgments which are not mediated are called immediate. To say that a judgment *M* was mediated by judgments *A*, *B*, *C*, *D*, . . . it is not enough that *A*, *B*, *C*, *D*, . . . were an occasion for the formation of judgment *M*; rather, their presence must be the complete cause for the occurrence of judgment *M*. "By a *complete* cause I understand not that nothing other than the presence of *A*, *B*, *C*, *D*, . . ., not even an activity of the mind, is required, but only that aside from the indicated judgments *A*, *B*, *C*, *D*, . . . no other judgment is necessary in order to bring about *M*. If, for example, the judgment was formed, that Caius is a scholar and it happens that later on the judgment 'scholars are often vain' comes to mind, we should not consider the latter as mediated by the former; it is conceivable that it was occasioned by that other judgment, in that the concept of a scholar brought about the concept of vanity as an associated concept and in this way finally the judgment 'scholars are often vain'; but the cause why we formed this judgment (why we have a suspicion that scholars are always vain) lies in judgments and experiences of an entirely different content than the judgment that Caius is one of the scholars." The relation of mediation holds, for example, between such pairs of judgments as 'Caius is a man' and 'all men are fallible' on the one hand, and the judgment 'Caius is fallible' on the other.

3. "From this and similar examples it becomes obvious that this relation of mediation does indeed hold between some of our judgments, and hence that there are mediated judgments. It is equally certain that there are also immediate judgments; for the existence of mediated ones can in the end be understood only because there are also immediate judgments."

4. "A judgment *M* which comes about by the mediation of *A*, *B*, *C*, *D*, . . . follows upon them in time, but in such a way that they have not

altogether disappeared when it comes about. The correctness of this assertion is confirmed in part by the observations which each of us can make with his own judgments, in part by the fact that between effect and cause there must be some simultaneity."

5. If certain judgments $A, B, C, D, \ldots$ have once mediated a judgment M, it does not follow that M will always be mediated by them.

6. We are often unable to tell whether one of our judgments was immediate, or whether it was mediated by other judgments, and by what judgments it was mediated. Thus "from the mere fact that we cannot tell by what judgments a given judgment was mediated, we must not conclude that it was immediate; we can do this only if closer observation shows that the given judgment is of such a nature that it can never be mediated, or mediated only if we are conscious of it.

7. In order to judge this matter, we must first find out what the different forms of mediation are. B. thinks that there are just three, namely (a) propositions $A, B, C, D, \ldots$ are all of them true and stand in the relation of *objective ground* to proposition M (cf. § 198) or (b) proposition M, if it is not a consequence of $A, B, C, D, \ldots$ must still be *deducible* from them in the sense of § 155; or finally, (c) proposition M has a certain degree of probability with respect to propositions $A, B, C, D, \ldots$ in the sense of § 161.

8. Nobody can deny that judgments can originate in the way described under (a) above, if he admits at all that truths can stand in the relation of ground and consequence to each other, and that men can sometimes recognize this relation. For if the truths, A, B, C, $D, \ldots$ are the ground for truth M, how can the recognition of M become more certain and complete than by recognizing that A, B, C, $D, \ldots$ are its ground?

9. But there are also cases which fall under (b); the acceptance of certain propositions $A, B, C, D, \ldots$ from which another proposition M follows with respect to certain ideas $i, j, \ldots$ sometimes, though not always, suffices to bring about the acceptance of proposition M. It may seem as if the acceptance of M can come about only if, in addition to the acceptance of propositions $A, B, C, D, \ldots$ the truth that M is deducible from $A, B, C, D, \ldots$ is also recognized. I admit that this is often the case. For example not everybody who accepts as true the propositions:

All P are M
Some S are not M

will progress to the judgment: 'Some S are not P'; perhaps this inference will be drawn only by those who remember the rule of inference called *baroco* in the schools. But this is not always the case. We sometimes advance immediately from the acceptance of certain propositions $A, B, C, D, \ldots$ to the acceptance of a deducible proposition M, without first thinking of the truth that a proposition of form *M* is deducible from propositions *A*, *B*, *C*, *D*, . . . or without even being aware that there is such a truth; I think that this can be proved, since otherwise not a single inference could be drawn, i.e. not a single judgment could be based upon other judgments as its cause. For, if the truth of proposition *M* cannot be immediately recognized from the accepted truth of propositions *A*, *B*, *C*, *D*, . . . but if recognition of the truth that from propositions *A*, *B*, *C*, *D*, . . . a proposition like *M* can be deduced is also required, then there are actually two insights which precede the recognition of *M*, namely 'every class of ideas whose substitution for *i*, *j*, . . . makes propositions *A*, *B*, *C*, *D*, . . . true also make proposition *M* true' and 'propositions *A*, *B*, *C*, *D*, . . ., as they stand, are true' or 'ideas *i*, *j*, . . . which originally occur in propositions *A*, *B*, *C*, *D*, . . . are a class of ideas which make *A*, *B*, *C*, *D*, . . . true'. Is it not obvious that the way in which judgment *M* follows from these two judgments is again only the recognition of a conclusion from its premises? Hence if no conclusion can be recognized from the truth of its premises without first recognizing the rule according to which this inference is formed, then the two judgments just formed still do not suffice to produce recognition of proposition *M*; this not only contradicts the just-mentioned supposition, but also shows that it is altogether impossible ever to derive a judgment from other judgments in this way, since an infinite number of judgments would be required. For, what I just said concerning the necessity to add a third judgment to the two given judgments (so that the proposition which we want to derive is really derivable as a conclusion) also holds when this third proposition is already added. Now a fourth proposition is required which shows that the three indicated propositions really are related as premises to the (desired) conclusion and so *ad infinitum*.

10. Finally, it cannot be denied that sometimes judgments are generated in the third manner mentioned under no. 7: the acceptance of certain propositions *A*, *B*, *C*, *D*, . . ., which merely bestow *probability* upon another proposition *M*, sometimes also leads to an acceptance of *M*. There are many examples where the premises *A*, *B*, *C*, *D*, . . . are accepted and are present in our mind, and where the conclusion is not merely that *M* is probable (this would be deducible from *A*, *B*, *C*, *D*, . . .,) but where the judgment *M* itself is formed. Thus we do not only say that it is probable that in the coming year the surface of the earth will be covered with plants and flowers, since this has happened a great many times, but we expect it, i.e. we form the judgment (with a greater or smaller degree of confidence) that this will indeed happen.

11. We must now prove that these three ways are the only ways of generating a judgment from other judgments. The most likely alternative is this: It might be argued that occasionally we proceed from certain judgments *A*, *B*, *C*, *D*, . . ., which we have formed, to a new judgment *M* merely because we *think* that it stands in one of the three indicated relations to *A*, *B*, *C*, *D*, . . ., while this is not actually the case. Do we not often perform *invalid* inferences and hence deduce propositions from others which are not in truth implied by them? My answer is this: occasionally we take an incorrect rule of inference for correct and use it in order to proceed from certain judgments *A*, *B*, *C*, *D* . . ., to a new judgment *M*, where *M* is not actually implied by them; this cannot be denied. Suppose that this advance to the new judgment *M* does not flow *immediately* from a consideration of propositions *A*, *B*, *C*, *D*, . . . but takes place only because we have mistaken an incorrect rule of inference for a correct one, i.e. because we made the mistake of thinking that a proposition like *M* is deducible from propositions like *A*, *B*, *C*, *D*, . . . alone. It follows that judgment *M* was not generated by *A*, *B*, *C*, *D*, . . . alone, but by them together with the tacitly assumed judgment that a proposition like *M* can be deduced from propositions like *A*, *B*, *C*, *D*. . . . In other words, the judgment *M* comes about because we form the following two judgments: 'if propositions *A*, *B*, *C*, *D*, . . . are true then proposition *M* is also true' and 'propositions *A*, *B*, *C*, *D*, . . . are true'. But from these two propositions, proposition *M* is not only apparently, but actually deducible; hence the way in which we here arrive at judgment *M* is no exception to the claim of no. 7, but is there described as one of the ways of generating a judgment. In order to find a true exception, we should have to claim that judgment *M* was immediately produced by a consideration of judgments *A*, *B*, *C*, *D*, . . . and that the fallacious rule of inference was not tacitly assumed. But to bring this about, it would be necessary that there is in our minds some kind of mechanism [*Einrichtung*] which brings it about that from judgments of the form *A*, *B*, *C*, *D*, . . . a judgment of the form *M* is generated although the latter does not stand to the former in a relation of ground and consequence, nor deducibility, nor even probability. But who can believe that such a mechanism is present in our mind? Its existence should be asserted only if it could be shown that certain judgments which we form with every care and precaution nonetheless contradict other judgments which we have formed just as cautiously. But nobody has been able to prove this. On the other hand, if we wanted to assume the existence of such an error-producing mechanism in our mind merely because

its *impossibility* has not been established, then, for the same reason, we should not trust a single one of our judgments, i.e. we should not form any judgments at all. But this would be a contradiction, since it is a consequence of trusting those judgments which determined that very decision.

12. Let me now adduce some *examples* of judgments which I take to be immediate: I assert of the following two forms of judgment, not that all judgments that fall under this form must be immediate, but that *some* of them are. One of these forms is 'I—have—the impression *A*'; the subject of all these judgments is the person who utters them (I) while their predicate indicates the possession of an impression which is just now present in that person, for example the presence of an idea, a just-formed judgment, a present sensation, an act of will, etc. The other form is this: 'this (what I just now observe)—is—an *A*'; in this case the subject-idea is an intuition which is just now present in the judging person, and is subsumed by this person under a certain concept *A*; this is the case, for example, when somebody says 'this (what I just now observe) is something red, a pleasant fragrance, etc.'. It is indeed true that judgments of this kind are sometimes the product of an inference, for example when we infer from the perception of one of our own actions that we must have had an idea of it, or when we judge of an intuition which we just now have, that it is the effect of an external object of such and such a kind. But it is impossible that all judgments of these two forms are mediated judgments, since every mediated judgment of this kind presupposes another one of the *same* kind. In order to infer from a present action of mine that I must have had an idea which was necessary to perform this action, I must first form the judgment 'I have performed action *A*'. And if the judgment 'the intuition *X* is an *A*' is not to be immediate, then it must be deduced from a pair of other judgments, namely 'the intuition *X* is a *B*', and 'all *B* are *A*'.

I wish to call all judgments which fall under one of these two forms *judgments of perception*, provided that they were formed immediately; it would be still more precise to call them *immediate judgments of perception*. All mediated judgments, if they contain an intuition, can reasonably be called *judgments of experience* in the narrower sense of this word.

13. It will become clear, upon some reflection, that there must be far more immediate judgments than we have just described. In particular, among purely conceptual judgments there must be some which are not mediated. For, if the judgments described in no. 12

were the only immediate judgments, then all the rest of our judgments would have to be generated either immediately or mediately from them. But it is not conceivable how this can be the case. Thus, from a proposition of the form 'I have intuition *A*' we can immediately draw the inference 'there must therefore be something real which has caused this intuition in me'; but the judgment 'every intuition which comes about either within me or within any other finite being presupposes the existence of a real object which generates it' can neither be inferred from the given judgment nor from any other kind of judgment mentioned under no. 12, nor from a combination of them, either immediately or mediately.

14. I have already mentioned in no. 6 that when we derive a judgment from other judgments we are often not conscious of the mode of derivation. This may be the case because the necessary inferences, by frequent repetition from early childhood, have become so familiar that we carry them out with a speed too rapid for our intuitive faculty. But whatever the reason, these judgments must be distinguished from those which are derived consciously, often with great effort. . . . To give an example: I do not think that we recognize immediately the existence of even a single (external) object, much less any of its properties or changes, e.g. if it is in motion just now. For example, the judgment that there is a bird flying over there I take to be inferred. But since we are not generally conscious of the premises upon which it rests, we tend to take it for an immediate perception and assert, for this reason, that we do not infer the flight of this bird, but see it immediately. I admit that there is a great difference between the manner in which we recognize the flight of this bird and, for example, the motion of the hour hand on a pocket watch. We recognize the latter when we realize that this hand is now in a different position than when we saw it a short time ago, i.e. we are conscious of the reasons from which we inferred its motion. To express this distinction, we say that we did not see the motion of the hand immediately but inferred it. Although these expressions are not exact, we can still use them provided that no misunderstanding can arise from them or we can use them with explanatory comments to avoid misunderstanding.

There follow three notes; one historical, one defending the position of no. 9 above, and one defending that of No. 11. None of them contain new arguments.

§ 301. *The Generation of Judgments by the Relation of Probability*

In this section, B. points out that it is not surprising that we should form judgments on the basis of the relations of deducibility and ground–consequence. He claims that when we recognize one proposition as a consequence or conclusion of others, we do so on the basis of our cognitive faculty [*Erkenntniskraft*]. A being with an infinite cognitive faculty would be in the possession of all truths, including propositions stating relations of ground–consequence and deducibility. Humans, who possess part of the cognitive faculty of such a being would therefore be in the possession of some truths stating relations of deducibility and ground–consequence. But the case of probability is entirely different. If a certain proposition *M* merely stands in a relation of probability to *A*, *B*, *C*, *D*, . . ., and we are induced to accept *M* because we have already accepted *A*, *B*, *C*, *D*, . . ., we cannot assume that our accepting of *M* is brought about by our *cognitive faculty*. "For *M* is not always a truth, and a being whose cognitive faculty is infinite does not judge in the same way as we. He does indeed form the judgment that proposition *M* has probability, but does not form judgment *M* itself. Hence the question arises how *we* arrive at this judgment? It is not appropriate to assume a special simple faculty which leads us to such judgments; otherwise we should have to say that the most perfect being lacks a faculty which we have. Rather, we must explain the origin of such judgments as a phenomenon which rests upon the *limitation* of our powers, and which could be brought about by a combination of several of them. The case could be similar to the faculty of *desire* which, in finite beings, divides into a faculty of *wishing* and of *willing*, the former of which is not found in the infinite being, since it results from a mere imperfection, namely from the limitation of our own well-being; so also with the faculty of *judgment*; in finite beings this faculty sometimes leads to a combination of ideas resulting not in certain, but merely in probable propositions. And just as wishing can be considered an imperfect willing, so the confidence with which a finite being forms a judgment based on a merely probable proposition is imperfect compared with the confidence with which it asserts something which is completely certain."

2. Concerning the actual nature of the limitations, B. makes the following suggestion: in propositions which assert the probability of a certain proposition *M*, where *M* is so highly probable that we have to act as if it were true, "the proposition itself is by far the most important thing, while the fact that it is not decidedly true but only probable is not nearly as important. Now if we are unable to keep in our mind all of the parts of *M* as well as the thought that this proposition is not completely certain, but merely so highly probable that we have to act upon it, then it seems natural that we omit the latter thought, since it is the less important, and concentrate our attention merely upon proposition *M* itself in order to grasp it completely and retain it. In this way it happens that the judgment which attributes only a certain degree of probability to proposition *M* gradually disappears from our clear consciousness (if it ever attained such a position), and judgment *M* alone remains."

§ 302. *How we Obtain our Immediate Judgments*

"The two most important kinds of immediate intuitive judgment were already introduced in § 300, no. 12. The first of these 'I—have—impression *X*' presupposes for its occurrence only the following two things: the judging person must have the capacity to undergo a change like impression *X* and must raise this impression to the level of clarity. If the cognizing being considers a present impression *A* (idea, sensation, etc.), then it is also in a position to form the judgment that it has such an impression. The origin of judgments of the form 'this (what I just now perceive) is an *A*' is much more difficult to explain, especially if idea *A* (as I think possible) is not simple, but very complex (cf. § 286, no. 8)." B. does not think that the mind will be able to form such a judgment on the first occasion, but he does not think it surprising that the mind directing its entire attention upon these changes will eventually be in a position to form more and more precise ideas of their peculiar properties, especially if changes take place in between which follow an entirely different law, because the contrast will facilitate the comprehension of the changes in question. "I do not consider it difficult to conceive that the mind gradually forms a concept of these specific attributes, even if they must be compounded from several parts. For the compounding of a concept from several fitting parts is a capacity which we must attribute to the mind, since we are in the possession of concepts of that sort. All this can be much more easily comprehended if we presuppose that in doing this the mind does not form the judgment that it does it, but that it performs these operations without being clearly conscious of them."

2. Concerning immediate conceptual judgments B. says that these judgments may be occasioned by the occurrence of certain other judgments about external objects, but that these things cannot be the complete ground why we judge, for example, that every object represented by a certain concept *A* has an attribute represented by concept *b*. If they were the complete ground, the judgment '*A* has *b*' would not be immediate. Rather, part of the complete ground for forming this judgment must lie in the specific characteristics of the two concepts *A* and *b* themselves.

§ 303. *How we Arrive at, or could Arrive at, our most General Judgments of Experience*

In contrast to immediate judgments, the origin of mediated judgments can be more precisely explained. In order not to become too prolix, I want to describe only a single class of such judgments, namely the judgments of experience, which we usually take to be immediate, since we are rarely conscious of the inferences which mediate them. It must be noted that it is not sufficient to indicate the modes of inference which we use in order to form these judgments (this has already been discussed in another place), rather, we must discuss these inferences themselves, i.e. we must indicate the most important propositions of which they are composed. Finally I do not merely want to point out how these judgments are actually formed, but also how they *could* and *should* be formed if a rational man wanted to justify them.

1. The greatest difficulty arises precisely where I think we must

make the start, namely when we have to explain how we form those judgments in which we determine *temporal relations* which hold between certain phenomena that take place within ourselves, e.g. with the judgment 'idea *A* precedes idea *B* within me', etc. It seems to me that judgments of this kind should not be considered immediate. We do not immediately perceive in our ideas, sensations, etc. which of them are earlier, which later, etc.; rather, we must infer this from the observation of certain of their properties or from other circumstances. But it is even more obvious that the temporal relations between mental changes are not always inferred from the perception of temporal relations among certain *external* changes. It is true that we *occasionally* determine the temporal order in which in our mind ideas and sensations follow each other by reference to the observed temporal order of certain external objects; but this can not always be the case; rather, the temporal order of external changes is, as a rule, inferred from temporal sequences that take place within ourselves. It is impossible for us to know what is earlier or later in the external world if we did not know beforehand what is earlier or later within ourselves. It is indeed possible that I am unable to say which of two objects, this chapel or that house, I have known longer, until I am told that the former was erected a few years after the latter; but generally things are reversed: only by perceiving internal temporal sequences can I recognize the temporal order of phenomena in the external world. From this it follows that if we want to explain the recognition of internal temporal sequences, we cannot take recourse to our recognition of *spatial* relations. The locations of things are no more than certain characteristics of their determinations which we must assume them to have so that we can explain why they act upon each other in just such and such temporal relations (given the forces which we recognize in them). It is obvious, therefore, that we can recognize spatial relations only after we have first recognized certain temporal relations. This is not contradicted by the fact that we use space (certain movements in it) in order to measure time, since closer consideration shows that the question whether a change in space took place at a certain time can in the end only be decided by reference to relations of simultaneity, which is an easily recognizable temporal relation. It follows from all this that temporal relations between individual mental phenomena (especially ideas, sensations, etc.) can at least sometimes be determined from their inner characteristics. The question is, how this is done. I think that it must be possible for us to recognize *immediately* that a given idea, sensation, or appearance is *at this moment present* in ourselves. For, if we cannot even recognize this immediately, then it is inconceivable how we could bring it out by inferences. For it is clear that no moment in time can be determined through concepts alone unless one such moment is given. But there is no reason whatever to consider a future or past moment as given rather than the present moment. But a moment is given by characterizing it as that moment at which a certain internal phenomenon takes place. Thus it must at least sometimes be possible to recognize immediately that an idea or sensation is presently occurring in us. Without doubt this will be possible only with those mental phenomena which have been raised to the level of clarity (§ 280).

2. Furthermore, it must sometimes be possible for us to recognize immediately that a certain idea or judgment which we form contains as parts certain other ideas *A*, *B*, *C*, *D*. . . . But if this is the case, then we can conclude that these several ideas *A*, *B*, *C*, *D*, . . . are in our mind at the same time. Thus here is a method with which we can recognize the *simultaneity* of certain ideas; it is evident that similar considerations also hold of judgments, sensations, and other phenomena within ourselves, if they appear as parts of a whole which can come about only through their interaction (hence their simultaneous presence).

3. It is also certain that we can often recognize the inconsistency of saying that two phenomena *A* and *B* are simultaneously in our mind. Sometimes we can see this immediately, sometimes for reasons quite independent from the present investigation. Thus it would be an obvious contradiction to say that a judgment *A* as well as its actual contradictory, the judgment '*A* is false', are formed simultaneously. Hence, if we are forced to admit the presence of certain contradictory phenomena in ourselves, we conclude that they could occur only at different times. Here then is a means by which we recognize that certain phenomena within ourselves are not simultaneous, but take place at *different* times.

4. Finally, if we recognize that a phenomenon *A* is just now present, then we know that its contradictory, *B*, which we also find in us, must have taken place in the past. For example, if we recognize that we just now formed the judgment *Neg.A*, then we know that the judgment *A*, if we have also formed it, must have been formed in the past. In this way we can determine the relation of priority between one time and another.

5. It is likely that there are other such methods. There are several phenomena in ourselves of which we can be certain that one of them must be earlier or later than the other. For example, a wish must be

in our mind earlier than the sensation of its fulfilment. Also, the individual ideas whose combination results in another idea must be earlier than the latter, etc. Now if we recognize by the criterion given in no. 2 that a certain mental phenomenon occurred at the same time as a wish, and another simultaneous with the feeling of its fulfilment, then we can form a judgment about the temporal relation between these two other phenomena.

6. If an internal phenomenon A is a part of several phenomena $M, N, O, \ldots$, or if we realize in some other way that it was simultaneous with them, and if we also know that phenomena $M, N, O \ldots$ contradict one another in some respect and hence cannot have been in the mind at the same time, or if we conclude this for some other reason, then we discover that phenomenon A must have been in the mind at different times, hence either continuously over a span of time, or repeatedly.

7. The same also holds when phenomenon A is composed of several individual phenomena, e.g. ideas, sensations, etc. Consequently we can recognize that several ideas (or phenomena) A, A', A'', . . . have repeatedly *occurred together*. In this way I become aware of the fact that I repeatedly had the idea of certain colours (namely the colours of roses) and the sensation of a certain odour (the odour of a rose) at one and the same time.

8. Now we already know several ways in which we can form judgments that assert that a certain mental idea or phenomenon belongs into an earlier time. But there are several other ways. For example we can conclude, at least with probability, that a certain idea must have been present in our mind earlier, perhaps even several times, if we can generate it more easily and with less attention than is usually the case with ideas of a similar composition, etc. In particular with respect to *judgments* we can surmise that the judgment 'A is B' must have been formed once before, if it appears in our consciousness and we are unable to indicate the reason why we have formed it, provided it does not belong to the class of judgments that are recognized immediately. Thus I can conclude, justifiably, that I have formed the judgment '$\sqrt{2} = 1.414 \ldots$' once before if it appears within me without calculation.

9. In order to generate a certain sum of ideas, judgments, or whatever other internal phenomena, a certain length of time is required; in order to form the same sum of ideas or judgments a second time, we need about the same time or somewhat less. The same holds if these phenomena form a certain *series*, i.e. if there is a certain order in which they follow each other, as, for example, the ideas 1, 2, 3, 4,

etc., when we count. Now let *a*, *b*, *c*, *d*, *e*, . . . *y*, *z*, be such a series of phenomena in our mind; let *A* be a certain other mental phenomenon which is simultaneous with *a*, and let *B* be a phenomenon simultaneous with *z*. In this case we know that the two phenomena *A* and *B* are separated by the span of time which is required to generate all of the ideas *b*, *c*, . . . *y*. If we repeat the series after having once run through it, in such a way that after the final idea *z* the idea *a* comes about for the second time and if phenomenon *C* occurs in the mind as soon as we reach *z* for the second time, then we know that phenomena *A* and *C* are about twice as far apart as *A* and *B*, etc. Here we have a method of estimating the relation between different lengths of time. We actually use this device if we count in order to measure the time lapses between phenomena *A*, *B*, *C*, *D*. . . .

10. If we have a present intuition *A*, then we must recognize immediately whether it belongs to the class of external or internal intuitions (cf. § 286), i.e. whether or not the change in our mind, which we presently observe, must be thought as the result of the influence of a certain external object. In this way we find out that there are also *external* objects, which have the power to produce intuitions in us.

11. If we have certain *similar* intuitions, i.e. if they all fall under the same concepts (we must at least sometimes be able to recognize this immediately, cf. § 300, no. 12), we conclude from this, if not with certainty then at least with probability, that the objects which caused them will also be similar. This inference is not certain, since dissimilar causes can sometimes have similar effects. Thus an intuition of yellow can sometimes be caused by the influence of a yellow object upon our eye, but sometimes by the juices of the eye itself, if they are coloured in an unnatural way, etc. But so long as no extraordinary circumstances obtain, it is more probable that similar effects proceed from similar causes.

12. If it happens repeatedly that we experience simultaneously certain intuitions which fall under the concepts *A*, *B*, *C*, *D*, . . ., and we rarely, if ever, experience one of them without the other, then we conclude with a good deal of probability that it is one and the same real object which causes these intuitions in us. We may therefore attribute several powers to it, namely the power to generate intuition *A*, to generate intuition *B*, etc. This method of inference allows us to discover, by and by, the different powers which surrounding objects have. For example, if we have repeatedly experienced a red colour together with a certain very pleasant fragrance, we conclude that it is one and the same object which caused both of

these intuitions in us or, what comes to the same thing, that the rose has a fragrance.

13. But now we must explain why the object whose presence we have just assumed does not always cause the indicated intuitions *A*, *B*, *C*, *D*, . . . in us. This can be explained in only one of three ways, namely, either this object changes from time to time, or there are certain other objects whose interference prevents this influence, or the spatial relations between this object and ourselves change.

14. When we experience that in many cases some things happen after we have previously *willed* them, we conclude that our will is the cause of their occurrence and that we have the power to bring about results of this kind. In this way we find out about several powers that reside *within ourselves*. In particular, we realize that under certain circumstances we have the power arbitrarily to produce within us intuitions of certain objects, and to remove these intuitions. For example, if we have a rose in our hands, we can produce in us the sensation of its fragrance, and we can remove that sensation. According to no. 13, this presupposes that a change takes place either in these objects themselves, or in the spatial relations between them and ourselves, or in one of the surrounding objects; thus we discover that we have the power to produce, at will, a great many changes in the external objects which surround us, or in the spatial relations between them and ourselves.

15. This imposition of our will upon external objects, which we experience in this way, is not always immediate; often external objects are influenced by changes which we bring about in others. But there must always be some external object upon which we act immediately. The totality of these is called the mental organ [*Seelenorgan*], and if we add to it everything that stands to it in an *organic* connection . . . it is called our *body*. We will soon see in what way we gather more and more information concerning the existence and attributes of this body.

16. If certain intuitions *M*, *N*, *O*, . . . (i.e. intuitions which fall under the concepts *M*, *N*, *O*, . . .) *never* occur unless certain other intuitions *A*, *B*, *C*, *D*, . . . preceded them in time, we conclude that things or changes which were necessary to bring about intuitions *A*, *B*, *C*, *D*, . . . are a *condition* for the occurrence of those things or changes which belong to intuitions *M*, *N*, *O*. . . . If intuitions *M*, *N*, *O*, . . . always follow as soon as intuitions *A*, *B*, *C*, *D*, . . . have taken place, we conclude that the things or changes which bring about intuitions *A*, *B*, *C*, *D*, . . . contain a sufficient ground or complete cause for the occurrence of those things or changes which bring

about intuitions *M*, *N*, *O*. . . . Conversely, if intuitions *M*, *N*, *O*, . . . never occur without *A*, *B*, *C*, *D*, . . . but the latter sometimes without the former, we realize that the particular changes which bring about *A*, *B*, *C*, *D*, . . . are not the complete ground, but a mere condition for the things or changes which bring about intuitions *M*, *N*, *O*. . . . In this way we can judge the *means* or *conditions* which lead to the origin of many objects.

17. If certain phenomena *M*, *N*, *O*, . . . which I want to bring about in my mind do not always occur, but only after I have succeeded in first producing *A*, *B*, *C*, *D*, . . ., I conclude that my mind cannot immediately bring about the changes which lead to the phenomena *M*, *N*, *O*, . . ., but only through *mediation* of those changes which lead to phenomena *A*, *B*, *C*, *D*, . . ., hence only after the latter are actualized. Thus I want to experience the taste of an apple. I bring this about only after first producing all those intuitions which follow upon looking at the apple, bringing it to my mouth, etc.; therefore I conclude that the appearance of that taste within me did not come about immediately but only by mediation. On the other hand, if there are certain phenomena *M*, *N*, *O*, . . ., in my mind which always, or almost always, come about when I wish, I conclude that the objects which are necessary for their production can be immediately influenced by my mind or are intimately connected with those that are so influenced. I therefore count them as part of my *body*.

18. I have intuitions of colours; I cannot arbitrarily create them but I can arbitrarily remove them (namely by shutting my eyes). I conclude from this that these intuitions do not arise within me immediately, but only through the mediation of an object which depends upon my will, i.e. a part of my body. This part I call my eyes. In a similar way I find out that I have organs through which I receive fragrances, intuitions of taste, intuitions of sounds, etc.

19. By the well-known laws of association of ideas, the sight of a rose in front of me awakens within me the memory of that pleasant fragrance which I have repeatedly experienced at the sight of a rose; occasionally, there is also a desire to repeat this sensation. But this desire is not satisfied until I have picked the rose (perhaps for no particular reason) and moved it to my nose. During this activity I *see* the changes which take place with my hand, i.e. several intuitions arise within me for whose explanation I must assume certain changes that take place in that part of my body. Since this change resulted in the satisfaction of that desire (to smell the rose), I conclude that that part of my body which brought about the various different intuitions

was through its change (motion) also the cause (partial cause) of the satisfaction which followed. In this fashion I learn more and more about the services which can be performed by my hands and the other members of my body.

20. I daily experience pleasant and unpleasant sensations which occur simultaneously with changing appearances of this or that part of my body, and which can become more vivid or more faint depending upon certain influences which are exercised upon this part. From this I conclude that the cause of these sensations lies in a change of these parts. Thus I can have a strong toothache without knowing which tooth contains the cause of this pain, until I find out that there is a change in the pain when one of my teeth is touched.

21. All these cognitions become much more determinate when we begin to assess the *spatial relations* in which external objects stand to ourselves and to each other. I imagine that this happens in the following way, or could happen in the following way if we proceeded in a particularly strict manner: Some of my intuitions refer to external objects. Whenever these intuitions change, I must assume that the cause of this change is either a change within myself, or in some of the external objects; the latter can be a change either in their spatial relations to me or in their inner attributes, or (what is probably more correct) in all of these things at once, only in varying degrees. Since my intuitions which refer to an external object are changes within me brought about by the influence of an external object, the character of these intuitions is determined by my attributes, the attributes of the external object, and the spatial relation between us. Without a change in one of these three parts there can be no change in those intuitions. But if one of these things changes then, because of the interaction between all things in the world, all other things change as well, except that the change in one or the other can be so insignificant that we can neglect it. Now, according to no. 11, if the same class of intuitions *A*, *B*, *C*, *D*, . . . occurs within me repeatedly, I conclude that it is one and the same object which brings them about. I do not know, so far, what happened to the object in the interim, when I did not have intuitions *A*, *B*, *C*, *D*, . . ., whether it was changed, or perhaps destroyed, and only regenerated at the moment when these intuitions recurred, or whether the disappearance and reappearance of its intuitions stemmed merely from a change in its spatial relation to me. If I now undertake several actions through which we can usually achieve that a certain sum of intuitions *A*, *B*, *C*, *D*, . . . disappears without returning, i.e. diverse activities through which objects are usually changed, but if no such change occurred

in this case, then I surmise that the disappearance and reappearance was not caused, in this case, by a change in the object in question, but only by a change in its spatial disposition relative to me, or by the interference of another object. In this case I must ascribe to the object a certain power to remain in a given condition (a certain connection of parts), i.e. a certain *solidity*. For example if I grasp a piece of wood with my hands and touch it in various ways, then the same intuitions recur from time to time (as often as this body is brought into the same position between my fingers); on the other hand, in the case of a soft or liquid body the sensations which it causes change constantly, and earlier ones do not recur. (In a similar way we convince ourselves that our hands themselves, or rather their parts, are solid bodies of this kind.)

22. Assume that two sums of intuitions *A*, *B*, *C*, *D*, . . . and *M*, *N*, *O*, *P*, . . . are given, where the first is taken as the effect of body *X*, the second as the effect of body *Y*. If these sums occur either simultaneously or in quick succession, either *A*, *B*, *C*, *D*, . . . after *M*, *N*, *O*, *P* . . . or vice versa, then I surmise that the locations of these two bodies are very close together, or that the bodies *touch* each other. This conclusion is not certain; for example, two stars which are always seen simultaneously can still be very far apart. But so long as there is no circumstance which indicates an exception, this assumption is probable, especially if the bodies do not only appear to the eye, but also to the other senses, i.e. touch, hearing, smell, etc. at almost the same time.

23. If a certain appearance which I ascribe to the influence of an external object becomes either stronger or weaker from moment to moment, then I may surmise that this object either *approaches* or *recedes*, as case may be.

24. Assume bodies, *M*, *N*, *O*, *P*, *Q*, . . . and *A* are all found to be solid. Furthermore body *A* has been found to have two different parts *a* and α, and now it is realized that at a given time *a* touches *M* and α touches *N*, at a later time *a* touches *N* and α touches *O*, then, later, *a* touches *O*, and α touches *P*, etc. I conclude that the distances between things *M* and *N*, *N* and *O*, *O* and *P*, etc. are approximately similar to the distance between parts *a* and α. The body *A* which I use for this purpose can, for example, be my own hand and parts *a* and α can be two specially chosen parts of it (as for example the thumb and the little finger).

25. If a body *A* moves from the vicinity of *M* into the vicinity of *N* and from there into the vicinity of *O*, etc., and the distances *MN* and *NO* are equal to each other and the times which pass are also

equal, then I know that bodies *A* and *M*, *N*, *O*, move uniformly relative to each other.

26. Assume that a body *B* has so far not changed its relation to certain other bodies *M*, *N*, *O*, *P*, . . . (among which there is my own body). Assume furthermore that another body *A* changed this relation and eventually came into contact with *B*: when this happened, the relation between *B* and those other bodies *M*, *N*, *O*, . . . also changed. I then conclude with probability that the contact between the moving body *A* and *B* (i.e. the *impact*) was the cause of the ensuing motion of *B*. In this way I discover that the members of my own body, for example my hands, can produce motion by impact.

27. I frequently notice that a body, as long as it is found to be in contact or connection with a certain other body, for example my hand (i.e. if it is carried or held by it), does not change its spatial relation to me; I also experience a certain effort on this occasion, which becomes more unpleasant the longer it lasts. I therefore conclude that this effort caused the body to remain in its position; from this it follows that it must have a certain propensity to change its location. In this way I discover that certain bodies *attract* or *repel* each other, and hence the general law of gravity.

28. Occasionally I notice that a certain body *A* (for example a ball) can touch another body *B* (for example a horizontal board), in a variety of different ways, although the latter stays at rest, and although the parts with which both bodies touch, at least in the case of *B*, change constantly. I furthermore find out that it takes about the same expenditure of effort and time to move body *A* from its initial position on *B* to any other position so long as they are equidistant. Now if body *A* is at rest and I give it a push with another body, for example my hand, in such a way that in the ensuing motion it continues to be in contact with *B* (is not propelled above *B*), then I can conclude that this motion, being caused by the impact alone, lies in a plane. Through my eye I can also comprehend the line which is described by *A*. And if I now find that a similar line is described no matter from what point and in what direction the motion is originated, it follows that the surface of body *B*, upon which all of these motions take place, can only be one of two things, either the surface of a sphere, or a plane; and the lines can either be segments of circles or straight lines. But my eye gives me a simple means to distinguish a straight line from any other line (actually, in this context, a line is a series of small bodies which are distinguished by their colour; and the eye can distinguish straight series of this kind from curved ones). The former has the property that, no matter from what vantage point

it is observed, it generates upon the retina a picture which coincides with a piece of that line generated at a different distance. Thus the ideas which are produced by a whole line, and the ideas which are produced by a piece of it (after the distance has been changed), must be equal to each other. Every line in which I ascertain this property, I recognize as straight. But if I have once seen several straight lines, I become familiar with the peculiarity of this phenomenon so that I can tell by sight whether a line should be reckoned among the straight ones or the curved ones. In this way I find in the above case that the lines are straight and that the surface of body *B* is a plane. Soon I learn from the mere illumination of a surface whether it is plane or curved. But once I can distinguish straight and curved things, I acquire the ability to measure distances; for if lines *mn*, *no* are straight, and are juxtaposed in such a way that *mo* is also straight, then the distance *mo* is the sum of the distances *mn* and *no*. Finally, since all spatial relations can be determined by distances, it follows that if I can estimate distances, there are no relations in space which I cannot determine.

29. But it should not be imagined that the just-indicated method of estimating distances is the only possible one, and that the sense of sight is indispensable for this matter. Even if we lack sight, we can determine these relations, although with more effort. We have seen from no. 21 how the sense of touch alone can tell us whether or not a given body is solid. Through the sense of touch we also find that there are bodies whose connected parts resist complete separation and destruction, though the parts change their spatial relations to each other in many ways and, in particular, can be brought closer together. Such a body we tend to call flexible. Now grasp parts of this body, which are not in immediate connection with each other (perhaps with both hands) and try to move one of them in a direction which it resists; we now know that the distance between these two parts is the largest there can be, unless the connection between them is severed. Now if the individual parts of this body are small, for example so that they can be easily grasped, while the whole body is much larger (longer), than the position in which we have now brought it represents a straight line (actually a thread), and by touching it we can learn how such a line is represented by the sense of touch.

Consider some other body which offers the same sensations, no matter where we touch it; we must conclude that its surface is either plane or cylindrical or spherical. Which of the three is the case can be decided in various ways. For example, if we also have a body which can be moulded into different shapes and retains them, then we can make an imprint of the surface in question upon this material, and if the surface which is generated by the impression is like the given surface (i.e. if it produces the same sensations when touched), then it is flat, in the other case it is either cylindrical or spherical, the former if we encounter different tactile sensations in different parts, the latter if this is not the case. If two plane surfaces of a body meet (an example is the edge of a prism), then the parts of the solid where this occurs (where a new sensation starts) form a straight line, and by a frequent and attentive touching of such a solid, we can learn how the peculiar properties of such straight lines are manifested to touch, etc.

30. Once we have become acquainted with the number of solids which have a fairly permanent shape and size, it will become easier every day to gauge the shape, magnitude, and distance of other bodies, especially if we can employ the sense of sight; in doing so we should particularly employ the following rules: every body which is kept out of sight by another body must be farther away than the covering body; every body which does not change its magnitude must be the farther away from us the smaller the angle of sight under which it appears (the angle of sight is measured by comparing the body in question with another body which changes neither distance nor size); every body which does not change either its size or shape must be the farther away the darker it appears to us and the less we are able to distinguish its parts, all other things remaining equal, etc.

31. It could be asked how a child or even animals arrive at a knowledge of all these relations in space, since it is quite obvious that they do not perform the inferences which I have here described. My reply is that the grasp of spatial relations which we find in children and animals can be explained with no more than the well-known law of the association of ideas, or the expectation of similar cases under similar circumstances which issues from it, especially if we assume that instinct has certain effects. We say that an animal has a grasp of relations in space if we notice that it avoids a stone which lies in its path, or if it wants to acquire a food object and turns either to the left or to the right depending whether the object really is at the left or the right. But all this can be explained satisfactorily from the mere association of ideas; it avoids the stone because several times when it had an intuition of the kind that it now has of this stone, its running was impeded; it turns right in order to snatch its prey since earlier on, when it had these kinds of intuitions, it also had to turn right in order to grasp it, etc. A child, when it is old

enough to clarify some of its concepts, can go a step further and use successful associations of ideas to make rules for future cases, and thus to act consciously.

So far I have considered only the most general points which had to be stated by way of introduction. However, some readers may welcome the discussion, in the following notes, of some points which arise only after a more detailed investigation of this subject.

NOTE 1: In no. 15 ff. it was shown, roughly, how we arrived at a knowledge of our body and its sensory organs. In a similar way we distinguish whether a certain change in our perceptions derives from a change in our body or in one of the other external objects. If it is a change in one of our sensory organs, then it must uniformly extend to all perceptions which derive from this organ. Thus, for example, I realize that the red spots which I have had before my eye for a few seconds (since I have cast a glance into the bright sun) stem from a change in the eye, not from a change in another external object; the reason is that I see the spots superimposed upon every object upon which I fix my eye. One could object that this rule would also attribute the darkening of all objects after sundown to a change in our eye or an extinction of vision, which would be absurd. But there are two matters to be considered: first of all the same blackout can be achieved during the daytime as often as we want, so long we bring about a certain external change, e.g. close the shutters; second, we can see at night as soon as we make certain external changes, e.g. light a fire. These circumstances demonstrate that the cause for this darkening does not lie in our eyes, but is to be found in some external object.

NOTE 2: Even simple people often ask the question whether and how they can assure themselves that things represent themselves to their senses in the same way as they represent themselves to the senses of other persons. However, we can conclude that another being observes things in about the same way we do if we notice that it reacts to them in the same way we do, i.e. that it, too, has a desire for impressions which we find pleasant, that it makes a distinction between objects between which we, too, differentiate, etc. On the other hand, if we notice that somebody differentiates things which we confuse, we justly ascribe to him sharper, or more perfect, senses than ours.

NOTE 3: There is another very interesting question, namely why we consider the location of some of our ideas and sensations to be in the mind, while in other cases their actual place is considered to be in some part of the body, and in still other cases we think neither of the mind, nor of the body, but only of certain external objects. My opinion is that we consider an idea or sensation as present only in ourselves, if we do not notice any external object as its immediate cause. Thus we say of a concept which we form by mere thought, e.g. the concept of a perpetual motion machine, that it is to be found only within ourselves (in our mind). On the other hand, if an idea is caused by a noticeable change in our body, e.g. a wound on our finger, then we locate this phenomenon in that specifically changed part of our body. I have already discussed in no. 20 why we assume that a certain change in the body is the cause of that mental phenomenon. Finally if we do not notice the change itself which an external object produces in the organs of our body, then we place the cause of this idea outside of our body It is for this reason that we place the sensations which are given us by the sense of touch, taste, or smell in special organs of the body, since the change which takes place in them whenever they give us these sensations, is also noticeable by other senses. If we touch something with the tips of our fingers, the impression which this body makes upon the fingers can be seen, etc. On the other hand, the ideas which we derive from the sense of hearing or sight are referred to bodies outside of our own body, since the changes suffered by these organs when they are externally influenced are not usually noticeable by any of the other senses. Only in rare cases, where a strong impression causes a sensation of pain (when our ears ring or our eyes burn, etc.), are we reminded of the organ which produced those ideas.

NOTE 4: Given these assumptions, several questions which were raised with respect to judgments derived from the sense of sight can be easily answered. For example, how we can make such quick and automatic decisions concerning the magnitude, distance, and position (left, right, up), of every visible object, and why such judgments appear to us to be immediate perceptions, so that we cannot disregard them even if other circumstances show them clearly to be mistaken, whence we say that the thing, though it *is* not really this way, at least *appears* to be this way. The origin of such judgments is all the more puzzling since (a) the pictures in the retina of the eye which (as was thought) are immediately seen by the mind, appear twice (once in each eye), while actually there is only one object; since (b) these pictures are usually much smaller than the external object which causes them, while we do not perceive this object to be diminished in size, but in its normal size; since (c) a small body at a close distance produces a larger picture than a large body at a greater distance, while the distances are very nearly alike in all these pictures, and especially since (d) their position is reversed so that what is to the left in the picture is to the right in reality, and what in reality is on top is at the bottom of the picture, etc. In order to overcome these difficulties some philosophers have assumed that the mind has an immediate sensation of the direction in which a light ray meets our eye; but aside from the fact that this assumption still does not explain everything, it is very hard even to understand. Rather, the correct explanation rests upon different grounds. We have to make no decision whatsoever whether it is the retina of the eye or some other part of our bodily machine which receives the light impressions in the first place: the only definite thing is that there is such a part and that the mind must perceive immediately the changes which take place in it, and must infer, merely from its attributes, the figure, magnitude, distance, and position of the visible objects in the following way: whenever the organ is changed in the same way, the mind infers a similar cause (unless some other circumstance inhibits this conclusion). Hence, if our eye (and indirectly also the actual mental organ) is stimulated, and we reach for the stimulating object, and find it above, then we expect an object in an upward position as often as the light ray meets that place, i.e. we see it above, and so in other cases. If we observe the change in our perceptions when a body, which changes neither size nor shape, approaches or recedes, if we notice what aspect it offers when we have measured its distance with our hands or in some other way, then we are in a position, the next time, to judge, merely from the way in which it is represented to our organs, whether it has the same, a larger, or a smaller distance. The fewer parts we can distinguish in it, the farther we take it to be away. Every object whose size we have learned by touch becomes a means for estimating either the distance or the size of other objects: the distance if we know its size, and the size if we know its distance. Assume that you see, at a certain distance, a person with a staff in his hand. Because he holds the staff in his hand, you know that they are at the same distance. If you find that one end of the staff coincides with the feet, the other with the head of the person, then you can conclude that the staff is about as long as the person. If we see a bird of known size pass in front of the sun or the moon, in such a way that these bodies are not altogether covered by the bird, we conclude that they are larger than the bird. Through practice we learn to draw such conclusions with great rapidity, so that we no longer observe them, and are no longer conscious of them. This explains why we mistake these judgments for immediate perceptions, and why we cannot discount them even in cases when they are incorrect and we are convinced of their falsity through inferences of a different sort. Thus, sun and moon appear larger in the horizon than when they are overhead, because they emit a dimmer light when they are in the horizon, and are therefore taken to be farther away. That this is the correct cause can be seen from the fact that the same bodies appear to be of various size even in the horizon, if they appear under different circumstances; they seem especially large if we see a considerable number of objects of known size, e.g. villages, rivers, valleys, etc. between ourselves and them, especially if objects of considerable size are seen next to them in the horizon, for example, trees or buildings, though they appear to us under the same or even a smaller visual angle. An additional factor is that from childhood on we found out nothing about these objects, except what we learned by sight. Hence we find it natural to explain the changes in their appearance from changes that take place in these bodies themselves, while we do not usually assume that a body with which we are more familiar, and of which we know that it does not quickly change its size (e.g. a man), becomes smaller whenever we see it from a larger distance. We are so used to giving it the same size, even though it takes up much less space on our retina, that we take this judgment to be an immediate perception and claim that it always appears to have the same size.

Heavenly bodies exhibit various degrees of brightness, depending how far they are above the horizon, and this explains why we assume that the skies have the form of a depressed sphere. But these considerations do not explain the specific height that this sphere seems to have, and the distance at which the stars appear to be. The experience that even the highest towers, mountains, clouds, and other objects sometimes obscure a star, shows only that the distance of the stars is still larger; but we do not know how large it is, nor have we found a limit within which they lie. Rather, it seems that people assume the heavenly sphere to be at various distances, depending on rather accidental circumstances. Those who have never lived in the open or are nearsighted tend to think of the sky as being lower, and the stars as being smaller, than those who have often climbed high mountains and have keen eyesight. Perhaps when we were children we looked after a bird that was flying away, and it seemed to us that it got lost in a cloud, and we thought that the cloud or, what comes to the same thing, that the stars are as high as the bird; but the height of this bird we estimated from the time it took for its flight, etc. Even astronomers do not see the stars as larger, and this is due to the fact that a judgment to which we became accustomed in childhood can be corrected in later years but cannot be replaced in such a way that it does not come to mind at all. And it is this coming to mind which we call *seeming* [*Scheinen*]. If these explanations are correct, then the difficulties mentioned above are overcome. For example, we assume that there is only one object, even though there are two parts of our body (both eyes) which are changed so as to indicate the presence of an object; this is a consequence of the fact that we know from experience that if such and such corresponding parts of our visual organ are simultaneously affected, there must be a single object which is the cause of this affection. Consequently, we see double only if our eyes change position in a non-uniform manner, for example, when we look up with one and down with the other. For now we do not just grasp those parts which we took in when there was only one object, but others: one eye sees parts which make the object appear in one position, and the other as if stood in another. Hence we see it double. The retinal images are very small, but we do not attribute the same small size to external objects; the reason is, simply, that we do not gauge the magnitude of external objects from these pictures at all (we do not even know that they are there), but mainly, we estimate size by touching. Similar points can be made about distances and position. Finally, it should be remarked that the question whether we see objects in their true magnitude, distance and position, has a reasonable sense only if it is applied to individual objects. If we wanted to ask whether or not perhaps *all* objects appear smaller or closer than they are in reality, or upside down, we should be asking something entirely absurd. For an object can be smaller, or closer, or reversed, only with respect to others, but we cannot assert this of all objects at the same time.

NOTE 5. In this section B. rank-orders the senses in this way: sight, hearing, touch, smell and taste. He ranks sight as the most perfect because most objects perceivable by any of the senses are also perceivable by sight, and because sight shows us objects at greater distances than the other senses, etc.

§ 304. *Other Opinions on this Subject*

B. gives a brief review of epistemological theories, contrasting "Descartes, Cudworth, Leibniz and many others" with "the Epicureans, Stoics, Campanella, Gassendi, Hobbes, Locke, and almost all English and French scholars to the present day." The former maintained that there are two different kinds of judgment; one begins with (i.e. is stimulated by) but is not grounded in, experience; the other consists of mere intuitions and perceptions. The latter held that all our judgments are grounded in experience. "To refute the empiricists," B. continues, "it is only necessary to give a precise explanation of what is meant by 'empirical origin'."

Nobody claims that we begin to judge before certain external objects have acted upon us, and have produced intuitions in us. The only question is which of the following two cases actually takes place:
a. All judgments which we form are either immediate judgments of perception (cf. § 300, no. 12), or can be deduced from a number of such judgments without the help of judgments of a different sort, namely judgments which follow from pure concepts.
b. We also form some conceptual judgments without having to deduce them from others.

Those who claim that all our judgments have empirical origin want to opt for the first alternative; but I believe that this is refuted by the mere fact that we form thousands of judgments whose origin cannot be explained in this way. For example, how could we explain in this manner the simple judgment that fire hurts? No matter how often we had simultaneously a perception of fire and an impression of pain, we could never have formed the judgment that the same object is the cause of both, unless we had first formed the judgment that a pair of phenomena which always happen at the same time have a common cause. I do not wish to claim that we must put this judgment in words, but at least we must have recognized it obscurely. Yet this judgment is purely conceptual. Indeed the ways in which the defenders of empiricism tried to explain the origin of these judgments are not satisfactory in the least. Locke, Reid, Beattie, Oswald, and others found it necessary to take recourse to certain instincts, i.e. they all confessed that the matter is inexplicable. Hume could explain the origin of the concepts of cause and effect only by reference to observations of the repeated succession of certain phenomena. But it seems obvious to me that such observation could only induce us to an *application* of these concepts; we must already have been convinced that everything that happens has a cause if the observation that phenomenon *A* always follows upon phenomenon *B* is to lead us to the conclusion that there is a causal connection between the two.

B. now gives a somewhat more detailed account of the doctrines of Berkeley, Tetens, de Tracy, and Herbart. The comments on Berkeley are the most important.

3. One of the most remarkable attempts was, without doubt, that

of Berkeley (in *Treatise Concerning the Principles of Human Knowledge* and other writings). The fact that this philosopher denied the existence of the external material world need not have any detrimental effect upon his explanation concerning the origin of mediated judgments of experience, since this mediation remains the same whether the intuitions which we take to be caused by external objects are really caused by them or (as Berkeley thought) directly by God. He noted quite correctly that the reason why we unite into one whole intuitions which come through several senses, and take them to be the effect of one and the same object, merely derives from the simultaneous presence of such intuitions. He also noted correctly that we infer a causal connection when certain phenomena always follow each other. He was not quite as successful in his explanation concerning the origin of our knowledge of spatial relations. He was quite correct in saying that the eye does not immediately give us any other ideas but those of colours, hence that we never recognize immediately the shape of a surface which is in front of us (as others had thought) but he went too far when he asserted that through sight alone we can never arrive at ideas of the relations in space; that the theorem of the three dimensions of space and other geometrical doctrines cannot be understood unless we have a sense of touch, and that for this very reason a man born blind who is suddenly made to see cannot, by sight alone, tell a cube from a sphere when they are put before him, and that he could not even notice whether or not a body suspended before his eyes is in motion. The last remark seems to be the one most obviously false. It is certain that the mind must notice immediately, and not only through the mediation of the sense of touch whether, during a period of time, its visual organ is affected in just one way, or in different ways (whether it receives one or several intuitions at that time); hence we must be justified, in the latter case, to conclude that a certain motion (in whatever body) is the cause of the changing ideas which are coming in from the external world. Concerning Molyneux' problem about the possibility of distinguishing a cube from a sphere, Leibniz gave what I take to be the correct answer (in *Nouv. Ess.* II, IX, 8), namely that this discrimination could be made after some thought. In fact, it can be seen from the preceding section that the sense of touch is not absolutely necessary in order to give an estimate of the spatial relations between external objects. For example, we are justified in surmising that a given body recedes from us, if the intuitions which it causes in the senses of sight, hearing, and smell (i.e. senses other than touch), become weaker and weaker, or disappear altogether. And if we are very attentive and notice how the sensory intuitions, e.g. of sight or hearing, change gradually, we are later on put in a position to infer from the nature of these intuitions the relative distance between the object and ourselves. The only difference is that now the scale of distance is no longer determined by the sense of touch, but will be distance as estimated by sight or hearing. We will no longer say 'the body is at arm's length' (i.e. we can barely touch it when we stretch out our hand), but we will say 'the body is as far away as a fly which we can barely discriminate' or 'the body is as far away as a man whom we can barely hear talking' or 'as far away as a rose which we can just smell', etc. But even if it were very difficult or even impossible, without the sense of touch, to estimate the mutual spatial relations of bodies that act upon us, we must never say that we are unable to grasp the truths of geometry. It is true enough that the ordinary expositions of this science take many properties of space as given, for example that it has three dimensions, and we should perhaps avoid this; it is also true that most of our proofs are carried out in a way that they convince only those persons who already have some acquaintance with most spatial relations by a mixture of touch and sight, but should it not be possible to proceed in an entirely different way? Should it be altogether impossible to demonstrate the truths of geometry from pure concepts, in such a way that even those must admit them who connect no pictorial ideas with the words 'line', 'surface', etc.?

§ 305. *The Doctrine of Critical Philosophy concerning this Point*

1. B. rejects Kant's doctrine of analytic judgments on the grounds that we cannot claim truth for a judgment '*A* which is *B*, is *A*' until we know that the subject idea '*A* which is *B*' has a referent.

2. He also takes Kant's explanation of empirical judgments to be faulty, since they explain only general judgments, such as 'all bodies are heavy' while, according to B. it is more important to explain the origin of such singular judgments as 'this body is heavy'. In this judgment the subject idea 'this body' does not contain the idea of heaviness, but only certain other intuitions, e.g. of colours, odours, etc.; hence it is of the form 'the object which is the cause of the colour which I see, of the odour which I smell, etc.' The judgment itself says that this very object is also heavy, i.e. it is the cause of certain other ideas which I have, namely of a pressure which I feel, etc. We can see from this that the formation even of this singular judgment requires several repeated perceptions. "There must have been a repeated co-occurrence of the intuitions that fall under the concepts 'colour', 'odour', etc. with the intuition that falls under the concept of that pressure,

if I am to be justified in the conclusion (even probable conclusion) that the same object which is the cause of one is also the cause of the other."

3. On the other hand, B. finds nothing problematical in synthetic judgments *a priori* "what justifies the understanding to attribute to a subject *A* a predicate *B* which does not lie in the concept *A*? Nothing I say but that the understanding *has* and *knows* the two concepts *A* and *B*. I think that we must be in a position to judge about certain concepts merely because we have them. For, to say that somebody has certain concepts *A*, *B*, *C*, . . . surely means that he knows them and can distinguish them. But to say that he knows them and can distinguish them means that he can assert something of one of them which he would not want to assert of the others, hence it means that he judges about them. Since this holds generally, it also holds in the case when these concepts are simple. But in this case, the judgments which we make about them are certainly synthetic. Without doubt, these synthetic judgments, whose subjects are the simple concepts *A*, *B*, *C*, *D*, . . ., put us in a position to form further synthetic judgments about compound concepts which result from the combination of *A*, *B*, *C*, *D*, . . . either by themselves, or together with still other concepts. Thus it seems certain to me that we must be in a position to form some synthetic judgment about any object, so long as we have a concept of it."

4 & 5. B. here takes exception to the so-called *a priori* intuitions by which Kant attempted to vindicate synthetic judgments *a priori*. In particular, he confesses not to be able to see the distinction between a so-called pure *schema* and a *genetic definition*: "if the schema of a circle is nothing but the idea of a method how an object can be provided for the concept of a circle, then a schema is nothing but an idea in which a circle is generated, i.e. nothing but the well-known definition of a circle, which is called a genetic definition, namely the concept of a line which is described by a point that moves in a plane in such a way that it always maintains the same distance to a given point." He further takes exception to Kant's contention "that we cannot think any relation in space without constructing it, that we cannot think any line, any circle, without first describing them, that we cannot think the three dimensions of space without putting three lines at right angles to each other in the same point . . .". B. admits that such constructions can increase vividness and comprehensibility "but that such constructions are not indispensable can be shown by many examples. Thus nobody will deny that even a beginner can understand me when I define a dodecahedron as a body which is bounded by twelve identical plane faces. From that moment on he will connect a concept with that word; he will think something by it, something that is correct. On the other hand, will he be able to construct such a dodecahedron? Does he already know whether the bounding faces are triangles or pentagons, etc.? The dispensability of such a construction is even more obvious in the following concept: 'a finite spatial object which is fully determined by two of its points, and each point of which stands in the same relation to the whole'. Perhaps several readers will not even know, for a moment or two, whether this spatial thing falls into the class of lines, surfaces, or bodies. How could they be in a position to produce a construction of it? Or are we supposed to believe that such an indistinct and obscure picture as our imagination can produce in this case can actually contribute something to the recognition of the properties of this object? But we certainly

think this object, and we are even in a position, by taking some thought, to determine what kind of thing it is, namely the surface of a sphere, where those two points form the poles."

6. B. quotes Kant on the addition of five and seven (*Critique of Pure Reason*, B 15) and a demonstration that the angles of a triangle add up to 180° (*ibid.*, B 744). B. continues "From these and similar examples I conclude that assumptions that are based upon certain obscure reasons, usually merely upon the testimony of the senses, especially the eye, were described by Kant to be effects of a special pure intuition. Although mathematicians have succeeded in defining their concepts more precisely, and giving stricter proofs than is the case in any other science, they still have not succeeded in defining their fundamental and most general concepts, and they introduce their initial propositions either without proof or with proof that is not truly scientific. Thus the usual text books of arithmetic do not give an exact definition of the concept of a sum. If this had been done, and if they had said that a sum is a class of objects where the order of parts is not taken into account, and where the parts of the parts are considered parts of the whole (§ 84) they would have found that the following analytic proposition follows immediately from this definition '$a+(b+c) = (a+b)+c$'. By an application of this theorem together with the definitions that $7+1 = 8$, $8+1 = 9$, etc. the proposition $7+5 = 12$ is produced as a purely analytic truth for which no intuition whatsoever is required. Even Kant must have been somewhat hard put to find the kind of intuition which allegedly occurs in this case. He concluded that, since the units of 5 are added to 7 *one after another*, it is an intuition of *time* which underlies these arguments. What a conclusion! One could say, for the very same reason, that every sorites depends upon an intuition of time, since we arrive at its conclusion only in time.

The other example is more difficult, since the concepts which are present in the proposition are much more complex than in the preceding case. I should have to write an entire text book of geometry if I wanted to show how this proof can be produced from pure concepts. I will therefore be satisfied with the following remarks. When a geometrician (according to Kant's description) begins his proof by drawing a triangle, either in imagination or on a piece of paper, we must ascribe these activities to the present imperfection of his science. We can think of a proof procedure which does not require an imaginary triangle, at least which requires it only to facilitate the argument, in the same way in which we need to write down the propositions when we go through a sorites, so that we don't forget what we began with and how far we have proceeded. It is said, quite properly, that the geometrician *sees* how the extension of one side produces an external angle, and how it is divided into two parts by a line parallel to the base, and how these two angles are equal to two angles in the triangle, etc. But it is not necessary, I say, that we must *see* these things; the testimony of the *eyes* is not required; the geometrician could conclude all of this from his *concepts*. If we consider that this kind of seeing is not an immediate conception, but is inferred from our immediate perception through the unconscious application of many geometrical truths, then we will hardly be inclined to elevate these judgments, which are formed without a clear consciousness of their ground, to a special source of knowledge under the name of *pure intuitions*."

§ 306. *Survey of the most Important Activities and States of Mind which Concern the Business of Judging*

In this section B. defines various mental terms: "If we form the judgment '*S* is *P*' whenever the question arises in the mind whether the subject *S* has the attribute *p* . . . we are said to be *constantly inclined* toward that judgment, namely as long as this relation between us and the judgment obtains. Since there is no word which designates sentences of this sort, and which imputes neither truth nor falsity to them nor implies that they are held with a certain confidence, I usually use the words 'opinion' [*Meinung*] or 'view' [*Ansicht*]. Hence by the opinions of a being I understand propositions which this being takes to be true, irrespective of whether they really are true, and whether this acceptance has a high or low degree of confidence." "If we raise the question whether proposition *M* is true, and wish to form one of the two judgments, *M* or *Neg.M*, but form neither *M* nor *Neg.M* although we have concentrated our attention upon [subject and predicate] *S* and *P*, then the state of our mind with respect to *M* is called doubt." B. then terms a judgment voluntary if it can be prevented by redirecting one's attention. Other judgments are called forced. "Thus the judgment that I now feel a certain warmth is forced [*abgedrungen*], while the judgment that the square over the hypotenuse, etc., like most mathematical and scientific cognitions, is voluntary." "If the propositions from which judgment *M* is deduced, as well as those from which the former follow down to the immediate judgments are all purely conceptual propositions, then judgment *M* can be called a judgment from pure concepts, or pure, or *a priori*. In all other cases it could be called "drawn from experience" or *a posteriori*.

PART III

Of the Relation between our Judgments and Truth

§ 307. *More Precise Determination of the Concepts: Knowledge, Ignorance, and Error*

1. Knowledge is that state of mind "where we have once formed a true judgment (whose content is called the content of our knowledge) [*Erkenntnis*] and where we can still remember this judgment and continue to agree with it."

2. "We will be justified in calling somebody *ignorant* of a certain truth *A*, if he does not have knowledge of it; hence, according to no. 1, if he has never formed the judgment *A*, or does not remember it, or, in any case, if, when we pose the question whether *A* is true, he does not at once, and on the basis of memory alone agree with *A*." An error is any false proposition that is believed by somebody.

3. If somebody knows certain truths from which *A* is deducible, he will not be said to know *A*, unless he has performed the deduction. Similarly, nobody will be said to maintain a certain error *A* merely because it is deducible from other errors to which he subscribes.

§ 308. *Grounds for the Possibility of Ignorance in Man*

Since ignorance is simply the failure to form a certain true judgment *A*, and since there are infinitely many true propositions, not all of them can be judged to be true. Hence ignorance is inevitable.

§ 309. *Grounds for the Possibility of Error, and Circumstances which Promote Error*

"It is much more difficult to explain how error arises. We can consider the difficulty which we encounter in this undertaking to be another proof for the magnitude of our ignorance. We make so many errors, and still, in many cases, we cannot explain how the error came about; worse than that, we can hardly explain how error is possible at all. All our judgments are either immediate, or derived from others. Hence the question is in which of these two cases error can occur. None of the judgments which are formed without the mediation of others can reasonably be suspected of being mistaken; for if we wanted to doubt one of them, we should have to doubt them all, since they all come about in the same way. But if we wanted to doubt all immediate judgments, we should also have to doubt all derived ones; hence we should have to doubt all our judgments, which is absurd. Thus we can reasonably look for error only in those judgments that are derived from others. But how can it occur even here? According to § 300, we should not assume that the mode of inference by which we derive some of our judgments from others is erroneous, unless it is a rule which we have formulated, i.e. a judgment. But if all our immediate judgments are correct, and if the modes of inference by which we derive mediated judgments from them are all valid, how do erroneous judgments occur?"

B.'s view is that "error arises whenever it happens that something which we have recognized as probable, following perfectly correct rules of inference, and which we expect and take to be true, turns out to be false". The error, B. continues, "does not lie in the proposition which states the degree of probability of such a judgment, but only in the judgment itself. We do not err when we estimate that the degree of probability that a black ball will appear is 99/100, but only when we *expect* the appearance of a black ball (on the basis of this estimate or for some other reason)" (he had previously described an appropriate experiment). B. maintains that herein lies the only source of error. It follows that only beings that must use probability inferences are capable of error, that error occurs only in those judgments which were inferred according to probability inferences, and finally, that every error is a proposition which stands in a certain relation of probability to the remaining propositions which are taken to be true by the person who is in error. It follows that error will be found mostly in judgments of experience, but can also sometimes occur in conceptual judgments. As particular causes of error, B. mentions (a) ignorance, (b) misdirection of attention, (c) inappropriate association of ideas, (d) faulty memory, (e) misdirected will, (f) tenacity, (g) the presence of an erroneous opinion from which we deduce other erroneous opinions, (h) inappropriate symbolism "since every symbol, since . . . it is connected with many other ideas, can also produce the latter and can therefore produce a sequence of ideas . . . which influence our thinking," (i) false testimony.

§ 310. *Other Treatments of the same Subject*

B. considers the theories of error of Descartes, Berkeley, Locke, Kant, and others.

§ 311. *Origin of True Judgments—Understanding and Reason*

In this section, B. maintains that all immediate judgments are true, likewise all those that follow from them by non-probabilistic inferences. Other true judgments are generated by probabilistic inferences from immediate judgments, and by inferences from false judgments.

In the sequel, B. defines *understanding* [*Verstand*] and *reason* [*Vernunft*] in the following way: "I do not think that we are too far removed from common usage if we define understanding as the capacity for empirical cognitions, which may require the mediation of purely conceptual truths, but where we do not have to be conscious of these truths. On the other hand, any truth which requires for its derivation the clear recognition of a purely conceptual truth is justly attributed to reason, even if it also requires a perceptual premise."

§ 312. *Whether we can Recognize a Truth without Recognizing its Grounds*

B. maintains that we rarely recognize the ground of a truth. All judgments of experience are based upon probability inferences; hence we infer all judgments of experience from premises which are not their ground. But many purely conceptual truths are also inferred from premises which are not their ground, though they follow deductively from them.

§ 313. *Grounds for the Recognition of Truth*

If a true judgment *M* is caused in us by other judgments *A*, *B*, *C*, *D*, . . .t then the latter are called the grounds for the recognition of the former. I, follows that some of our judgments (immediate judgments) do not have grounds; On other occasions, all or some of *A*, *B*, *C*, *D*, . . ., may be false, and *M* true. Hence B. distinguishes two cases: If *A*, *B*, *C*, *D*, . . . are the ground of *M*, and the recognition of *A*, *B*, *C*, *D*, . . . is the ground of the *recognition* of *M*, then *A*, *B*, *C*, *D*, . . . are called the objective grounds for the *recognition* of *M*, otherwise its subjective grounds.

§ 314. *Whether there is a Definite Limit to our Knowledge*

In B.'s view the boundary which limits our knowledge [*Erkenntnisvermögen*] can be properly drawn only if we determine completely the class of truths we can recognize, as well as the class of those which we cannot. Now it would be absurd to enumerate all those truths which we cannot attain, nor is it very informative to say that we can recognize all and only those truths which are useful. Moreover, it is not sufficient to show that we have *so far* been unable to recognize a certain truth; we may be able to do so in the future.

We might be inclined to think that the best way of showing that there are limits to our knowledge is to indicate that there are *objects* of which we can know nothing. Hence it would seem that we can determine the limit by saying 'every object which falls under idea *A* is beyond human knowledge' or (what comes to the same thing) 'no truth of the form '*A* is *X*' can be recognized by man'. "But it is obvious that such an assertion is absurd. We know something of every object, at least what it has in common, *qua* object, with all other objects. Moreover, the assertion that we cannot know anything about this object is itself a judgment about it; hence it is actually a contradiction to say of an object that we cannot know it at all, i.e. that we cannot form a single true judgment about it. One might reply that in saying that we do not know this object, we do not wish to assert that we are ignorant of absolutely all its properties and that we cannot even form a single judgment about it. I accept this, but I should like to ask now for a more definite identification of the attributes that must be unrecognizable if the object is to be called unknown. . . ."

It might be thought that another way of indicating the limits of our knowledge is to identify certain propositions by name, and to claim that we shall never be able to find out whether they are true. But how could one possibly prove such an assertion? "The fact that we have so far not been able to decide whether attribute *b* belongs to the objects which fall under the idea *A* does not entitle us to conclude that we will never find grounds for such a decision". B. now considers the case where *A* and *b* are pure concepts. In this case it cannot be a future experience which decides whether it is true that *A* is *b*. "But who could want to claim that even a very long-continued consideration of the two concepts *A* and *b*, through more and more precise and many-sided comparisons between them and related concepts, etc. could not lead to the recognition that the two concepts can or cannot be combined into a judgment?" On the other hand, if the ideas *A* and *b* are mixed concepts, then mere contemplation cannot decide the truth of '*A* is *b*' but experience is necessary. However, from the fact that we have not had these experiences, it certainly does not follow that we will not have them in the future.

NOTES: "Hence I confess to be of the opinion that it is hardly possible to make a determination of the limit of our knowledge [*Erkenntnisvermögen*], at least not the kind that is generally desired. It might be argued that this assertion contains a contradiction, similar to the one pointed out above; for this impossibility itself seems to form a limit for our faculty of knowledge. I reply, in the first place, that I do not assert that such a limit can *never* be given, but only that *I* do not know it; furthermore that the assertion of this impossibility can be considered an indication of an eternally unknowable truth only if it is presupposed that such a limit actually exists even though we cannot indicate it. But I do not say this; I rather think that we are unable to indicate this limit because it does not exist, so that the sum of human knowledge is capable of infinite increment."

§ 315. *The Doctrine of Critical Philosophy on this Subject*

B. here gives an extended criticism of several of Kant's doctrines. His main criticism is directed against Kant's view that we make true synthetic judgments only about those objects which we can intuit empirically, or which fall under one of the two pure intuitions, namely time and space. B. takes it for granted that the falsity of this doctrine has been sufficiently established. He then begins a critical examination of some of the facets of Kant's system.

1. Kant's account is directed mainly toward logic, mathematics, physics, and metaphysics, and does not give a satisfactory explanation of the origin of practical judgments, which must also be synthetic, and of the judgments of certain other theoretical sciences, such as aesthetics.

2. Kant assumed that logic (i.e. general logic) consisted of nothing but analytic judgments. B. takes this to be false, counting among the propositions of logic such assertions as 'there are ideas', 'there are simple and complex ideas', 'there are intuitions and concepts', etc. Moreover, while "the proposition 'if all men are mortal, and Caius is a man, then Caius is mortal' might reasonably be called analytic in the wider sense of § 148, the rule itself, namely that from two propositions of the form '*A* is *B*' and '*B* is *C*' a third proposition of the form '*A* is *C*' follows, is a synthetic truth."

3. B. now reiterates some of his criticisms against Kant's theory of

mathematics. In particular he holds that there is no such thing as *a priori* intuition.

4. Kant thought that he could sufficiently explain, from his theories, that the four sciences logic, arithmetic, geometry and pure physics are in the undisputed possession of generally acknowledged truths, the reason for this being that in the case of merely analytic propositions, as they are found in logic, as well as with propositions which can be tested by intuition, errors cannot easily creep in or maintain themselves.

I do not wish to deny that the indicated sciences have a marked advantage over others in the reliability of their doctrines. But nobody who has accepted even the larger part of the foregoing arguments can hold that the reasons for this are to be found in the circumstances cited by Kant. Indeed, I see no reason why we should forsake the explanation which was given long before Kant. It has always been maintained that these sciences enjoy such a high degree of certainty only because they have the advantage that their most important doctrines can be easily and variously tested by experience, and have been so tested, and that those doctrines which cannot be immediately tested are deducible by arguments which have been tested many times and have always been found valid, and finally, that the results which are obtained in these sciences do not infringe upon the human passions; hence most of these investigations were begun and finished without bias, and with suitable leisure and peace. The only reason why we are so certain that the rules *barbara*, *celarent*, etc. are valid is because they have been confirmed in thousands of arguments in which we have applied them. This also is the true reason why we are so confident, in mathematics, that factors in a different order give the same product, or that the sum of the angles in a triangle is equal to two right angles, or that the forces on a lever are in equilibrium when they stand in the inverse relation of their distances from the fulcrum, etc. But that $\sqrt{2} = 1.414\ldots$, that the content of a sphere is exactly two-thirds of the circumscribed cylinder, that in each body there are three free axes of revolution, etc. we assert mainly because they follow from propositions of the first kind by arguments which others have conducted hundreds of times and have found valid; an additional factor is that in all these matters we do not have the slightest advantage if the thing turns out to be otherwise. That the reason for our confidence really lies in these circumstances can be seen most clearly from the fact that our confidence rises and falls as these circumstances dictate. If we have not tested the truth of a proposition either by experiment, or by repeated checking of its derivation, we do not give it unqualified assent, if we are at all sensible, no matter what Critical Philosophy may say about the infallibility of pure intuition.... Does not experience teach us that we make mistakes in mathematical judgments, and that we make these mistakes more easily the more we trust what that philosophy calls by the high-sounding name of pure intuition? The geometrician who thought that a pair of solids have the same content if they are bounded by similar and equal sides made such a gross mistake certainly only because he trusted his intuition, i.e. mere appearance, too much....

5. B. objects most strongly to Kant's doctrine that we cannot make any synthetic judgment about objects that we cannot perceive, so that we can make synthetic judgments neither about God, nor about our own soul, its immortality, about freedom or other super-sensible objects. But if sensory or perceivable objects are defined as objects which under certain circumstances can act upon the organs of our body, then it turns out that these objects are always *bodies*, i.e. "classes of an infinite number of simple, finite substances which are capable of change, in such a way that their class at any moment fills a certain finite three-dimensional space." But if a body is defined in this way, then it follows that the substances of which it is composed are not themselves sensible objects, hence they are super-sensible. "But we cannot say of them that they are neither in space nor in time; on the contrary, they are in both, although it is true that a single one of them fills only a single point in space." B. maintains that any number of true synthetic propositions can be found concerning these simple substances. Similarly also about the human mind (soul), etc.

6. Since it is metaphysics which deals with these super-sensible objects, it follows that Kant's indictment of metaphysics is premature. There are a great number of synthetic propositions from metaphysics of which we can be certain, for example, 'there is a God', 'this God is immutable', 'all substances in the world influence each other', 'no simple substance can perish in time', 'ideas can exist only in simple substances', and many others.

7. The most remarkable fact is that Kant, not satisfied with denying the ability of human reason to judge super-sensible objects, went on to say that in the domain of metaphysics there are four pairs of contradictory propositions (antinomies), each of which can be proved with reasons that do not violate any of the rules of logic. In the *Critique of Pure Reason* (B 454 ff.) these eight propositions are given with their alleged proofs.

I believe that I can find several mistakes in these proofs; but I will be satisfied with giving merely a few indications:

a. The proof of the proposition that the world has a beginning in time is based upon the claim that an infinite past sequence is a contradiction; the reason for this is that the infinity of a sequence is said to consist in the fact that it can never be completed through a successive synthesis. This I believe to be quiet false, for the infinity of a sequence does not consist in the fact that it cannot be completed through a successive synthesis (i.e. through a synthesis which is carried out gradually in time). The concept of time does not belong to the concept of an infinite sequence at all, since there are sequences of things, finite as well as infinite, which are not in time at all. One is the infinite sequence of natural numbers 1, 2, 3, 4, 5, . . ., or the following: . . ., −5, −4, −2, −1, 0, +1, +2, +3, +4, . . ., the first of which goes to infinity only in one direction, the second in both directions. But if there are infinite sequences of numbers, and magnitudes in general, then there must also be infinite sequences in time, indeed in both temporal directions, into the future as well as into the past. And Kant himself did not find anything objectionable in the possibility of an infinite sequence into the future; he should therefore also have admitted the possibility of such a sequence into the past, i.e. the possibility of an already past infinite sequence, since both directions in time are completely similar to each other.

b. The proof of the proposition that the world must also be limited in *space* is supported by the preceding proof. Kant says "in order to think a world that fills all spaces as a single whole the successive synthesis of parts of an infinite world must be considered as completed, i.e. an infinite time must be thought to have passed in counting all existing things, which is impossible." We have already seen that this impossibility is a sheer invention. An additional error is that world which consists of infinitely many parts can only be thought by means of a successive synthesis of parts, while we already think such a world when we simply speak about it.

c. In the second antinomy I take the thesis "every compound substance in the world consists of simple parts, and nothing exists except the simple or what is compounded from the simple" to be true, only with the proviso that there must be infinitely many simple parts out of which compound objects consist. I find the proof of the antithesis, or the proposition that no compound thing in the world consists of simple parts to be completely untenable. This proof proceeds from the assertion that space does not consist of simple parts, but of spaces, i.e. of manifolds which are external to each other. This has frequently been said by mathematicians and is indeed true if taken in a certain sense, but if it is taken in another sense, the one required for Kant's proof, it is false. In compound things we can often distinguish two kinds of parts, namely homogeneous parts, i.e. parts which fall under the same concept as the whole, and heterogeneous parts for which this is not the case. Thus we show that a piece of saltpetre has homogeneous parts, if we break it into pieces which are themselves pieces of saltpetre; heterogeneous parts appear when we show that saltpetre consists of hydrochloric acid and potash. And if we want to speak merely of homogeneous parts, then it is quite correct to say that every extended space consists only of other extended spaces, every line only of lines, etc. But if we want to speak of parts in general, so that heterogeneous parts are also included, then we should not hesitate to admit that every extended space, whether line, surface, or body, also consists of simple parts, namely points; it is nothing but a certain class of such points. To make this clear I note, firstly, that it is a generally accepted doctrine that in every line, surface, or body there are, or as they say, there *lie* points. But what can this being or lying of points in a spatial object be, if not that they are parts of these objects? It will be objected that these parts are not integral parts, since it is impossible to produce a body out of a number of points, no matter how large. I freely admit this if this number is finite; I also admit that not every infinite number of points forms an extended object. But as soon as we think of a class of points of such a nature that for every one of them, and for every distance, no matter how small, there are one or several points in the class, which have this distance, then we have a true *continuum* (which will be either a line, a surface, or a body). Thus we think a surface of a sphere if we think the class of all points which are equidistant from a given point. That there is nothing in this spherical surface than this infinite set of points which is represented by the concept of the totality of points which have the same distance from a given point follows from the fact that it is quite impossible to indicate anything that belongs to the surface of this sphere which is not such a point or a class of such points. Hence it is a false assertion that space does not consist of simple parts in general this holds only of homogeneous parts. . . .

8. . . . I must admit that I do not find the method of "practical postulation" very satisfactory. I think we can speak of a duty, or a necessity, of believing something only where there are decisive grounds for the truth of the respective proposition. If there are no such grounds for our freedom and immortality, or for the existence of God, then nobody has the duty, nor is it even possible, to believe in these objects. It seems to me a perverse procedure to deduce, as Kant did, from a postulate of reason the possibility of its satisfaction. Strictly speaking, reason never demands anything unconditionally, but only if a being exists which has the power to do a good thing, then reason demands of this being that it should bring about this good. For example, reason does not demand unconditionally that a sick person should be healed; it only demands from a physician, and

from nobody else that, if he is present, and if he can heal the sick person, then he should do so. Hence if, as Kant asserted, the realization of the highest good is possible only if there is a God and if we are free and immortal, then it follows that we are justified in asserting that reason demands the realization of the highest good if the existence of God, freedom and immortality are already proved. Hence these three truths cannot be deduced from that postulate; on the contrary, the postulate can be proved only if these three truths are presupposed.

§ 316. *Survey of the most Important Distinctions between Judgments, Based on their Truth Values*

In addition to true and false judgments, B. distinguishes judgments which are based upon a true (or false) ground, and judgments of whose (true) ground one is conscious (these are called clear cognitions). True judgments for which we know the grounds of the ground, etc. are called insights, or scientific knowledge; others common cognitions.

PART IV

Of Certainty, Probability, and Confidence in Judgments

§ 317. *Definition of the Concepts of Certainty and Probability with Respect to Thinking Beings*

1. The relation between our judgments and truth has been sufficiently discussed, and it is now necessary to consider their *certainty* or *probability* and the degree of *confidence* which depends upon them. If the theory concerning the origin of error which was maintained in § 309 is correct, then we can say of every fallible thinking being that it commits errors only for the following reason: it takes propositions to be true which stand in a real or fancied relation of probability (cf. § 161) to certain other propositions, which this being, rightly or wrongly, assumes to be true. In the case of judgments which are not mediated, likewise with judgments which follow from immediate judgments by non-probabilistic arguments . . . no error can occur; all such judgments are true.

2. Consequently, judgments which are generated in a way that makes error impossible are, for this reason, called *certain*, or *secure*. Thus, certainty in this sense must be distinguished from the concept defined in § 161, no. 3. The latter is merely a relation which holds between propositions, where one of them is deducible from the others, and where it is a matter of indifference whether these propositions are in themselves true or false or whether there is a thinking being which takes them to be true or even thinks them. Certainty in the present sense is a property which belongs only to judgments, and belongs to them only in relation to a thinking being which presently forms them, provided that they are generated in a way which makes error impossible. The context will make clear in which sense the word 'certainty' is taken in a given case; if it does not, I will add the phrase, 'for a certain thinking being'. It follows from this explication that a proposition which has certainty for a being must be considered true by that being, or (what comes to the same thing) is one of its judgments. This is also in conformity with common usage; it would be strange if we wanted to call a proposition *certain* for a person even if he does not take it to be true, perhaps only because he could be easily convinced of it. Furthermore only propositions which are true

in themselves can be called certain with respect to a thinking being; for the judgment which contains this proposition must come about in a way which makes error impossible, hence it must be true. Since true judgments are also called cognitions [knowledge, *Erkenntnisse*], propositions which are certain can all be called cognitions. One and the same proposition will have to be considered sometimes certain, sometimes not, depending on the being to which it is related. Thus, for God everything which is true is certain, since He recognizes all truth and recognizes it without possibility of error. With men, the class of certain propositions is restricted to truths which are recognized without mediation, or which follow from immediate judgments by a non-probabilistic deduction.

3. On the other hand, propositions about which we may be mistaken are, according to no. 1, only those that stand in a real or fancied relation of probability (cf. § 161) to other propositions that have already been accepted as true by the being in question. Hence the following cases only can arise: (a) the premises *A*, *B*, *C*, *D*, . . ., which are taken to be true by the being are indeed true and proposition *M* stands to them in the relation of probability (cf. § 161); or (b) the propositions *A*, *B*, *C*, *D*, . . . are true, but the proposition *M* does not really stand in the relation of probability to them, but the being mistakenly thinks that it does; or (c) the premises *A*, *B*, *C*, *D*, . . . are not all true but a real relation of probability obtains between *M* and them; or, finally, (d) neither the propositions *A*, *B*, *C*, *D*, . . . are all true nor is there a real relation of probability between *M* and them. A moment's thought shows that cases (b) and (d) can be reduced to (c). Since reason never errs in the judgment that a proposition is deducible from certain other propositions, unless we think a rule *R* according to which this deduction is to proceed, we can always add this erroneous rule *R* to the propositions *A*, *B*, *C*, *D*, . . . which are accepted as premises by the being in question. But then the premises *A*, *B*, *C*, *D*, . . . and *R* stand not only in an imagined, but in a real, relation of probability. Hence we can say that error arises only if a being accepts as true propositions which stand in the relation of probability to certain other propositions which it has already accepted as true. It is very important to draw attention to these propositions, and to give them a name; let us call them propositions which have *probability for* this being. Hence we must distinguish probability in this sense of the word from probability as it was defined in § 161 .There we meant by probability a relation which holds between propositions as such, no matter whether they are in themselves true or false or whether somebody takes them

to be true or false. Here, on the other hand, probability is an attribute which a proposition can have only in relation to a certain thinking being, and only if that being accepts as true the propositions relative to which that proposition has probability. Hence, if we say that a proposition *M* has probability for a certain being, with respect to premises *A*, *B*, *C*, *D*, . . . and ideas *i*, *j*, . . ., then propositions *A*, *B*, *C*, *D*, . . . must be considered true by that being, and proposition *M* must stand to *A*, *B*, *C*, *D*, . . . in the relation of probability (cf. § 161) with respect to ideas *i*, *j*. . . . But this definition does not require that the being should also take proposition *M* itself for true, nor does this seem to be required by common usage. Could it not be the case, for example, that there is a proposition which follows with probability from certain other propositions which are accepted by somebody, although he does not know this because he is not paying attention to the matter? It would seem reasonable, in this case, to call this proposition probable for him (though he is not conscious of this fact). It follows from this that probability in this sense can be divided into *recognized* and *unrecognized*. Concerning proposition *M* itself, it is not necessary that a thinking being must accept *M* even if it is aware of the relation of probability which holds between *M* and the accepted propositions *A*, *B*, *C*, *D*, . . .; this, indeed, never happens if the recognized probability of the proposition is smaller than one-half, since in this case the judgment will tend toward *Neg.M*. Hence we must distinguish probable propositions which are *accepted as true* and others where this is not the case. The premises *A*, *B*, *C*, *D*, . . . which are accepted by the thinking being, and to which *M* stands in a real relation of probability are either all true or some of them are false. In the first case the probability of *M* can be called a *real* or *objective*, in the second a merely *apparent* or *subjective* probability. In the case where the probability of *M* with respect to *A*, *B*, *C*, *D*, . . . is objective for a certain being, i.e. if the propositions *A*, *B*, *C*, *D*, . . . are all truths, they are occasionally called *grounds* of that probability, sometimes even the grounds (though not sufficient grounds) of proposition *M* itself. According to the sense of the word *ground* given in § 198, we could only say of these propositions that, taken together, they are not the ground of *M* itself—but the ground of the proposition which states its *probability* for that being. . . .

Consider all propositions which are accepted by a thinking being and which stand in the relation of probability (§ 161) to a given proposition *M*; the probability which accrues to *M* on the basis of all of these propositions is called the *complete* or *absolute* probability of *M* for this being, a probability which accrues to *M* only on the basis of part of this class is called a *relative* probability of *M* for that being. Thus, for example, the probability of the proposition that the earth turns around its axis merely on the basis that such revolutions are found in several other heavenly bodies is only a relative probability of this proposition; on the other hand, the probability which accrues to it from the consideration of all circumstances known to me which speak for or against such revolutions, is the absolute probability which this proposition has for me. We can see from § 161, no. 18 that the absolute probability of a proposition does not always have to be larger than every relative probability. For, among the totality of propositions which are accepted by somebody as true, there may be some which, taken in themselves, give a probability less than one-half to it; such propositions will diminish the degree of probability that accrue to it from other considerations. . . . If the absolute probability of a proposition for a certain being is very large and if there are circumstances which make it foolish or illicit to consider the possibility of the opposite and to act on it, then I call these propositions *trustworthy*, *secure*, *reliable* for that being. In ordinary life these propositions are often called *certain*, but in science they are merely called *morally* or *sufficiently* certain in contrast to proper certainty (cf. no. 2). Hence I call it a trustworthy proposition that the ceiling of this room will not fall down. This calamity, though I do not find it impossible, has such a low degree of absolute probability that it would be foolish to consider it in my actions and, for example, leave the room, since the danger of an accident if I remain here is not larger than the dangers encountered when I leave it and spend the night outside. Hence trustworthiness has several degrees; and one and the same degree of absolute probability, which makes one proposition trustworthy for a given being, may be too low for another proposition. For it is not the low degree of probability alone which makes it foolish or illicit to think of the possible opposite of proposition *M* and to prepare for it, but certain other circumstances also play a role: Consider the damage which would arise if the opposite of *M* were true and we did not prepare for it. The product of this damage and the degree of probability of *Neg.M* is *the danger* to which we are subject when we do not prepare for *Neg.M*. What makes it foolish or illicit then, to prepare for or consider the opposite of a given proposition *M* is that the danger in neglecting *Neg.M* is smaller than the danger in neglecting *M*. Hence, whether or not a proposition is morally certain depends on the magnitude of damage that would result if its opposite came about and we were unprepared for it, and on other circumstances as well. Thus, the probability that somebody who is in good health will not live to see the next day is small enough to be neglected in connection with business that does not greatly suffer from his death; hence his survival for one day is credible; but where great disadvantages accrue, it is certainly one's duty to think of the possibility of sudden death, and make appropriate preparations; hence in this respect it is not certain enough that he will live to see the following day etc.

NOTE 1: Kant (*Logik*, A 90) and several others defined certainty as a belief [*Fürwahrhalten*] which is connected with a consciousness of necessity. Presumably the necessity under discussion consists in a certain relation to the judging person, i.e. a judgment is certain whenever the judging person feels that his judgment could not be otherwise and that he is necessitated to judge in this way; against this I should like to call to mind that all judging is in a certain sense necessary: it is not our arbitrary decision to form a judgment in this or in some other way; on the contrary, after we have directed our attention to certain objects, we feel necessitated in each of our judgments. If I am told that out of a hundred balls most are black and only a few are white, and I see somebody choosing one at random, I cannot decide to expect that a black ball will be drawn; I am necessitated in this expectation although I know that it could be mistaken and that a white ball could be drawn. On the other hand, if the notion of necessity is related to the proposition itself which is the content of the judgment, I must reply that necessity in the proper sense of the word is an attribute which belongs only to real things, but not to objects which do not have reality, for example propositions. In an improper sense one may attribute necessity to a certain kind of proposition, namely those that express purely conceptual truths; but the class of certain propositions comprises more than these; it also includes so-called empirical truth, and Kant himself accordingly distinguishes *rational* and *empirical* certainty. . . .

3. Some will take exception to the fact that on the one hand I connect the concept of probability with certain thinking beings, and on the other hand I distinguish an *objective* from a *subjective* probability; They probably object because we call something objective if it can be thought without relation to a subject. I wish to reply that my distinction between these two kinds of probability depends upon the kind of premises which this thinking being accepts as true. If they are objectively true, then the probability can also be called objective; if they are in themselves false, but are taken to be true by the subject, then we ought to call the probability subjective. . . .

Hoffbauer, in his logic (*Anfangsgründe der Logik*, Halle 1794, § 419) points out that in games of chance "the possibility of all cases can be determined *a priori*, but nobody can give reasons why one outcome rather than another should be the case." I must confess that I cannot understand him. For the assertion that the possibility of all cases can be determined *a priori* can only mean that we can determine *a priori* that each of several enumerated cases has the same probability. But we can attribute this equal probability to several cases only if we ignore all dissimilarities (e.g. that one ball is heavier than the others, etc.). If we assume that these cases are *in actual fact* equal, then not only is it impossible for any person, but even for God to give a reason why "this rather than that outcome will be the case". Actually, nothing would happen, since the exact equality of circumstances would give no preference to one case over all others. But this is not the case; rather, in each of several cases which have an equal probability there are certain peculiar circumstances which make it happen that only one of them comes about, and which makes the occurrence of all the others impossible. Even we are capable of ascertaining, under certain circumstances, some of these dissimilarities, for example that some balls are heavier or smoother, etc., and we can make estimates which case is more likely to occur. . . .

Often *mathematical* and *dynamic* or *philosophical* probability are distinguished; the former is said to occur where all so-called grounds are similar and can only be counted; the latter, where this is not the case. I, too, believe that there are probabilities which we can calculate, and others which we cannot calculate, but I think that this distinction is very subjective and vacillating. What seems more important to me is the distinction, already touched upon in § 161, no. 8 between probabilities which are *determined* by their premises and those that are not. Perhaps it was only this distinction which they had in mind without being aware of it.

Some logicians, for example Kiesewetter [*Grundriss einer allgemeinen Logik nach Kantischen Grundsätzen*, Pt. I, Berlin 1802, § 297) distinguish between *real* and *logical* probability. The former is said to be the probability of a *thing* the latter of a *judgment*; the former is said to be calculable "since the reasons are envisaged as similar, the latter not, since the reasons for and against an opinion are dissimilar" (*ibid.*, p. 468). I should like to reply that, strictly speaking, probability is an attribute which belongs only to propositions, hence to judgments only in so far as they are propositions, but which does not belong to any other object. Hence, if we frequently use expressions such as 'this event is probable', etc., all we want to say is that the proposition which states this event is probable. Just as it is improper to call an event *true* (at most this could mean that the *proposition* which states it is true) so it is also improper to call it *probable*. If this is true then this distinction and all the consequences that follow from it require no further counter argument.

§ 318. *Circumstances which Determine the Confidence of our Judgments*

... Nothing can *immediately* determine the degree of our confidence in a judgment *M* except the following two things: (a) the degree of absolute probability or complete certainty which accrues to proposition *M* with respect to the totality of judgments *A*, *B*, *C*, *D*, ... which are presently in our mind, and which provide evidence for or against *M*; and (b) the degrees of confidence of the judgments *A*, *B*, *C*, *D*, ... themselves. What besides these two things could it be that determines the confidence of our judgment? Judgments which we have formed earlier, but have forgotten, or which are not presently in our mind, either clearly or obscurely, cannot act, simply because they are not there. The greater or smaller degree of attention with which we test proposition *M*, the strength of our wish that it should turn out true or false: such matters certainly have an influence. They determine to some extent how our judgment turns out (whether it will be affirmative or negative) and in particular they will also determine the degree of our confidence in it. But it seems to me that all these circumstances act only indirectly, not immediately, and only in so far as they stimulate judgments that we have formed earlier, and on whose presence in the mind the degree of probability of proposition *M* depends. It is also obvious that the degree of our confidence does not depend upon the relative degree of probability which proposition *M* has in relation to a *part* of our judgments, but it depends upon the degree of absolute probability of this proposition, provided that we mean by absolute probability the probability which accrues to it with respect to all judgments presently in our mind. It is similarly obvious that these judgments can influence the degree of confidence of judgment *M* only in proportion to the confidence that they themselves have. But there is a twofold difficulty connected with this view. (a) Since the degree of confidence with which we judge is always real and finite, it must in each case be completely determined. However, according to § 161, no. 8, it is reasonable to suppose that there are premises which leave the degree of probability of a certain proposition completely undetermined.

How then can the degree of confidence with which we judge have a determinate degree, if it depends upon something that is itself undetermined? (b) Is it not incomprehensible how a degree of probability which we do not especially *notice* can have effects in our mind such that it determines the degree of confidence with which we form a judgment? I should like to make the following reply. Concerning (a): if the judgments which are presently in the mind, *A*, *B*, *C*, *D*, . . ., and which function as premises to the proposition *M*, do not produce a determinate degree of probability for *M*, we will neither form judgment *M* nor judgment *Neg.M*, just as we do not form this judgment when the premises bestow a probability of one-half upon proposition *M*. Concerning (b): we must envisage judgments *A*, *B*, *C*, *D*, . . . from which a certain probability accrues to proposition *M*, as forces which generate judgment *M* with its determinate degree of probability. Their force is larger the larger the degree of their own confidence, and judgment *M*, which is generated by them, will become stronger, i.e. will enjoy greater confidence the greater the force of its generators. It is not surprising that all this should happen without our *awareness* of the magnitude of the forces of *A*, *B*, *C*, *D*, . . . or of the strength which judgment *M* receives from them. This is no more surprising than the fact that we bring forth thousands of other effects without being aware of the laws which bring them about.

§ 319. *The most Important Levels of our Confidence*

According to B., the highest degree of confidence is bestowed upon immediate judgments and those that follow from them by non-probabilistic deduction. Their degree of probability is 1, and confidence in them is called *perfect* or *complete*. The confidence in a judgment which is morally certain is called *moral confidence* or *conviction*. Lower degrees of confidence are indicated by such words as 'opinion', 'surmise', etc.

§ 320. *How the Various Degrees of Confidence can be Represented Numerically*

1. It will rarely be very important to represent the degree of confidence with which we judge numerically. Still, it might be a useful intellectual exercise to consider the question how these degrees should be calculated; this motivates the present discussion. If the degree of probability of a proposition *M* with respect to the premises *A*, *B*, *C*, *D*, . . . (which are certain for us) = is one-half, i.e. if this proposition is just as probable as its negation, or the proposition *Neg.M*, and if we notice this, then it is obvious that we will not form judgment *M*. For, if we form that judgment then, for the same reason, we should

also have to form judgment *Neg.M*, and it is certain that we cannot do both at the same time. Hence, we do neither, i.e. we do not judge at all. In this case the degree of confidence is called doubtfulness, possibly because we do not judge but doubt, whenever the probability of a proposition = is one-half.

2. On the other hand, if the degree of probability that judgment *M* has for us with respect to the (certain) premises *A*, *B*, *C*, *D*, . . . is larger than one-half, then it is understandable that we form judgment *M*, if premises *A*, *B*, *C*, *D*, . . . are then present in our mind. Likewise, the degree of confidence with which we form this judgment will be proportionate to the surplus of the probability of proposition *M* over the probability of its Negation *Neg.M*, since it is nothing but this surplus which determines us to judge. Let us use μ to designate the probability of proposition *M*; then the probability of *Neg.M* equals $1-\mu$. Hence the degree of confidence with which *M* is formed will be proportionate to the magnitude $\mu-(1-\mu) = 2\mu-1$, and can be expressed by $(2\mu-1)\,C$, where *C* is a constant. Now since the degree of confidence which has no higher degree above it is finite, we can take it as a measure for the rest and use it as the unit. Hence the expression $(2\mu-1)\,C$ should equal the unit if $\mu = 1$. For if the probability of a proposition turns into certainty, then the confidence of our judgment is perfect, and there can be no higher degree. Hence the value of the constant $C = 1$, and the degree of confidence with which we form a judgment which has probability μ is simply $2\mu-1$.

μ can be represented as $\dfrac{m}{m+n}$ (cf. § 161), where $m+n$ represents the number of mutually exclusive cases where *A*, *B*, *C*, *D*, . . . all become true, while m represents the number of cases where *A*, *B*, *C*, *D*, . . . together with proposition *M* become true. Hence the probability of $Neg.M = \dfrac{n}{m+n}$ and the difference between them or $2\mu-1 = \dfrac{m-n}{m+n}$. "Hence we can say that the degree of confidence with which we form a judgment can be found by calculating the differences between the number of mutually exclusive cases in which the judgment becomes true and the number of cases in which it becomes false, divided by their sum."

3. This mode of calculation gives a value of zero for the confidence of a proposition whose probability equals one-half, which is as it should be. Furthermore the degree of confidence of the judgment *Neg.M* turns out to be $\dfrac{n-m}{m+n}$. Since $\dfrac{n-m}{m+n}$ equals $\dfrac{-m-n}{m+n}$, a negative value for the degree of confidence simply shows "that we no longer form judgment *M* but judgment *Neg.M* with the same, but positive, degree of confidence."

4. If a proposition *M* is deducible from premises *A*, *B*, *C*, *D*, . . .

and if we have these premises presently in mind, and consider them true then, according to § 161, the degree of probability *M* for us should never be smaller but somewhat larger than the product $\alpha \times \beta \times \gamma \times \delta \times \ldots$. If we are afraid that we might be mistaken in the judgment that *M* is deducible from *A*, *B*, *C*, *D*, . . ., then the degree of probability will be somewhat lower. Hence on balance the probability of a conclusion will rarely be much larger, rather it will usually be somewhat smaller than the product of the probabilities of all its premises. Only if this product is larger than one-half will the judgment *M* actually be formed. If this product is equal to one-half, we will not judge at all, and if it is smaller than one-half we will form the judgment *Neg.M*.

5. If everything is as above, but *M* is not deducible from *A*, *B*, *C*, *D*, . . . but only receives the degree of probability μ from them, then we will form judgment *M* with a degree of confidence which is proportionate to the degree of probability of $\mu \times \alpha \times \beta \times \gamma \times \delta \times \ldots$.

6. Consider a number of partial reasons, each of which bestows a certain probability upon proposition *M*. No matter how large the number of these partial reasons is, no perfect confidence for judgment *M* can result from them, though a moral confidence can result, which can approximate perfect confidence as closely as desired. For, the degree of confidence with which we form judgment *M* depends upon the degree of absolute probability which this proposition has for us. But, according to § 161, this will never approach perfect certainty, i.e. unity, though it can come as close to unity as desired. But if μ cannot equal 1, though it can approximate 1 as closely as desired, $2\mu - 1$, i.e. the degree of confidence, can also never be equal to 1, although it can approximate this value as closely as desired.

NOTE: What has been said will suffice to make clear the difference between the concepts of *probability*, including the sense defined in § 318, and the concept of *confidence*. Probability is always an attribute which belongs to propositions as such, whether or not they are taken for true or even mentally represented; hence it can be divided into known and unknown probability. On the other hand, confidence is always an attribute of judgments that have been formed. Hence, with respect to one and the same proposition *M* and to one and the same thinking being we can distinguish several kinds and degrees of probability, but not several kinds and degrees of confidence. The necessity of distinguishing the confidence with which we form a judgment, and the probability of the proposition itself, and of assigning different values to their respective degrees is particularly apparent in cases where the probability equals or is smaller than one-half. Assume that we are told that in an urn there are ten black, ten white, and ten red balls. It will certainly not occur to us to form the judgment that somebody who picks a ball at random will draw a black one. Similarly, we will not want to judge that he will pick a white one or a red one. Whatever reason we might have for forming one judgment, we have the same reason for forming another; nonetheless, it is obvious that we cannot assert all three of them, since they are contrary propositions. Hence we form none of them. Calculation shows that the degree of probability for each of them is one-third. Now if we did not distinguish between the degree of probability of a proposition and the degree of confidence with which we form the corresponding judgment, we would have to form all three judgments, each of them with a confidence of one-third, which is quite absurd. Given the concept of confidence that I have introduced here and its mode of calculation, anything is explained properly. The degree of confidence for each of the three judgments is $2 \times 1/3 - 1 = -1/3$, which shows that we form none of these judgments, but only the three negations (which are compatible with each other, each with a degree of confidence which equals $+1/3$). But it could be objected that nonetheless the drawing of a black (likewise of a white or a red) ball is not impossible, but expected; consequently, if somebody were to assert the opposite, namely that no black ball will be drawn, if he wanted to wager more than two to one on it, we would contradict him and perhaps accept his offer of a wager in the hope of a gain, all of which might be said to prove that the judgment that a black ball will be drawn has not been completely given up. I reply that to contradict somebody who does not want to admit that a black ball will be drawn, and to engage in a wager with him, does not require that we should believe that such a ball will be drawn; rather, it will suffice for this that we form the judgment that the indicated proposition (namely that a black ball will be drawn) has a degree of probability of one-third. But this judgment is quite different from the judgment that a black ball will be drawn. The former not merely has probability, but it is deducible from the premises by means of a perfect deduction, and therefore has as much certainty as they have. As I become more and more clear about this distinction between probability and confidence, I find it more and more surprising that all previous writers on the subject have never clearly made this distinction.

§ 321. *Knowledge and Belief*

There are two important states of our mind with respect to our own judgments, states which (if I understand them correctly) have to do with a consideration of the confidence with which we are attached to a given judgment; they do not depend upon the *degree* of this confidence, but rather on whether or not we are able to destroy this confidence. Given that a truth *M*, either immediately or after we pay attention to its grounds, has become so obvious to us that we assume we could not persuade ourselves of the opposite even if we wanted. In other words, we think that it is not in our power to destroy the confidence with which we are attached to judgment *M*. In this case I should like to call truth *M* an item of *knowledge* [*Wissen*]. Such knowledge we receive, for example, of the truth of the Pythagorean theorem once we have been exposed to its proofs. For now we know the truth of this proposition in a way that assures us that we could not persuade ourselves of its falsity even if we tried. On the other hand, consider a proposition *M* which we take to be true, but of which we do not have knowledge. In this case it does not seem impossible that we could come to form the opposite judgment, *Neg.M*, by concentrating our attention on true or apparent reasons against *M*. In such a case I call the relation between our mind and proposition *M* a belief in this proposition, provided we intend to continue to pay attention to the grounds for the truth of *M*. For example, I believe that the heavenly bodies are populated, since I have decisive reasons to form this judgment, although I do not doubt that, if I wanted, it would be possible for me to persuade myself of the contrary. Hence, *knowledge* is a relation of our faculty of judgment to a given proposition which is initially, but not permanently, dependent upon our decisions; belief on the other hand does not only depend upon our will at its inception, but permanently; hence it also designates a relation of our *attitude* to the proposition in question. Usually, the degree of confidence is higher in the case of knowledge than in the case of belief, but there are also cases where the confidence of a belief does not fall behind that of knowledge. This is usually the case with our belief in God. A belief which has a lower degree of confidence is usually called an opinion, or a taking to be true. Moreover, with respect to the firmness of our resolution to keep the grounds for our proposition in mind, we can distinguish a solid and a wavering belief, with respect to the ground which produced this resolution, we can distinguish a reasonable, moral, as well as a foolish or immoral belief and, finally, with respect to its object (i.e. whether it is a conceptual or an empirical proposition) a theoretical and a historical belief.

The concepts which I have here assigned to '*knowledge*' and '*belief*' seem to me not only important and worthy of a special name, but also adequate to the meaning which common usage connects with these words. Kant's definition is of course quite different since it describes opining, believing, and knowing as three essentially different modes of taking something to be true. He says that the ground of an opinion is neither objectively nor subjectively adequate, that of a belief subjectively, but not objectively, and that of an item of knowledge is both objectively and subjectively, adequate. I should like to call grounds merely subjective if they appear to be grounds to the person forming the judgment, irrespective of whether or not he followed correct rules of thought, but objective if they still appear to be grounds in cases where the rules employed were correct. But then it is strange to say that "what I believe I do not hold to be objectively, but only subjectively necessary (for me)." It seems to me that if I become aware that I accepted something on the basis of incorrect rules of thought, then I can no longer believe it. Other passages (e.g. *Logic* A 97) lead us to believe that grounds are called subjective if they are "derived from the nature and interest of the subject". But then it would have to be permissible to assert that we do not know, but merely opine or believe, that we are hungry and thirsty; for is it not obvious that the grounds upon which we base such judgments are derived from the nature and the interest of the person judging?...

Fries (*System der Logik*, Heidelberg 1811, p. 451) says that "Belief in its logical sense is the acceptance of an opinion only because I am moved by interest to form a judgment with respect to it, e.g. when the physician has to minister even if the case is doubtful". I deny that we are ever necessitated to decide for or against a doubtful proposition *M* merely in order to be able to act. I say that in order to act, and act rationally, nothing is required but a deliberation whether the probability of *M* is large enough to take certain measures appropriate under the circumstances. If this is the case, and if we take these measures, then it is said, improperly, that we have decided in favour of *M*, whereas a rational person takes special care before he makes such a decision. The example of the physician shows this clearly: though he makes arrangements for one possible case, he also prepares for other possible outcomes.

In the metaphysical sense Fries distinguishes knowledge, belief and sentiment [*Ahnung*] as three modes of conviction of which the first is said to derive from intuition, the second without intuition, and the third from feelings without definite concept. It is obvious why I cannot agree with these definitions. It is generally acknowledged that mathematical cognitions can become an object of knowledge, yet in my view they do not rest upon intuition. Furthermore, I do not think (the reader already knows why) that there are cognitions which proceed from feelings in such a way that it is altogether impossible to reduce them to clear concepts. . . .

BOOK FOUR

HEURETIC

A number of sections of a heuristic nature follow. Only two of them are included in the present edition.

§ 329. *Tentative Acceptance or Indirect Procedure*

B. here considers the case where the truth of a certain proposition *M* is to be ascertained, and where the method employed for this purpose is the indirect method, i.e. *M* is accepted tentatively, or as an hypothesis. B. points out that two things are required when this method is employed namely (a) the propositions in question must be suitably chosen, and (b) they must be tested in every conceivable way. Concerning the choice of suitable hypotheses he has little to say beyond the assertion that a preliminary investigation will usually reduce the number of testworthy hypotheses to a manageable size. For the testing of hypotheses he makes the following suggestions: First of all we should try to think the proposition in question as clearly as possible. Often, the truth or falsity of the proposition will become apparent at this stage.

"If the mere clear representation of a proposition *M* does not lead to a judgment about it, or if this judgment does not appear reliable enough, the next stage in its testing is that we attempt to deduce, either from *M* alone, or from *M* together with other already known premises, several consequences and from these further consequences, etc. If we recognize any one of these consequences as a false proposition then we are entitled to the conclusion that our tentative hypothesis *M* itself is false. Conversely, if all consequences drawn from the proposition *M* are true, we ought to do the same thing we just did with *M* now also with its negation, the proposition *Neg.M*, i.e. we should collect all consequences which follow from *Neg.M* either immediately or in conjunction with known truths. If we can find one which is decidedly false, then it is shown that *Neg.M* is false and hence that the initial hypothesis *M* itself is true. This procedure of showing the truth of proposition *M*, and thus of solving the indicated problem, is generally called the *reduction to absurdity*, or apagogic procedure. Examples are common in the mathematical sciences so that I need not mention any at this time."

If neither the assumption of *M* nor that of *Neg.M* leads to an absurdity then the above method will not usually lead to a decision except, perhaps, to the assertion that *M* has a certain probability.

As a further method for testing hypotheses the finding of known premises from which they follow is suggested.

B. next considers the testing of propositions of the form 'all *A* are *b*'. "If, in a case like this, we find only one *A* which lacks attribute *b*, then the falsity of *M* is decided. . . . But if we want to recognize the truth of proposition *M* in this way, and if we want to have certainty, then we can obviously not restrict our investigation simply to a great number of *A*'s, but we must extend it to all of them. Now if *A* comprises many, perhaps even infinitely many, objects, then this becomes possible only by finding certain concepts A', A'' . . ., which fall under the concept *A* in such a way that their domains together exhaust the domain of *A*; moreover they must be such that the truth of the propositions 'every A' has *b*', 'every A'' has *b*', . . . can be more easily established than the proposition 'every *A* in general has *b*'. . . . It is clearly not necessary that the ideas A', A'', should be exclusive, but only that they together exhaust the domain of *A* (§ 95). If we cannot establish a sufficiently large number of propositions of the form 'A' has *b*', 'A'' has *b*', . . . to conclude decisively that all *A* in general have *b*, since the ideas A', A'', . . . taken together do not completely exhaust the domain of *A*, then the propositions already established still entitle us to conclude with a certain degree of probability that the attribute *b*, which belongs to all the *A*'s already investigated, also belongs to the remaining ones and hence to all *A*. In particular, assume that the number of all *A* is $m+n$, and the number of those already investigated and found to have attribute *b* is *m*; also assume that besides our observation we have no other reason whatever to suppose that property *b* should belong only to the *A*'s which we have investigated rather than to the others; assume furthermore that concept *A* itself does not furnish any reason why it is more likely than not that the attribute *b* should belong to the objects falling under *A*; and, lastly, assume that we have no reasons other than those produced by our observations to suppose that the *A*'s should be all alike with respect to an attribute of the kind of *b*: if all these conditions are fulfilled, then the degree of probability with which we can assume that the remaining *A*'s, and hence all *A*'s, have attribute *b* is $\frac{m+1}{m+n+1}$. (The calculations leading to this expression can be found in the literature.) Since $\frac{m+1}{m+n+1}$ either equals or is smaller than one-half, whenever $m<n$, and since we cannot accept propositions whose probability is smaller than one-half, it follows that the truth of a proposition *M* cannot be assumed unless we have tested at least half of all *A* and have found them to have property *b*. It is, of course, a different matter if we have independent reasons to suppose that the *A*'s are uniform. For example, if we go into a small section of a large library and take a few books from the top shelf, and a few books from the bottom shelf, and we find them to be theology books, we can conclude that all the books in this section are on theology, since it is reasonable to suppose that the books in the library are ordered according to subject."

§ 379. *The Discovery of Causes for Given Effects*

One of the most common and important tasks for human inquiry is the discovery of causes which produce given effects. According to § 168 I call an object *A* the cause (either the complete or partial cause) of another object *B* if the proposition that *A* exists forms the (complete or partial) ground of the proposition that *B* exists. It follows that, in this sense, every cause is a real thing; properly speaking, a cause is not a substance, but an attribute or force which resides in a single substance or a class of them. But in an extended sense, we give the name of a cause to an individual or a substance or a class of them if they possess a certain force. It is my view that two kinds of cause must be distinguished, namely those which do not require any time in order to produce an effect of finite magnitude, and those which can produce such an effect only in a finite period of time. An example of the first kind is the creative force of God; for God needs no time, no matter how small, in order to produce a substance; rather, such a substance will exist from eternity if God wants that it should so exist. An example of the second kind is every moving force which can produce a finite speed only if it has acted through a finite time. Existence as well as the attributes of causes of the first kind can be demonstrated from concepts alone; hence they are discussed in a special science, namely metaphysics. It is not necessary that logic should give special instructions how such causes can be discovered, since the rules are the same as for the discovery of purely conceptual truths in general. But the same does not hold with respect to causes of the second kind, which act in time, and merely produce changes. One of the reasons is that these causes have a definite magnitude and character only at a certain time; hence they cannot be recognized *a priori*, but only through perception. We know the nature of the mutual influence of heavenly bodies only from perception; this influence depends upon their mass and spatial relations, which can and must be different at different times. The identification of causes which can be recognized only through perception is of extreme importance. Hence it is reasonable to indicate, in broad outline, how we must proceed in this matter. But if we are to investigate the cause of a given change, we must never doubt that there is such a cause. Common sense itself insists that for each change there must be a cause which has brought it about, and only very few scholars have dared to assume the possibility of certain changes without the existence of a cause which determines them completely; they have done this only in the single case where those changes are the acts of will of a sentient and rational being. Hence these philosophers allow us to assume that every perception has a cause, but we cannot simply suppose of this cause that it has another cause and that the second cause has a third and so to infinity. For if we are forced to explain a certain phenomenon by claiming that a free act of will of a sentient and rational being was a near or remote cause, the sequence of causes finds its end at this point since we have arrived at a cause which itself has no (fully determining) further cause. But in all other cases, especially where we do not encounter acts of will, we must assume that every perception has a cause, and that the latter has again a cause, etc., even if we subscribe to this system of indeterminism. But before we consider how to discover these causes, we must clearly determine what is wanted when the discovery of such a cause is desired. Obviously nothing can be meant but that we should *determine* this cause, i.e. that we should enumerate its various attributes, especially those which belong exclusively to it, or at least which it does not have in common with every other cause. . . . When we consider how we recognize these attributes, we can again make the following distinction: some of them we can deduce with the aid of pure concepts from others, already found, while we are led to the discovery of others merely through *perception*. Thus, if we know that the cause of a certain change is a rational being, we can determine from pure concepts that this cause is a simple being, that it has had a rational purpose in its activity, etc. . . . But whether this purpose is good or evil cannot be decided from the indicated attribute alone; this would require further perceptions. But I need not give any instructions as to how one can find attributes which are deducible from those already found, or from pure concepts, or both. Hence I have to consider only those attributes whose presence is found through perception. But all attributes of an object whose presence is deduced from mere perceptions consist in nothing but certain *forces*, namely those in whose presence we make the assertion that the object in question is capable of causing certain perceptible changes in us and in others. . . . . Hence it follows that a closer determination of a cause . . . always consists in showing that one and the same object, which we assume to be the cause of a certain change, has also produced other changes of such and such a kind under certain circumstances. . . . Finally, concerning the effects themselves to which we want to find a cause, two essentially different cases must be distinguished. The effect can be a *single* determinate object, and it is desired that we should indicate the determinate forces which have brought it about; or it is a certain *kind* of phenomenon and it is desired that we should indicate the *kind* of force which can bring about such phenomena. An example of the first kind is the task of finding the cause of the eclipse of the sun described in St. Matthew 27 : 45. An example of the second kind is the task of finding the causes of eclipses of the sun in general.

1. In the first case, when the object whose cause we wish to find is a single fully determinate real object, it follows from the well-

known truth that God is the Creator and Governor of the whole world, and from the law of the universal reciprocal influence of created objects that the complete cause of the existence and the attributes of our object can only be the totality of *all* existing objects. For each of them has some influence upon it, no matter how small, so if that other thing did not exist, our object would be different. It follows from this that the task of finding the complete cause of something is impossible for us. But that is not usually what is wanted when we are asked to find the cause of an object or effect. Normally only an indication of objects which have a humanly discernible influence upon the fact that the object has such and such observed attributes is desired. Moreover, we usually allow ourselves to neglect all those circumstances whose presence is understood, as well as those which are not important for the purpose of the investigation in question. The latter explains why one and the same event, depending on the point of view from which it is considered, and depending on the reason why we want to know its cause, can truly be said to have very different causes. Thus the death of Clitus can be attributed, by a doctor to bleeding, by a moralist to rage, by a judge to a murder, and by others to other objects. Hence the various designations *causa principalis, concomitativa, effectiva, instrumentalis, physica, moralis,* etc. If it were already known that effects of kind E could be brought about only by the particular kind of object C, then no further investigation is required to show that C was present in this case. On the other hand (by far the most common case), if there are several objects C, C', C'', . . . each of which is suitable for producing an effect of kind E, then, if we already know this, we only have to investigate which of these various causes, C, C', C'', . . . was present in this particular case. We will find this by investigating the phenomena simultaneous with E or slightly before E and by considering whether they indicate the presence of C or C' or C''. . . . From this we can see that the first of the indicated problems can be solved if only the second is solved, i.e. that we can find the cause of the given effect easily enough if we only know which kind of cause produces what kind of effect. But how do we arrive at this knowledge?

2. B. now concerns himself with the question how the cause or causes of a given effect can be found. For example if phenomena a, a', a'', . . . or b, b', b'', . . . or c, c', c'', . . . always take place whenever E takes place (so far as we know), then it is reasonable to suppose that objects A, B, C, . . . which respectively bring about these phenomena, are present whenever an effect such as E occurs. Now the degree of our confidence increases as the number of observed cases of this sort becomes larger. But saying that E never occurs without a, a', a'', . . . or b, b', b'', . . . c, c', c'', . . . does not

mean that we are entitled to say that the objects A, B, C, . . . are the *cause* of effect E. The presence of phenomena a, a', etc. can also be explained by saying that A, B, C, . . . are not a complete but only a partial cause or merely a condition of E, or even an effect of E. If in some cases only a, a', a'', . . . were present, in other cases only b, b', b'', . . ., etc. then there is no doubt that neither of them is the effect of E alone, but E could be a partial cause of them. But this is ruled out as soon as we discover that a, a', etc. occur all of them earlier than E, since it is obvious that the effect can never be earlier than its cause, though it can occur later if we mean by *effect* that which remains after the force of the cause is already spent. Now if a, a', a'', . . . always occur somewhat earlier than E, then we are entitled to assume that A is either a complete or a partial cause of E. If A, B, C, . . . are not only to be a partial, but a complete cause of E, then, wherever there is E, there must be A, B, C, . . . and conversely, wherever A, B, C, . . . are, there must be E. "The larger the number of cases is in which this occurs, and the greater the differences in all other respects, the greater the confidence with which we may assume that the object A or one of the several A, B, C, . . . is a complete cause of E. For, the greater the differences between the remaining circumstances under which E took place . . . the smaller the danger that some other, hidden, object A' or B' was a partial cause in the production of E. Thus somebody can be all the more confident that he gets a headache from drinking coffee, the more often he realizes that he gets this headache when he drinks coffee and the greater the differences in the rest of his activities on these days."

If we find that the complete cause of an effect E lies in certain objects A, or B, etc. we may ask whether *all the parts* of these objects are required to produce this effect, i.e. whether it could not be the case that the complete cause of W can be found in a simpler object than the one which brings about the changes a, a', a'', a''' . . . or b, b', b'', b''', . . . etc. We have no other reason for this surmise than that E never occurred unless these phenomena also occurred together. "Hence if it is within our power to bring about these phenomena singly or partially, then it would be reasonable to try whether an object which produces only a, a''', or only b and b''', is sufficient to produce E. For example, people would have thought that the propagation of sound requires air, if they had not made experiments which showed that one can propagate sound without air, for example in water; from this it followed that the true cause of sound lies only in those properties of air which it shares with other elastic bodies."

3. The difficulties which stand in the way of the just-indicated procedure are greater than might be thought. The reason is that the number of circumstances which accompany every event, including the event E which is to be explained, is infinite. Hence it happens only too frequently that in spite of careful attention we fail to see just those circumstances which are required to find the true cause. Among the large number of perceptions it is difficult enough to find those which occurred in all or most cases where E took place. It is even more difficult to remember precisely whether these perceptions

have been found together on other occasions although E did not follow. . . . Given these difficulties, it is desirable to know some rules which would shorten and facilitate this work somewhat.

a. One of these rules is the following: if there are a great number of cases when E occurred, where E is the event to be explained, then we should first concentrate on those cases which differ most widely from each other, i.e. which have very few common features outside of the presence of effect E. In these cases it would be permissible to assume only one of two things, namely either that the cause of the effect to be explained was different in these two cases, hence that there are several causes which can bring about E, or that this cause lies in one of the few circumstances which the two cases have in common. Now since it is not very probable that so many altogether different causes have one and the same effect, it is reasonable to begin with an examination of the second assumption especially since we can expect that this examination will not take a lot of time since there are only very few circumstances which need to be examined.

b. It often happens that whenever a certain event E took place a considerable number of similar circumstances occurred which are not all necessary for the occurrence of E. In order to find out whether this is the case, the quickest method is to look for cases where most of these circumstances were present without the occurrence of E. From this we find out at once that none of these circumstances contains the complete cause of E; hence we do not have to make a lot of guesses which we must eventually discard.

c. If the effect E is of a kind which admits of more or less, or (as we say) which admits of *degrees*, then the discovery of its cause may be greatly facilitated if we first concentrate upon phenomena which increase and decrease concomitantly. For, wherever the effect has a degree, the cause also must have a degree, and the larger the latter the larger must be the former. Hence it is reasonable to suppose that the phenomena which increase and decrease concomitantly with the effect belong to its cause, unless, indeed, we discover that they are its effect; the latter case could be discovered if we found that these phenomena occur somewhat later than the event we wish to explain. Thus we conclude that the light of the sun causes warmth on earth because we notice that a spot becomes warmer the larger the number of sun rays which fall upon it.

d. Let E be an object which has a mathematical opposite, and let C be its cause; then we may assume that the opposite of C is the cause of the opposite of E. Hence we investigate whether among the circumstances which accompany E there are some which are the opposite of those which accompany the opposite of E. If we find any, then we may assume that they are, if not the complete, then at least a partial, cause of E. For example, if we notice that the volume of a body becomes larger if it becomes warmer, and becomes smaller if we remove heat from it, we conclude that the cause of its expansion lies in the increased heat.

e. If we wish to find the cause of a phenomenon which is found to accompany many objects only *occasionally*, then it is reasonable, above all, to see if there are not some of them which display this phenomenon *all the time* and others which never display it. In the former case there must be something which causes this phenomenon, in the latter something which prevents it. If we therefore consider the attributes which belong exclusively to the former class, or at least to the former class in contrast with the latter, then we may hope to find at least a partial ground of the event. For example, transparency is an attribute which is found in all bodies that have a high degree of fluidity and in all expansible fluids, but not in solids that have a non-homogeneous structure. Therefore, we surmise that the similarity of parts of a body is one of the main conditions for its transparency.

4. B. next considers the question how we can determine which parts of a cause are responsible for what parts of an effect, given that both cause and effect consist of parts. He concludes that if certain parts of C change, but no change occurs in E, then those parts of C which have changed are not responsible for the occurrence of E. Furthermore if E becomes smaller if certain parts of C are put in a different position, we may conclude that these parts are in part responsible for the occurrence of E. According to B. the safest way of determining which parts cause what aspects of the effect is to show that c (a part of C) causes e (a part of E), etc.

5. If we claim that a certain object E is caused by several objects A, A', A'', then we must also establish the spatial and temporal relations between these parts of the cause. For example if we wish to explain a multiple echo, it will not suffice to state that there were several elastic surfaces, but the unequal distances between these surfaces must also be indicated.

6. If we know that a given effect E can be caused by one or more of A, B, C, . . ., and E has taken place, but we have not witnessed any of A, B, C, . . ., we must first decide whether E could have been caused by some further object. In case this is less probable than the presence of either A, or B, or C, or . . ., we must decide which of these might have been the cause. This will generally depend upon the number of cases in which E is known to have been caused by A, by B, etc. "But if the number of cases in which one of the causes, e.g. A, was found to be operating is not larger than half of the total number of observations, then the probability of the assumption that E was caused by A is not large enough to warrant the assertion that it really was that way. For, given that we have observed n cases, and a was found to

have occurred α times, then the probability of the proposition that *A* was present in this case equals $\frac{\alpha+1}{n+2}$ and this figure becomes larger than one-half only if α is larger than $\frac{n}{2}$. Unless the circumstances of the present case speak for each of *A*, *B*, *C*, . . . with equal strength, we must also include into our calculation of probability the relevant conditions of the present case in addition to the considerations of former cases. Finally we must favour that assumption which has the highest degree of probability over all the others."

7. According to B. we should not postulate the presence of an unknown force which brings about *E*, until we are convinced that every attempt to explain *E* from the cooperation of several known forces is either illicit or rests upon assumptions which have an even greater improbability than the postulation of that unknown force. To judge this we must consider two things, namely (a) "whether the several objects, whose combined operation would explain *E*, are indeed compatible; whether one of them might not have a force which cancels the presence or efficacy of another. If the mode of operation of these objects is unknown to us, if we have no experience that they can act together, then it is very uncertain that they can together produce the sum of what they can produce individually." (b) Secondly, we must consider whether it is probable that the number of objects required can actually be found in the necessary spatial and temporal relations. The larger we assume the number of these objects to be, the smaller the probability that they will be found together, unless there is some particular cause that can bring about this disposition of objects.

8. The most difficult case arises when a phenomenon is to be explained that has been observed only once, i.e. a phenomenon which can be brought under a concept *E* which does not fit any past observation. In my view, there are only two methods with which we can determine more closely the cause of such a phenomenon. (a) One of them is to find out whether there were not some unique circumstances accompanying or immediately preceding the phenomenon. We would then assume that they belong to the cause of *E*, since it is certain that an unusual effect can only be brought about by an unusual cause. (b) No matter how peculiar the given phenomenon is, it will still have certain similarities with other phenomena which we have often seen and whose cause we know. Now since similar causes bring about similar effects, we can conversely argue, at least tentatively, from the similarities of effects to a certain similarity of causes. In this way we can estimate not only several characteristics of the unknown cause of E, but if we find an object with these characteristics in the vicinity of *E* we can often strengthen our assumption by a closer investigation of this object, provided only that we find several further properties of the cause which is to be determined. It is understood that this

latter method can also be used when we have to explain a phenomenon which we have seen several times, provided it is done carefully enough. Thus from the similarity between the phenomena of light and of sound, we can argue for a similarity of causes for both of them, and by suitable experiments, which are based upon this assumption, we can arrive at a more detailed knowledge of the nature of light. A repeated application of this argument from similarity will often lead to at least a partial, if not complete, knowledge of the cause of the phenomenon which was quite inexplicable in the beginning. It can happen that repeated comparisons of many objects lead to a whole series of phenomena E, E_1, E_2, E_3, . . . E_n, which are such that each of them has great similarities to the next one, although members that are farther apart, especially E and E_n, have almost no similarity at all with each other. Finally, it may well be the case that we have a reasonably close knowledge of cause C_n of phenomenon E_n. Now the great similarity of E_{n-1} and E_n allows us to assume that phenomenon E_{n-1} has a cause C_{n-1} which is not exactly like C_n, but has a certain property c in common with it. In the same way the similarity between E_{n-1} and E_{n-2} can be of a kind which justifies us to assume also for E_{n-2} that it has the attribute c and so on, until we are finally justified to assume of the given phenomenon E, that its cause C, no matter how different from C_n, still has the attribute c, which was held in common by all the other causes as well.

As an example B. gives the claim that the motion of micro organisms is due to their being alive. He does not think that we would make this claim if it were not for a larger number of intermediate living beings between micro organisms and ourselves. . . .

9. From all of this it becomes obvious that in looking for the cause of a given effect we almost always have the choice among several assumptions, where the task is to decide which of them is the most probable. Now in order to calculate or even estimate the degree of probability of each of these assumptions, we must, first of all, know of which and how many *simple* (i.e. not further decomposable) and independent premises each of them is composed. Next we must determine the degree of probability of each of these individual premises, if not by precise calculation, then at least by a rough estimation, if only from an obscure feeling, and finally we must form the product of these degrees. Lastly, the assumption which shows the greatest value for the product of the degrees of probability of its individual premises is the most probable of all. But it can also be the case that the individual premises, into which two assumptions A and B can be decomposed, or at least those premises in which they differ,

have the same degree of probability. . . . It is fairly obvious that we should, in this case, declare for most probable that assumption which consists of the smallest number of presuppositions, even if we do not know the degree of their probability. Hence, it is generally maintained that, other things being equal, a hypothesis is the more acceptable, the smaller the number of the independent premises of which it consists. Thus, if the phenomenon *E* can be explained in two ways, namely by the accidental conjunction of three objects *a*, *b*, *c*, and also by the accidental conjunction of four objects *d*, *e*, *f*, *g*, and if these premises are of equal probability, then the first explanation must be preferred to the second.

NOTE: Many philosophers have claimed that the *true* causes of phenomena, or the powers of things, are forever hidden; the reason is that we can know only their *effects*, since our perceptions are always effect, never cause. These claims are occasioned by the discovery that the cause of a given effect can never be determined through experience in any other way than by showing that this cause also produces certain other phenomena. Now this provides a splendid opportunity for loud lamentations about the limitation of human knowledge. I gladly agree that human knowledge is limited, but it seems to me that this sort of complaint is improper. We cannot properly say that we do not know the powers of things since we perceive only their effects, for to know the powers of things means to know their effects. There cannot be any way of knowing the power of a thing, either for men or for anybody else, except through a recognition of its effects, since a power cannot be anything but a cause, or a ground for the generation of certain effects. What I would acknowledge as a limitation of our knowledge, and the knowledge of all other created beings, is that we can perceive the powers of things only in so far as they act upon us, either mediately or immediately, and that we know so very few of their powers and that of these we only have probable and quite unreliable knowledge.

It is usually maintained that the discovery of the cause of a given phenomenon does not always, but only sometimes, require the introduction of certain hypotheses. But if a hypothesis is nothing but a proposition which we introduce in order to give an explanation, and of which we are not fully certain, but which we maintain only with probability, then every determination of a cause to a given effect which is taken from experience requires hypotheses. The reason is that each such determination can be given only through probability arguments. The only difference is that these hypotheses may have a larger or smaller degree of probability; hence it is a mere difference of quantity and should be described as such. I think that we should issue a specific warning to novices not to overlook this fact, i.e. not to assume an opposition between hypotheses on one hand and propositions which are in ordinary life called *established* experiential truths; it would be false to assume that the latter are found in a different way than the former while in fact their greater reliability rests only on the greater number of observations.

Sometimes, the reason why a simpler hypothesis is to be preferred to a more complex one (cf. no. 9) is sought in the circumstance that *we may presuppose that nature always follows the simplest rules.* In my opinion, no special presuppositions are needed for this, rather it follows from the rules of the probability calculus itself; cf. § 253, note.

§ 384. *The Discovery of Effects of Given Causes*

In this section B. applies the methods of § 379 to the inverse problem. He gives comparative anatomy as an example for the method described in § 379 no. 8; it led, for example, to the discovery of the auditory organs of fishes.

BOOK FIVE

THEORY OF SCIENCE PROPER

The rest of the *Wissenschaftslehre* is concerned chiefly with the actual composition of handbooks. B. goes to considerable detail in such matters as how books ought to be divided into chapters, how chapter headings are to be composed, how a suitable class of readers is chosen, and he cautions the writer against undue prolixity. In addition to numerous passages of incidental interest, and others that shed light on doctrines discussed earlier, the sections on definitions (*Erklärungen*, §§ 554–559) and criteria (determinations, *Bestimmungen*, §§ 500–509) are of importance. In the present selections, we reproduce only B.'s chapter on indirect proofs.

§ 530. *Proofs by Reduction to Absurdity*

1. We have already seen in § 329 that a very good method of persuading ourselves of the truth of a given proposition is to assume its negation, and to try to deduce from it an obviously false proposition. For, if this can be done, then this negation is obviously false, and the negated proposition itself true. Proofs which employ such a procedure are usually called proofs by *reduction to absurdity*, also *apagogical*, or *indirect* proofs. All others are called *direct* or *ostensive* proofs. The question here is whether the use of indirect proofs should be avoided in a treatise. What we ask of a reader in such a proof is no more than that he should infer the truth of a certain proposition *M* because we have deduced a conclusion *Neg.A* from *Neg.M*, either by itself or together with certain other acknowledged true propositions *B*, *C*, *D*, . . ., and because he recognizes *Neg.A* to be false. Hence we want him to form the following deduction: 'if *M* were a false, hence *Neg.M* a true, proposition, then *Neg.A* would also be a truth. But *Neg.A* is false, hence *Neg.M* is also false and *M* is a truth.' Since this inference is quite persuasive, it is understandable that the degree of certainty which proposition *M* will acquire from an apagogical proof depends only on the degree of confidence with which the reader accepts its two premises. If these are certain enough, he will find proposition *M* also sufficiently certain. If the proposition *Neg.A*, *the absurdity* to which we reduce our deduction, is a proposition whose falsity is obvious, i.e., if proposition *A* is an indubitable truth, then everything depends on the degree of certainty with which the reader apprehends the major premise, i.e., the

proposition 'if *Neg.M* is true then *Neg.A* must also be true'. If this proposition is also obvious, and if the number of propositions *B*, *C*, *D*, . . ., through whose mediation we prove it is not too large, and if they are all sufficiently certain, and if the inferences which we draw from them are not probability inferences, but perfect arguments, then our apagogical proof will produce sufficient confidence for proposition *M*. Hence, if our only purpose is *persuasion*, there is no reason why the apagogical mode of proof should be avoided in a treatise; and the fact that such proofs often come easily to mind justifies the supposition that they will appear simple and natural to the reader. Nonetheless, I think that I should voice the following objections to this kind of proof: if from a proposition *Neg.M* which is false the false proposition *Neg.A* is to be deducible through the mediation of the true propositions *B*, *C*, *D*, . . ., then the true proposition *M* must be deducible from the true propositions *B*, *C*, *D*, . . ., and *A*. We can see immediately from this that it is more straightforward to prove *M* directly from *B*, *C*, *D*, . . ., and *A*, than first to prove *Neg.A* from *B*, *C*, *D*, . . ., and *Neg.M*, and then to conclude from the obvious falsity of *Neg.A* that there must be a false proposition among the others, and since propositions *B*, *C*, *D*, . . . are all certain to be true, that this false proposition must be *Neg.M*, and that *M* must therefore be true. Moreover, if the opinions voiced in § 221 about the inner connection between truths is not in error, it follows that the propositions upon which a given proposition rests in an apagogical proof can never be its objective ground in its pure form. It is certain that the objective ground of a truth cannot lie in the larger number of propositions from which it is deduced in this mode of proof, if it is also deducible from a smaller number.

2. I am of the opinion that every apagogical proof, if it contains no other error, can easily be changed so as to avoid the disagreeable consideration of false propositions. I want to show this in a general way, and to do this I should like to suggest, as a rule, that we use apagogical proof only with propositions whose predicate idea is negative. Hence I assume that the proposition for which we want to give an apagogical proof has the following form:

Every *A* is a non-*B* (1)

I hope that everybody will be in a position to apply to other cases what I say for this case. The apagogical arguer considers the contradictory of the proposition he wants to prove, hence the proposition that there is some *A* which is a *B*, or, more properly, the proposition

The idea of an *A* which is *B* has a referent (2)

He then shows that the acceptance of this proposition contradicts a proven and known truth, namely

Every R is an S (3)

i.e., that the following is deducible from proposition (2):

The idea of an R which is not an S has a referent (4)

But the only correct way to get to this conclusion is by deriving from the assumed proposition (2), in conjunction only with true and proven propositions, three propositions of the following form:

The idea X has a referent (5)

$$\left.\begin{array}{l}\text{Every } X \text{ is an } R\\ \text{Every } X \text{ is a non-}S\end{array}\right\} \quad (6)$$

The reason is this: the only way of showing that the idea of an R which is not an S has a referent (4) is to indicate an object which is both an R as well as a non-S. Obviously we cannot indicate such an object except by finding an idea X which refers exclusively to it and by proving, *first*, that there really is an object for this idea X, i.e. that it has a referent (5), and, *second*, that an object that falls under this idea, i.e. every X, is both an R as well as a non-S. It could be objected that a contradiction to (4) could be produced if one of the two propositions (6) were particular instead of universal. I wish to call to mind, however, that in this case, too, there exists a somewhat narrower idea X for which the general propositions (6) still hold, since the expression 'every X' does not necessarily mean that the idea X has *several* objects in its extension. If these three propositions, hence also the first of them (5), are to be deducible from the assumed proposition (2), then the idea whose reference is assumed in (2) must stand to X in the relation of inclusion in the sense of § 108, i.e., the latter must either be equivalent or wider than the idea in (2). It is obvious that we can conclude that an idea has reference from the fact that some other idea has reference only if the former is either equivalent or wider but not if it is narrower than the latter. Hence the idea X, if we look more closely, will be of the following form:

An A' which is a B' (7)

where A′ and B′ are a pair of ideas which are formed from A and B by the omission of certain components, and which are therefore wider, or at least not narrower than A and B. After these preliminaries, it is easy to show how every apagogical proof can be so transformed that we do not have to proceed from a false premise. Since it holds that every *R* is *S* (3), it is certain that not all three of the propositions (5) and (6) are true. Hence, either the first of them is false, i.e., the idea *X* does not refer since, though simpler than the idea in (2), it still contains too many characteristics and among them some contradictory ones; or, if *X* has a referent, then one of the two propositions (6) is false. In each of these cases it will be possible, by a suitable rejection of certain characteristics contained in the idea (7) to form a pair of ideas X' and X'', both of which have reference and to generate the following true propositions:

$$\left.\begin{array}{l}\text{Every } X' \text{ is an } R\\ \text{Every } X'' \text{ is a non-}S\end{array}\right\} \qquad (8)$$

But it must be possible to show the truth of these propositions merely by using the same premises which were required to prove propositions (5) and (6). For, no conclusion which was necessary for the proof of these propositions could have been taken from those characteristics of idea *X* which made it non-referring, and which were left out in X' and X''. But once we succeed in this, i.e., once we have a pair of propositions as in (8) and have proved them, the goal is achieved. For the connection of them with truth (3) entitles at once to the inference that X' and X'' are a pair of exclusive ideas (§ 103) or (what comes to the same thing), that the idea of something which is X' and X'' does not refer. Hence I am all the more certain that the even narrower (more complex) idea of an *A* which is also a *B* likewise does not refer. But from this truth the desired conclusion follows (as soon as we know that the idea A has a referent):

Every *A* is a non-*B*.

3. As an example to illustrate the above point, let me use Euclid's well-known proof (*Elements* I, 19) that in each triangle *acb* the longer side ($cb > ca$) is opposite the larger angle ($a > b$). The proof goes essentially as follows: if it were false that $cb > ca$, then either $cb = ca$, or $cb < ca$. If $cb = ca$, then, according to *El.* I, 5, the opposite angles would have to be alike ($a = b$); but this violates the assumption. If it were the case that $cb < ca$, then, according to *El.* I, 18, the opposing angles would have to stand in the relation $a < b$, which again violates the assumption. Hence the only remaining possibility is that $ca < cb$.

We see that this argument contains two apagogical proofs. Both of the following propositions are proved apagogically:

In a $\triangle$ *acb*, in which $a > b$, it is false that $cb = ca$
In a $\triangle$ *acb*, in which $a > b$, it is false that $cb < ca$

After they are established, the desired conclusion is proved directly. Hence it should be satisfactory if we use the method of no. 2 to show how the apagogical argument form can be avoided for the first of the above propositions; the procedure for the second would be exactly similar. The proposition to be proved is

In a △ *acb*, in which $a > b$, it is false that $cb = ca$ (1)

If we consider it in relation to form (1) introduced in no. 2, we see that *A* in (1) is here replaced by the idea of a △ *acb*, in which $a > b$, while *B* is replaced by the idea of a △ *acb* in which $cb = ca$. The false assumption from which we start in an apagogical proof, according to no. 2, is the proposition:

The idea of a triangle in which $a > b$ and $cb = ca$ has a referent. (2)

In Euclid, this proposition is tacitly contained in the hypothetical proposition 'if in △ *acb* it were the case that $cb = ca$, then it would have to be the case that $a = b$'. Proposition (3), whose truth is presupposed in this proof, is the proposition

In every △ *acb* in which $cb = ca$, $a = c$ (3)

Hence the idea *R* is the idea of a △ *acb*, in which $cb = ca$, and *S* is the idea of a △ *acb* in which $a = b$. Hence (4) takes on the following form:

The idea of a △ *acb*, in which $cb = ca$, and $a \neq b$, has a referent (4)

We arrive at this conclusion from the assumption (2) without any further premises simply by replacing the idea

'A △ *acb*, in which $a > b$, and $cb = ca$' by the wider idea
'A △ *acb* in which $a \neq b$, and $cb = ca$'.

This is now the *X*, or the idea (7) of an *A'* which is a *B'*, so that in this special case *B'* and *B* (and *R*) are the same, and only *A'* was generated by extending *A*, where $a > b$ was replaced by the more general $a \neq b$. Hence (5) and (6) are as follows:

The idea of a △ *acb*, in which $cb = ca$, but $a \neq b$ has a referent (5)

a △ *acb*, in which $cb = ca$, but $a \neq b$, has $a = b$
a △ *acb*, in which $cb = ca$, but $a \neq b$, does not have $cb = ca$ (6)

Now in order to avoid the apagogical form, all we have to do, according to no. 2, is to transform idea X, as the idea of something which is both A and B, into a pair of referring ideas X' and X'' by omission of suitable parts. In this way we will obtain the two true propositions (8). We find at once

$$\left.\begin{array}{l}\text{A } \triangle\ acb\text{, in which } ca = cb\text{, has } a = b \\ \text{A } \triangle\ acb \text{ in which } a \neq b\text{, does not have } ca = cb\end{array}\right\} \quad (8)$$

The proof of these two propositions should not be difficult. The first is *El.* I, 5, the second is generated from the first by transposition, from which we can see that in this special case we do not even need both of them to arrive at the proposition

> The idea of a $\triangle$ *acb* in which $a \neq b$, but $ca = cb$, does not have a referent.

From this we obtain the desired conclusion

> The idea of a $\triangle$ *acb*, in which $a > b$, but $ca = cb$, does not have a referent, or
> a $\triangle$ *acb* in which $a > b$ does not have $ca = cb$

Hence the entire proof can be briefly stated in this way:

> a $\triangle$ *acb*, in which $ca = cb$, has $a = b$, hence
> a $\triangle$ *acb* in which $a \neq b$ does not have $ca = cb$, consequently
> a $\triangle$ *acb* in which $a > b$ does not have $ca = cb$.

4. We can see that the only thing required to transform an apagogical proof in such a way that no false propositions have to be considered is to express some of its propositions in a *simpler* way, such that we delete from their subject ideas certain characteristics which make them non-referring. In the usual exposition of this subject such propositions are put in hypothetical form, such as the following: "If something which has attributes α, β, γ, . . . had the attribute μ, then it would also have non-γ, where non-γ designates an attribute which is incompatible with γ. But in such a proposition the attribute γ of the subject idea of the antecedent can obviously be deleted, and this simplification will not make the proof of this hypothetical proposition nor the deduction of other propositions from it any more difficult. From this it follows that the apagogical mode of proof should never be used where a really clear understanding of the grounds (if only the subjective grounds) of a given truth is intended. For it obviously does not meet this purpose, and through a few alterations in the propositions, which are actually simplifications, and which reduce the number of inferences, it is always

possible to generate a proof which does not include a reduction to absurdity. But I do not wish to say that it is not permissible to proceed in the apagogical manner where the point is not so much to generate a clear insight into the ground, but to produce conviction in a familiar and succinct manner. For, although in this case the *thoughts* are more complex, they can still be more familiar to the reader, and their linguistic expressions can often be shorter, since frequently a more complex idea can be more briefly expressed than a simpler one.

If it seems advisable to retain the apagogical proof procedure, then I should like to see at least the following rules obeyed:

Our mode of presentation alone should make it obvious, even to the inattentive reader, that the propositions which we deduce from the assumed negation of proposition *M*, up to and including the proposition *Neg.A*, are not asserted by us, but are merely introduced for the sake of argument.

The proposition *Neg.A*, whose absurdity is taken for granted, must be chosen very carefully. Since it assumes a rather prominent position in the proof, it always attracts the attention of the reader above all others; hence, if they are not sufficiently convinced of its falsity the whole proof would be without effect.

NOTE: B. considers a case where a false major premise, together with one true minor premise, generates a proposition which contradicts the major premise. He writes:

"Is it not possible to think of a proposition which is in itself false, which, in connection with certain true propositions, and with respect to certain variable ideas *i*, *j*, . . . produces a conclusion which, with respect to certain other variable ideas, stands in the relation of contradiction to the first premise?" He uses the following example:

"Every proposition is false. That every proposition is false is itself a proposition. Therefore it is false that every proposition is false'. I cannot see how it can be denied that in this case we deduce from a false major premise through mediation of a true minor premise, and in conformity with the most correct mode of inference a conclusion which negates the major premise. It must also be admitted that this major premise shows its falsity simply because it produces a conclusion which is its negate. Thus the possibility of this kind of apogogical proof is demonstrated by a single example. However, since I have stated that all apogogical proofs can be turned into ostensive proofs, the reader will wish to know whether this proof, too, can be stated ostensively; this question is all the more reasonable as I have used this proof in § 31 for a very important purpose, namely, in order to show that there are truths at all. In order to produce the conclusion, the major premise which we assumed there is not really required; all we need is the following undeniably true proposition: 'if the assertion that no thing of a certain kind

A has attribute *b* is itself a thing of this kind, then this assertion itself lacks attribute *b*'. If we add to this major premise the minor premise: 'the assertion that no proposition has truth is itself a proposition' then the desired conclusion follows, namely 'the assertion that no proposition has truth lacks truth'."

INDICES

The numbers refer to the sections of the book.

I. Special Symbols, Phrases and Sentence Forms (cf. §§ 169 ff.)

II. Index of Subjects

III. Index of Persons

IV. Names Omitted

Bolzano's *Wissenschaftslehre* is an admirable source book for, and commentary upon, the history of logical theory. The following is the list of persons to whom reference had to be omitted in the present edition. Bolzano's observations upon these authors can be traced only through the edition of 1929/31 since the other editions do not contain an index of persons.

Abel, Jak. Friedrich; Acontius, Jacob; Adelung, Johann Christoph; Agricola, Rudolf; Alefeld, Johann Ludwig; Ampère, Andre-Marie; Archytas of Tarentum; Averrhoes; Avicenna; Bacon, Francis; Baumeister, Friedrich Christian; Baumgarten, Alexander Gottlieb; Becher, Johann Joachim; Berger, Chr. Gottlieb; Bilfinger, Georg Bernhard; Boethius; Boerhave; Bonnet, Charles de; Bretschneider, Karl Gottlieb; Bruno, Giordano; Buhle, Joh. Gottlieb; Cauchy, August Louis; Channing, William Ellery; Combe, George; Cousin, Victor; Creuz, Friedrich Karl Casimir; Develey, Em.; Diodorus; Diogenes Laertius; Eubulides; Feder, Joh. Georg Heinrich; Fermat, Pierre; Fischer, Johann Carl; Fischer, Er. Gf.; Flatt, Joh. Fr. v.; Galenus, Claudius; Gengler, A.; Gerard, J. M. A.; Geßner, Joh. Anton Wilhelm; Gravesande, Willem Jacob Storm van's; Hartley, David; Heraclitus; Hermes, Georg; Hillebrand, Jos.; Hinrichs, Herman Friedr. Wilh.; Hoffman, Adolf Friedrich; Jäsche, Gottlob Benjamin; Jussieu; Kalmar, G.; Kircher, Athanasius; Klugel, Georg Sigmund; Knutzen, Martin K.; Krause, Karl Christian Friedrich; Lagrange, J. L.; Lambert, Joh. Heinrich; Lange, Samuel Gottlieb; Le Clerc, Jean v. Amsterdam; Linné; Lullus, Raymundus; Maczek, J.; Manzoni, Alessandro; Marheineke, Ph. Konr.; Mayer, Josef Ernst; Megarians; Meier, Georg Friedrich; Meilinger, Florian; Menagius; Mendelssohn, Moses; Mendoza, Pedro Hurtado de; Menedemus; More, Thomas; Neumann, Karl Georg; Nußlein, Franz Anton; Newton, Isaac; Ohlert, A. L. J.; Ohm, Martin; Oken, Lor.; Oswald, James; Philetas of Kos; Pichler, Karoline; Ploucquet, Gf.; Quinctilianus, Marcus Fabius; Rajus; Reimarus, Hermann Samuel; Reinhart, Franz Volkmar; Reinhold, Karl Leonhard; Reuß, Matern; Richter, Jean Paul; Ritter, Heinrich; Rohling, Joh. Ephr.; Rosenkranz, Karl; Rüdiger, Andreas; Sagner, Kaspar; Salzmann, Chr. Gottl.; Schaumann, Joh. Christian Gottlieb; Schlegel, Friedr.; Schmid, Heinrich; Schwab, Joh. Ephraim;

Scott, Sir Walter; Selle, Chr. Gottlieb; Seneca, Lucius Annaeus; Snell, Fr. Wilhelm; Snellius, Rudolphus; Solbrig, Karl Fr.; Spinoza, Baruch; Stattler, Benedikt; Staudlin, Karl Fr.; Steinbart, Gotthelf Samuel; Stiedenroth, Ernst; Storchenau, Sigmund v.; Suabedissen, David Theodor August; Tacquet, Andreas; Tournefort, Joseph Pitton de; Ueberwasser, Fd.; Vetter, Karl Wilh.; Villaume, Peter; Wagner, Lor. Hnr.; Wegscheider, Jul. Aug. Ludw.; Weib, Christian; Weibe, C. H.; Wilkins, John; Wyttenbach, Dn.; Xenophanes; Zabarella, Jacobus.

V. Translation of Key Terms

attribute	*Beschaffenheit*
being	*Dasein*
cognition, knowledge	*Erkenntnis*
compatible	*Verträglich*
concept	*Begriff*
definition	*Erklärung*
extension	*Umfang*
idea	*Vorstellung*
intuition	*Anschauung*
non-referring, without referent	*Gegenstandslos*
proposition	*Satz an sich*
reference	*Gegenständlichkeit*
referent, object	*Gegenstand*
relation of ground and consequence	*Abfolge*
satisfiability	*Gültigkeit*
scope	*Weite*
supporting truths	*Hilfswahrheiten*
theory of science	*Wissenschaftslehre*
treatise	*Lehrbuch*

www.ingramcontent.com/pod-product-compliance
Ingram Content Group UK Ltd.
Pitfield, Milton Keynes, MK11 3LW, UK
UKHW041845031026
15587UKWH00004B/1029

9 780520 367555